T0181149

Operational Amplifiers

Johan Huijsing

Operational Amplifiers

Theory and Design

Third Edition

 Springer

Johan Huijsing
Faculty of Electrical Engineering
Mathematics and Computer Sciences (EEMCS)
Delft University of Technology
Delft, The Netherlands

ISBN 978-3-319-80277-0   ISBN 978-3-319-28127-8   (eBook)
DOI 10.1007/978-3-319-28127-8

Printed on acid-free paper

This Springer imprint is published by Springer Nature
The registered company is Springer International Publishing AG Switzerland

To my dear wife Willeke and children Hans, Adriaan, Mirjam, and Reineke, who have given me love and support.

Acknowledgements

Simon Middelhoek stimulated me to write this overview book.
Wim van Nimwegen drew pictures in a very clear way.
Ovidiu Bajdechi wrote the problems and simulation exercises.
Anja de Koning and Mary van den Berg typed the original manuscript.
Wendy Sturrock and many students helped in correcting the manuscript.
Maureen Meekel provided computer support.
Thanks to all.

Above all, thanks to God through Jesus Christ, who is the Lord of my life.

Contents

Summary

This third edition has completed Chap. 10 on systematic design of µV-offset operational amplifiers and precision instrumentation amplifiers by applying chopping, auto-zeroing, and dynamic element-matching techniques. Wide-band and fast-settling capacitive-coupled operational and instrumentation amplifiers are added. The associated designs of floating input choppers are presented, that facilitate beyond-the-rails CM input voltage ranges. Furthermore, many improvements have been made and errors corrected.

A systematic circuit design of operational amplifiers is presented. It is shown that the topology of all operational amplifiers can be divided in nine main overall configurations. These configurations range from one gain stage up to four or more gain stages. Many famous designs are completely evaluated.

High-frequency compensation techniques are presented for all nine configurations even at high capacitive loads. Special focus is on low-power low-voltage architectures with rail-to-rail input and output ranges.

The design of fully differential operational amplifiers and operational floating amplifiers is being developed. Also, the characterization of operational amplifiers by macromodels and error matrices is presented, together with measurement techniques for their parameters.

Problems and simulation exercises have been supplied for self-evaluation.

Introduction

The goal of this book is to equip the circuit designer with a proper understanding of the theory and design of operational amplifiers (OpAmps). The core of the book presents the systematic design of operational amplifiers. All operational amplifiers can be classified into a periodic system of nine main overall configurations. This division enables the designer to quickly recognize, understand, and choose optimal configurations.

Chapter 1 defines four basic types of operational amplifiers on the basis of the external ground connections of the input and output port; and which port needs to be isolated from the ground has a big impact on the circuit design.

A complete set of linear parameters, by which each of the above four basic types of operational amplifiers can be quantified, is given in Chap. 2. This chapter also presents macromodels and measurement techniques for OpAmp parameters.

A systematic treatment of sources of errors in important applications of the above four basic types of operational amplifiers is presented in Chap. 3.

Input stages are evaluated in Chap. 4. Important aspects such as bias, offset, noise, and common-mode rejection are considered. Low-voltage input stages with a rail-to-rail input voltage range are extensively discussed.

A classification of push–pull output stages is presented in Chap. 5. Three possible topologies are explored: voltage follower stages, compound stages, and rail-to-rail general amplifier stages. Emphasis is on voltage and current efficiency.

A classification of operational amplifiers into nine main overall configurations is presented in Chap. 6. The classification consists of 2 two-stage OpAmps, 6 three-stage OpAmps, and 1 four- or multi-stage OpAmp. High-frequency compensation techniques are developed for all configurations. Methods are presented for obtaining a maximum bandwidth over power ratio for certain high capacitive load conditions. Slew-rate and distortion are also considered.

Chapter 7 presents design examples of each of the nine main configurations. Many well-known OpAmps are fully elaborated. Among them are simple CMOS OpAmps, high-frequency bipolar OpAmps, Low-voltage CMOS and bipolar OpAmps.

The design of fully differential operational amplifiers with common-mode feedback is developed in Chap. 8. Special focus is on low-voltage architectures.

The design of the most universal active network element: the operational floating amplifier (OFA) is presented in Chap. 9. It has both the output and input port isolated from ground. The concept of this OFA gives the designer the freedom to work with current signals as well as voltage signals.

An additional Chap. 10 has been added on the systematic design of µV-offset operational amplifiers and precision instrumentation amplifiers by applying chopping, auto-zeroing, and dynamic element-matching techniques. Capacitive coupling at the input gives these chopper amplifiers input CM ranges that reach far beyond the supply-rail voltages. The design of associated floating choppers is presented.

Problems and simulation exercises have been supplied for most of the chapters to facilitate self-evaluation of the understanding and design skills of the user of this book.

Notation

OpAmp	Operational amplifier
OA	Operational amplifier
OIA	Operational inverting amplifier
OVA	Operational voltage amplifier
OCA	Operational current amplifier
OFA	Operational floating amplifier
GA	General amplifier stage
VF	Voltage follower stage
CF	Current follower stage
CM	Current mirror stage
IA	Instrumentation amplifier
a	Temperature coefficient
A_v	Voltage gain
A_{vo}	DC voltage gain
β	Current gain of bipolar transistor
B_v	Voltage attenuation of feedback network
C	Capacitor value
Ch	Chopper
C_{ox}	Specific capacitance of gate oxide
C_M	Miller capacitor value
C_P	Parallel capacitor value
D	Distortion
f	Frequency
f_T	Transit frequency of a transistor
f_o	Zero-dB frequency
g_m	Transconductance of a transistor
i	Small-signal current
I	Current
I_B	Bias current

(continued)

I_C	Collector current
I_D	Drain current
I_E	Emitter current
I_S	Supply current
I_Q	Quiescent current
k	Boltzman's constant $K = \mu C_{ox} W/L$
L	Length of gate in MOS transistors
M	CMOS transistor
R	Resistor value
S	Signal
S	Switch
S_r	Slew rate
T	Generalized transistor
Q	Bipolar transistor
v	Small-signal voltage
V	Voltage
V_B	Bias voltage
V_{CC}	Positive supply voltage with bipolar transistors
V_{DD}	Positive supply voltage with MOS transistors
V_{EE}	Negative supply voltage with bipolar transistors
V_G	Generator voltage
V_{GS}	Gate-source voltage
V_{GT}	Active gate-source voltage $(V_{GS} - V_{TH})$
V_S	Total-supply voltage
V_{SN}	Negative supply voltage
V_{SP}	Positive supply voltage
V_{SS}	Negative supply voltage with MOS transistors
V_T	Thermal voltage kT/q
V_{TH}	Threshold voltage of MOS device
W	Width of gate in MOS transistors
μ	Mobility of change carriers

Extrinsic device parameters

R_L
C_L
C_M
$R_D \, R_C$
$R_G \, R_B$
$R_S \, R_E$

Intrinsic small-signal transistor parameters

$r_{ds} \, r_{ce} \, r_o$
$r_{gs} \, r_{be}$
$r_s \, r_e$
$c_{ds} \, c_{ce}$
$c_{gs} \, c_{be}$
$g_m \, g_m$
$\mu_n \, \mu_p$
$\beta_n \, \beta_p$

Chapter 1
Definition of Operational Amplifiers

Abstract In 1954, Tellegen introduced the concept of a universal active network element under the name of "ideal amplifier" (Rend. Sem. Mat. Fis. Milano 25:134–144, 1954). In this chapter a classification is given of ideal amplifiers on the basis of which ports are grounded.

In 1954, Tellegen introduced the concept of a universal active network element under the name of "ideal amplifier" [1]. The name "nullor," generally accepted now, was given to it by Carlin in 1964 [2]. The symbol of a nullor is shown in Fig. 1.1.

The nullor is defined as a two-port network element whose ports are called input and output ports and whose input voltage V_i and input current I_i are both zero, so:

$$V_i = 0,$$
$$I_i = 0 \tag{1.1}$$

The nullor concept only has significance if a passive network external to the nullor provides for a feedback from the output port into the input port [3]. The output voltage V_o and the output current I_o will be determined by the passive network elements in such a way that the input requirements $V_i = 0$, $I_i = 0$ are satisfied.

An accurate signal transfer requires, firstly, accurate passive components and secondly, a practical nullor realization which approximates $V_i = 0$, $I_i = 0$.

This implies that the nullor realization should have a high gain, a low input noise, and low offset voltage and current (see Sect. 2.1). All linear and nonlinear analog transfer functions can be implemented with nullor realizations and passive components.

We will now classify four nullor types on the basis of the number of ports which are floating, beginning with both ports grounded and ending with both ports floating. There are two main reasons for this kind of classification. Firstly, the larger the number of ports which are grounded the simpler the construction of the active device will be. Secondly, the larger the number of grounded ports the lower the number of possible feedback topologies will be allowed.

J. Huijsing, *Operational Amplifiers*, DOI 10.1007/978-3-319-28127-8_1

Fig. 1.1 A two-port
network composed of a
passive network and a
nullor

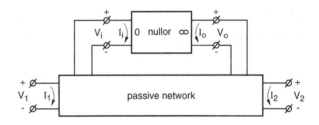

We will give each of the four nullor types a name which will be explained
later. The first one with two grounded ports will be called operational inverting
amplifier (OIA). The second one with the input port floating and output port
grounded will be called operational voltage amplifier (OVA). The third one with
the input port grounded and output port floating will be called operational current
amplifier (OCA). Finally, the fourth one with both ports floating will be called
operational floating amplifier (OFA). The adjective "operational" was coined by
John R. Ragazzini and his colleagues in a paper [4] published in 1947. This paper
described the basic properties of an OIA used with linear and nonlinear feedback.
The adjectives "inverting" (I), "voltage" (V), "current" (C), or "floating" (F), are
given by the present author to distinguish the four types of Operational Amplifiers according to their most striking attribute, as we will see in the next sections
of this chapter. The most popular one, the OVA will be shortened to OpAmp in
most parts of this book, where the distinction between the different types is not
needed.

1.1 Operational Inverting Amplifier

A practical approximation of a nullor having both ports grounded will be called an
"operational inverting amplifier" (OIA). The grounded input port makes the construction of the input stage relatively easy, because it only needs to function at one
voltage level [4, 5]. Similarly, the grounded output port makes it relatively easy to
construct an output stage having a high power efficiency, because the current return
path can be directly connected to the grounded supply voltages. The negative sign
(inverting) of the amplification factor makes it possible to obtain stable negative
feedback with passive components connected directly from the output to the input
port. The parallel connection of the feedback circuit at the input and output of the
amplifier results in a low virtual entrance impedance (see Sect. 3.1), suitable for
accurate current sensing at virtual zero input power, and a low exit impedance,
suitable for obtaining an accurate output voltage. The simplest realization is a
differential transistor pair.

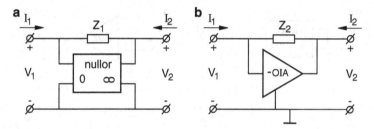

Fig. 1.2 Transimpedance amplifier (**a**) with a nullor symbol and (**b**) with an OIA symbol

1.1.1 Current-to-Voltage Converter

The most simple application of an OIA is the transimpedance amplifier or current-to-voltage transactor. This circuit is shown in Fig. 1.2a with a symbol of a nullor and in Fig. 1.2b with a practical symbol of an OIA.

The current-to-voltage transfer factor

$$Z_t = V_2/I_1 = -Z_1, \quad \text{at} \quad V_1 = 0 \tag{1.2}$$

can be accurately determined by $-Z_1$ if the OIA satisfies two requirements: firstly, a high gain, and secondly, a low input offset voltage and offset current. A high gain also assures low entrance and exit impedances.

1.2 Operational Voltage Amplifier

A practical nullor approach having only the output port grounded and the input port floating [6, 7] will be called "operational voltage amplifier" (OVA) or OpAmp. Currently, it is the most widely applied universal active device. The floating character of the input port imposes special demands on the construction of the input circuit, as will be discussed in Sects. 4.3 and 4.4. The floating input port allows series coupling of negative feedback.

This results in a high entrance impedance suitable for accurate voltage sensing at virtual zero input power. The parallel coupling of the feedback network with the grounded output port assures a low exit impedance.

1.2.1 Non-inverting Voltage Amplifier

The most essential application of the OVA is the non-inverting voltage amplifier or voltage-to-voltage transactor. The circuit is drawn in Fig. 1.3a with a nullor symbol and in Fig. 1.3b with a practical amplifier symbol for an OVA.

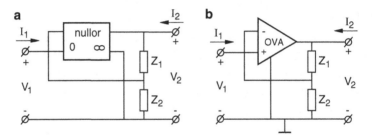

Fig. 1.3 Voltage amplifier (**a**) with a nullor symbol and (**b**) with an OVA symbol

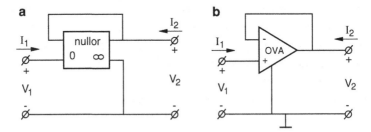

Fig. 1.4 Voltage follower (**a**) with a nullor symbol and (**b**) with an OVA symbol

The voltage amplification factor

$$A_u = V_2/V_1 = (Z_1 + Z_2)/Z_2, \quad \text{at} \quad I_1 = 0 \tag{1.3}$$

can be accurately determined by the impedance ratio $(Z_1+Z_2)/Z_2$ if the OVA satisfies the requirements: a high gain, a low input offset voltage and current, independent of the common-mode voltage of the input port, and a low input bias current. A high gain assures a high entrance impedance and a low exit impedance.

1.2.2 Voltage Follower

A special situation occurs if the OVA has its negative input terminal connected with the output terminal. We will call such a device a "voltage follower" (VF), because the exit voltage follows the entrance voltage. The construction of a universal active device with this connection may be simpler than without this connection, because no voltage shifting is required between the input and output. The VF circuit is given in Fig. 1.4b with an OVA symbol.

The voltage follower has the unique property that the voltage amplification factor

$$A_u = V_2/V_1 = 1 \quad \text{at} \quad I_1 = 0 \tag{1.4}$$

precisely equals plus unity, independently of any passive components, if the amplifier satisfies the three requirements: high gain, low input offset voltage and current, and a low input bias current. The accuracy of the plus-unity voltage transfer is not limited by the tolerances of any passive components. Note that the accuracy of the minus-unity voltage transfer of a voltage inverter does depend on the tolerance of a ratio of two impedances, as shown in Sect. 3.1.2, Fig. 3.2.

The voltage follower uses the most important attribute of a floating input port, viz. that the potential at one input terminal precisely follows the potential at the other input terminal.

1.3   Operational Current Amplifier

A nullor approximation which has only the input port grounded and the output port floating [8] will be called an "operational current amplifier" (OCA). An output port with a floating character is difficult to construct, as we will see in Chap. 9. However, this labor is rewarded for applications requiring a high output impedance by using feedback in series coupling with the output port. This series feedback results in an exit with a current-source character, while the grounded input port with parallel feedback assures a low entrance impedance.

1.3.1   Current Amplifier

The most elementary application of the OCA is a current amplifier, whose circuit is shown in Fig. 1.5b with an OCA symbol. The amplifier is the current dualogon of the voltage amplifier of Fig. 1.5b. The amplification factor

$$A_1 = -I_2/I_1 = -(Y_1 + Y_2)/Y_1, \quad \text{at} \quad V_1 = 0, \tag{1.5}$$

is accurately determined by the admittance ratio $(Y_1 + Y_2)/Y_1$ if the amplifier satisfies: a high gain, a low input offset voltage and current, and an output port with a

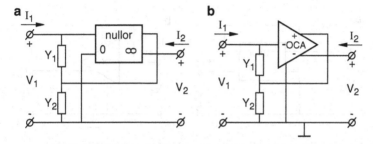

Fig. 1.5 Current amplifier (**a**) with a nullor symbol and (**b**) with an OCA symbol

low output bias current, because this current is directly added to the output. Note, that the minus sign merely results from the choice of the opposite current notations of I_1 and I_2. The low entrance impedance allows current sensing at a low entrance voltage V_1.

The current source character at the exit yields an accurate current transfer independently of the load impedance.

1.3.2   Current Follower

A special situation occurs if the negative input terminal of the OCA is connected with the output terminal. We will call such a configuration a "current follower" (CF), because the exit current follows the entrance current. This circuit is the current dualogon of the voltage follower. The circuit is drawn in Fig. 1.6a with a nullor symbol and in Fig. 1.6b with an OCA symbol.

The current follower has the unique attribute that the current amplification factor

$$A_i = -I_2/I_1 = 1, \quad \text{at} \quad V_1 = 0, \tag{1.6}$$

precisely equals plus unity, independently of any passive component values, if the gain is high, the input offset voltage and current is low, and if the output port has a low bias current. In contrast, the current-amplification factor of a current mirror, which nominally is minus unity, does depend on the matching of two passive elements (see Sect. 3.4).

Note that the minus sign in Eq. 1.6 is needed because the output current I_2 is defined in the opposite direction regarding I_2, when the current is being transferred through the CF.

The current-follower action reveals the most important attribute of a floating output port, namely that the current which flows into one output terminal is precisely followed by the current which flows out of the other output terminal. This attribute is the very dualogon of the voltage-follower action of a floating input port.

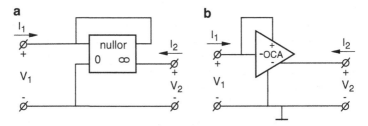

Fig. 1.6 Current follower (**a**) with a nullor symbol and (**b**) with an OCA symbol

1.4   Operational Floating Amplifier

A nullor approximation which has both the input and the output ports floating [9] will be called an "operational floating amplifier" (OFA) [8, 9]. Earlier it was called a "monolithic nullor" [10], or second generation current conveyer [11]. The construction of such a universal active device combines the demands of both floating input and output ports.

The OFA provides the maximum freedom for composing feedback configurations. With simple passive components it is possible to apply negative feedback in series with input and output ports, which results in both a high entrance and exit impedance.

1.4.1   Voltage-to-Current Converter

A specific application of the OFA is the voltage-to-current converter or transadmittance amplifier. Such a circuit is shown in Fig. 1.7a with a nullor symbol and in Fig. 1.7b with a practical OFA symbol. The voltage-to-current transfer factor

$$Y_t = I_2/V_1 = Y_1, \quad \text{at} \quad I_1 = 0 \tag{1.7}$$

will be accurately determined by one admittance $-Y_1$ if the amplifier satisfies four requirements: high gain, low input offset voltage and current, low input bias current, and low output bias current. The negative feedback in series with both ports ensures a high entrance impedance and a high exit impedance, which gives the transactor a voltage-sensing entrance and a current-source exit character.

1.4.2   Voltage and Current Follower

In fact, the transadmittance amplifier of Fig. 1.7a, b does not apply all potentialities of the OFA. It is applied in the special case in which the lower terminal of the input

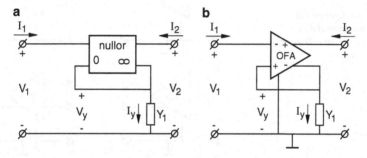

Fig. 1.7 Transadmittance amplifier (**a**) with a nullor symbol and (**b**) with an OFA symbol

port is connected with the lower terminal of the output port. This connection may simplify the construction of the OFA, because no voltage level shifter is needed between the input circuit and one output terminal, as we will see in Sect. 9.2. An OFA with this connection can be called a "voltage and current follower" (VCF).

A nullor which has this connection is also called a "three-terminal nullor" or a "unitor" [12]. In Fig. 1.7b the VCF firstly acts like a voltage follower, accurately transferring the entrance voltage V_1 towards the voltage V_y on the upper side of the admittance Y_1, and secondly like a current follower, accurately transferring the current I_y through the admittance Y_1 towards the current I_2 at the upper exit terminal.

1.5 Conclusion

A classification of universal active devices has been given on the basis of the number of ports which are connected to ground or to each other. The more ports not internally connected, the more freedom there is in the choice of the feedback configuration although this creates more complications with the construction of the device. Figure 1.8 presents an overview of the four types of active devices with different grounding schemes while Fig. 1.9 gives the three types of followers with one interconnection between the ports.

Four transfer functions are particularly suited to accurate signal transfer. They can be implemented with the four basic types of active devices:

Current-to-voltage converter with an OIA
Voltage-to-voltage converter with an OVA
Current-to-current converter with an OCA
Voltage-to-current converter with an OFA

In the ideal case, their signal transfer is independent of the source and load impedances. Moreover, the signal transfer depends on the theoretical minimum number of passive components.

Fig. 1.8 (a) Operational inverting amplifier (OIA), (b) operational voltage amplifier (OVA) or OpAmp, (c) operational current amplifier (OCA), and (d) operational floating amplifier (OFA)

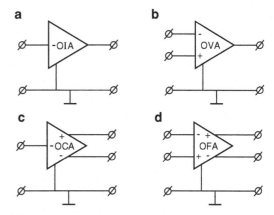

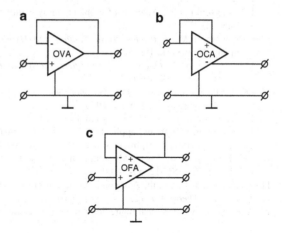

Fig. 1.9 (a) Voltage follower (VF) with an OVA symbol, (b) current follower (CF) with an OCA symbol, and (c) voltage and current follower (VCF) with an OFA symbol

Three transfer functions have the unique attribute that their accuracy is independent of any passive components. These types are:

Voltage follower (VF) with an OVA
Current follower (CF) with an OCA
Voltage and current follower (VCF) with an OFA

If each of these seven types of functions were realized with nullor approximations which do not have the right ports floating, a larger number of passive components and/or active devices is needed. Such realizations are less accurate and usually more expensive than the realizations with the right kind of active devices.

The requirements to be satisfied by the universal active devices are:

1. High gain, with a stable high-frequency close-loop feedback behavior.
2. Low input offset and noise voltage and current.
3. When a floating input port is needed, a low input bias current.
4. When a floating output is needed, a low output bias current.

It is remarkable that the grades of perfection of the four requirements mentioned have no absolute limits. This means that there is a large area of technical exploration present for the design of active electronic building blocks.

References

1. B.D.H. Tellegen, La recherche pour une serie complete d'elements de circuit ideaux non-lineaires. Rend. Sem. Mat. Fis. Milano **25**, 134–144 (1954)
2. H.J. Carlin, Singular network elements. IEEE Trans. Circ. Theory **CT-11**, 67–72 (1964)
3. B.D.H. Tellegen, On nullators and norators. IEEE Trans. Circ. Theory **CT-13**, 466–469 (1966)

4. J.R. Ragazzini et al., Analysis of problems in dynamics by electronic circuits. Proc. IRE **35**, 444–452 (1947)
5. G.A. Korn, F.M. Korn, *Electronic Analog and Hybrid Computers* (McGraw Hill Book Company, New York, 1964)
6. J.G. Graeme et al., *Operational Amplifiers, Design and Applications* (McGraw Hill Book Company, New York, 1971)
7. J.G. Graeme, *Applications of Operational Amplifiers, Third Generation Techniques* (McGraw Hill Book Company, New York, 1973)
8. J.H. Huijsing, Operational floating amplifier. IEE Proc. **137**(2), 131–136 (1990)
9. J.H. Huijsing, Design and applications of the operational floating amplifier (OFA): the most universal operational amplifier. Analog Integr. Circ. Sig. Process **4**, 115–129 (1993)
10. J.H. Huijsing, J. de Korte, Monolithic nullor – a universal active network element. IEEE J. Solid-St. Circ. **SC-12**, 59–64 (1977)
11. A.S. Sedra, K.C. Smith, A second generation current conveyer and its applications. IEEE Trans. Circ. Theory **CT-17**, 132–134 (1970)
12. A.W. Keen, A topological nonreciprocal network element. Proc. IRE **47**, 1148–1150 (1959)

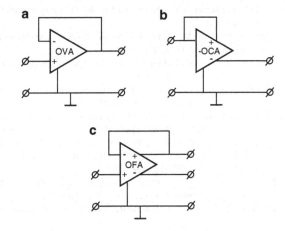

Chapter 2
Macromodels

Abstract The qualities of the universal active devices mentioned in Chap. 1 can be specified by their macromodels or equivalent circuits and by transfer matrices. These representations should contain all elements for quantifying the four qualities of gain, offset, and if applicable, the bias current of input and output ports. Macromodels may also include the HF parameters and nonlinear effects.

The qualities of the universal active devices mentioned in Chap. 1 can be specified by their macromodels or equivalent circuits and by transfer matrices. These representations should contain all elements for quantifying the four qualities of gain, offset, and if applicable, the bias current of input and output ports. Macromodels may also include the HF parameters and nonlinear effects.

The first four sections of this chapter contain only a linear representation of the macromodels of the four types of OpAmps. The elements may include a complex description to include the HF behavior.

Nonlinear behavior will be represented by SPICE macromodels for OpAmps in Sect. 2.5.

Measurement techniques for operational amplifiers are discussed in Sect. 2.6.

2.1 Operational Inverting Amplifier

The operational inverting amplifier is a three-terminal network in which one terminal is grounded. The equivalent circuit of Fig. 2.1 contains all elements of a unilateral driven source. The simplest unidirectional realization is a single transistor.

© Springer International Publishing Switzerland 2017
J. Huijsing, *Operational Amplifiers*, DOI 10.1007/978-3-319-28127-8_2

Fig. 2.1 Equivalent circuit of an operational inverting amplifier (OIA)

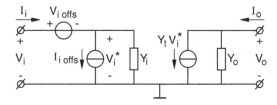

2.1.1 Definition of Offset Voltage and Current, Input and Output Impedance, Transconductance

For the purpose of standardization, all elements are chosen such that the currents at the terminals can be expressed in the voltages at the terminals. Thus, most elements are admittances. The main element is a voltage-controlled current source with a transadmittance Y_t.

Together with the input admittance Y_i and output admittance Y_o, these three admittances form the essential ingredients of an active device. The admittances may be taken as a Laplace transform to represent the high-frequency response. No elements are added to represent internal feedback because of the reasonable simplification that a universal active device with more than one internal cascaded amplifier stage is unilateral. And if there is internal feedback, for instance by heat transfer on the chip, this should be considered separately.

An offset voltage source V_{ioffs} and an offset current source I_{ioffs} have been added on the input side. They represent all additive DC errors of the device. In addition, they may include all noise quantities. The input voltage after subtraction of the offset voltage has been denoted by V_i^* (with asterisk).

The errors of an active device application directly follow from the deviations from the input requirements $V_i = 0$ and $V_i = 0$. For that reason, we will choose a mathematical description of the device which directly produces the deviation from the ideal behavior. Such a description for each of the different kinds of devices will be called an error matrix. In the case of an OIA, being a two-port element, the error matrix is equal to the following chain matrix:

$$\begin{vmatrix} V_i \\ I_i \end{vmatrix} = \begin{vmatrix} 1/Y_t & Y_o/Y_t \\ Y_i/Y_t & Y_iY_o/Y_t \end{vmatrix} \begin{vmatrix} I_o \\ V_o \end{vmatrix} + \begin{vmatrix} V_{ioffs} \\ I_{ioffs} \end{vmatrix} \tag{2.1}$$

The error matrix clearly shows the requirements to be met by the OIA.

Firstly, a high transadmittance Y_t is needed. This also includes a high voltage amplification factor $A_v = -Y_t/Y_o$ and a high current amplification factor $A_i = Y_t/Y_i$. Secondly, the input offset voltage V_{ioffs} and current I_{ioffs} should be low, and so should be the spectral input noise voltage V_{in} and current I_{in}, which denote the noise components of the offset quantities.

2.2 Operational Voltage Amplifier

The operational voltage amplifier is a four-terminal network of which one terminal is grounded. An equivalent circuit is drawn in Fig. 2.2. For an adequate description of the floating character of the input port we should distinguish the differential-mode (DM) input voltage $V_{id} = V_{i1} - V_{i2}$ and current $I_{id} = (I_{i1} - I_{i2})/2$ from the common-mode (CM) input voltage $V_{ic} = (V_{i1} + V_{i2})/2$ and current $I_{ic} = (I_1 + I_2)/2$. The simplest realization is a differential transistor pair with a common tail current source.

2.2.1 Definition of Input Bias Current, Input Common-Mode Rejection Ratio

The basic three elements, the admittances Y_t, Y_{idd}, and Y_o, are the normal ones in each active device. They have already been discussed along with the equivalent circuit of the OIA. In addition, there are the common-mode input admittances Y_{ic1} and Y_{ic2} and two common-mode input bias-current sources with an equal value I_{ibias}. Furthermore, the CM input voltage V_{ic}^* causes an output current $V_{ic}^* Y_t/H_i$ in addition to the current $V_{id}^* Y_t$ of the main voltage-controlled current source. This is the result of a crosstalk of the CM input voltage V_{ic}^* on the DM input voltage V_{id}^*. The factor H_i is called the "common-mode rejection ratio" (CMRR) $H_i = (\delta V_{ic}^*/\delta V_{id}^*)_{1o,Vo}$. The reciprocal factor $1/H_i$ can be called the "common-mode crosstalk ratio" (CMCR). All these additional elements show the nonideal floating character of the input port.

The main errors of an OVA are quantified by the following error matrix (Eq. 2.2):

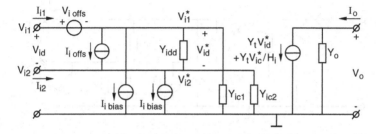

Fig. 2.2 Equivalent circuit of an operational voltage amplifier (OVA)

$$\begin{vmatrix} V_{id} \\ I_{id} \\ I_{ic} \end{vmatrix} = \begin{vmatrix} 1/Y_t & Y_o/Y_t & 1/H_i \\ Y_{id}/Y_t & Y_{id}Y_o/Y_t & Y_{id}/H_i \\ \cdot & \cdot & Y_{ic} \end{vmatrix} \begin{vmatrix} I_o \\ -V_o \\ V_{ic}^* \end{vmatrix} + \begin{vmatrix} V_{ioffs} \\ I_{ioffs} \\ I_{ibias} \end{vmatrix} \qquad (2.2)$$

with : $V_{id} = V_{i1} - V_{i2}$
$V_{ic} = (V_{i1} + V_{i2})/2$
$I_{id} = (I_{i1} - I_{i2})/2$
$I_{ic} = (I_{i1} + I_{i2})/2$
$Y_{id} \simeq Y_{idd}$
$Y_{ic} = (Y_{ic1} + Y_{ic2})/2$
$V_{ic}^* = V_{ic} - V_{ioffs}/2 \simeq V_{ic}$

The four elements in the upper-left part of the matrix represent the basic description of any controlled source, as discussed along with the OIA. The right-hand part of the matrix contains all errors caused by the nonideal floating character of the input port. An OVA should have a high value of the CMRR H_i and a low input admittance Y_{ic}. The dots in the matrix represent normally negligible effects.

Finally, the input bias current I_{ibias}, which is also a result of the nonideal isolating or floating character of the input port should be low and is placed in the separate column of additive error sources.

2.3   Operational Current Amplifier

The operational current amplifier is a four-terminal device in which one input terminal is grounded. An equivalent circuit is shown in Fig. 2.3. The floating character of the output port can best be emphasized by distinguishing the differential-mode output voltage $V_{od} = V_{o1} - V_{o2}$ and current $I_{od} = I_{o1} - I_{o2}$ from the common-mode output voltage $V_{oc} = (V_{o1} + V_{o2})/2$ and current $I_{oc} = (I_{o1} + I_{o2})/2$. The simplest realization is a differential transistor pair with a common tail current source.

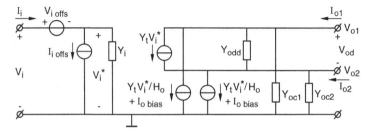

Fig. 2.3 Equivalent circuit of an operational current amplifier (OCA)

2.3.1 Definition of Output Bias Current, Output Common-Mode Current Rejection Ratio

In addition to the elements of an OIA, the equivalent circuit contains the elements which express the parasitic liaisons between the output port and ground. Firstly, there are the common-mode output admittances Y_{oc1} and Y_{oc2}. Secondly, we have two output bias sources with an equal value I_{obias}. Thirdly, these bias currents are modulated as a function of the input voltage V_i^* with a transconductance Y_t/H_o.

The quantity H_o will be defined as the output common-mode current rejection ratio (CMCRR) for the output currents. The reciprocal quantity $1/H_o$ will be called the output common-mode current crosstalk ratio (CMCCR). It describes the current crosstalk of a DM output current $I_{od} = V_i^* Y_t$ on the CM output current $I_{oc} = V_i Y_t/H_o$ at $V_{od} = V_{oc} = 0$. The CMCRR for output currents is dual in regard to the CMRR for input voltages.

The main errors are quantified in the following error matrix (Eq. 2.3):

$$\begin{vmatrix} V_i \\ I_i \\ I_{oc} \end{vmatrix} = \begin{vmatrix} 1/Y_t & Y_{od}/Y_t & \cdot \\ Y_i/Y_t & Y_i Y_{od}/Y_t & \cdot \\ 1/H & Y_{od}/H_o & Y_{oc} \end{vmatrix} \cdot \begin{vmatrix} I_{od} \\ -V_{od} \\ V_{oc} \end{vmatrix} + \begin{vmatrix} V_{ioffs} \\ I_{ioffs} \\ I_{ibias} \end{vmatrix} \tag{2.3}$$

with : $V_{od} = V_{o1} - V_{o2}$
$I_{od} = I_{o1} - I_{o2}$
$V_{oc} = (V_{o1} + V_{o2})/2$
$I_{oc} = (I_{o1} + I_{o2})/2$
$Y_{od} \simeq Y_{odd}$
$Y_{oc} = (Y_{oc1} + Y_{oc2})/2$

Again, the four elements in the upper-left part of the matrix represent the errors of any controlled source, as described along with the OIA. The elements of the lower row represent the nonideal floating character of the output port. An OCA should have a high CMCRR H_o and a low CM output admittance Y_{oc}. Moreover, the output bias current I_{obias} must be placed in the column of additive error sources. The dots in the matrix stand for negligible effects.

2.4 Operational Floating Amplifier

The operational floating amplifier is a five-terminal network in which one terminal is grounded. An equivalent circuit is shown in Fig. 2.4. The floating character of the input and output ports can best be expressed by distinguishing DM and CM input and output voltages and currents, as was done with the OVA and OCA.

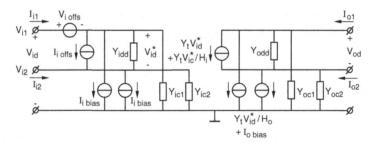

Fig. 2.4 Equivalent circuit of an operational floating amplifier (OFA)

2.4.1 Using All Definitions

The equivalent circuit contains all elements of a driven source:
$Y_t V_{id1}^*$, Y_{idd} and Y_{odd}, the offset sources V_{ioffs} and I_{ioffs}; all elements which represent the liaisons of the input port with the common ground: $Y_{ic1}, Y_{ic2}, I_{ibias}(2x)$ and $V_{ic}^* Y_t/H_i$; and all elements which have the same function for the output port: $Y_{oc1}, Y_{oc2}, I_{obias}(2x)$ and $V_{id}^* Y_t/H_o(2x)$. These elements have been discussed with the OIA, OVA, and OCA. The simplest realization is a differential transistor pair with a common tail current source.

$$
\begin{vmatrix} V_{id} \\ I_{id} \\ I_{ic} \\ I_{oc} \end{vmatrix} = \begin{vmatrix} 1/Y_t & Y_{od}/Y_t & 1/H_i & \cdot \\ Y_{id}/Y_t & Y_{id}Y_{od}/Y_t & Y_{id}/H_i & \cdot \\ \cdot & \cdot & Y_{ic} & \cdot \\ 1/H_o & Y_{od}/H_o & \cdot & Y_{oc} \end{vmatrix} \begin{vmatrix} I_{od} \\ -V_{od} \\ V_{ic}^* \\ V_{oc} \end{vmatrix} + \begin{vmatrix} V_{ioffs} \\ I_{ioffs} \\ I_{ibias} \\ I_{obias} \end{vmatrix} \quad (2.4)
$$

The errors of the OFA are quantified by the error matrix (Eq. 2.4):

The four elements in the upper-left part of the matrix represent the errors of any controlled source as discussed along with the OIA. The elements of the third column represent the nonideal floating character of the input port, described along with the OVA, while the elements of the fourth row do the same with regard to the output port, described along with the OCA. The elements in the additional row represent all offset and bias sources. The dots in the matrix stand for negligible effects.

2.5 Macromodels in SPICE

In Sects. 2.1–2.4 we have presented linear circuit models for the four operational amplifier types OIA, OVA, OCA, and OFA. These are theoretically correct but not always practical. The main practical shortcoming is the lack of nonlinear behavior description. It is desirable for shortening the simulation time in SPICE simulations of large systems with many operational amplifiers, to have relatively simple macromodels, which nevertheless do take into account the nonlinear behavior, saturation effects, and slew rate.

2.5.1 Macromodel Mathematical

In some cases, like the use of OpAmps in switched capacitance circuits, it is often sufficient to have only the nonlinear behavior of the input stage modeled. In some SPICE programs this can be entered by a formula description. An example of such a description is given by Lin et al. in Ref. [1] of a circuit like Fig. 6.14b and shown in Fig. 2.5.

The description consists of two blocks. The first block describes in a first-order approximation the nonlinear behavior of the input stage. The second block describes the frequency response with two-poles. With this model the slewing and settling behavior can be modeled in first-order approximation. For further details see [1].

2.5.2 Macromodel Miller-Compensated

If we need to take a larger number of nonidealities and nonlinearities into account, such as input offset and noise and saturation effects we need to use a more extended macromodel.

As an example the SPICE macromodel of Boyle et al. [2] is shown in Fig. 2.6. It models the input stage by using the differential pair M_1 and M_2, diodes D_3 and D_4 to

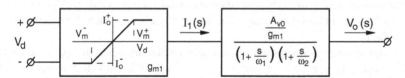

Fig. 2.5 A two-pole model taking into account the slew rate limitation. The maximum available currents of the input stage are given by I_0^+ and I_0^-. The transfer has the characteristic slope of g_{ml}

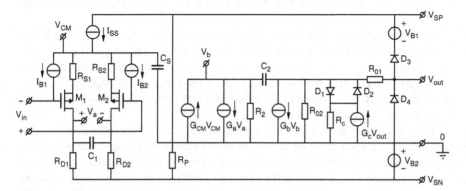

Fig. 2.6 SPICE macromodel of a Miller-compensated two-stage CMOS operational amplifier according to Boyle et al. The input stage is modeled by a differential transistor pair, while the output stage is built up by driven sources and saturation diodes

model saturation effects at the output, and a Miller (Chap. 6) compensation with C_2 across the output stage. The parasitic currents of the protection diodes at the input are modeled by means of current sources I_{B1} and I_{B2}. The output current is limited by the diodes D_1 and D_2 together with the series resistor R_{01} and a reproduction of the output voltage across $R_C G_C V_{out}$. The same model can be used with a bipolar input stage for a bipolar amplifier. The input bias current sources may be left out in that case, because they are already built in the bipolar transistor models.

The macromodel can also be used when a voltage follower is added at the output as a third stage like in the μA741 of Fig. 7.21.

More precise macromodels have been presented by Mark Alexander and Derek Bowers [3], and others.

2.5.3 Macromodel Nested-Miller-Compensated

The effects of common-mode-depending bias currents and saturation become even more serious in the application of low-voltage OpAmps with rail-to-rail input and output voltage ranges, such as the bipolar OpAmp NE 5234, as explained with Fig. 7.41. A SPICE macro-model should incorporate these effects. An example of such a model is given by Feyes et al. [4] using controlled sources and diodes. However, a simpler model is shaped if we just equip the input and output stages with transistors using strongly simplified transistor models, while we use controlled sources for the intermediate stage similar to Boyle's approach. This is shown in Fig. 2.7a, b.

The input stage is composed of two complementary transistor pairs and a tail current selector, as explained with Fig. 4.21. Two different offset sources can be inserted, one for each pair. The output stage is modeled together with its R-R saturation properties by two complementary bipolar transistors Q_{11} and Q_{12}, and a translinear class-AB loop through D_{11}, D_{12}, and a floating supply source replica $V_{SP} - V_{SN}$, as explained with Fig. 5.19b. The diodes D_{13} and D_{14} prevent internal overdriving.

The intermediate stage is linearly modeled by a simplified transistor model. The total macromodel has three poles: one at the output determined by the load capacitance, one at the input of the output stage determined by the diffusion capacitors of the output transistors, and one at the input of the intermediate stage determined by $R_{22} C_{22}$. These three poles are handled by nested-Miller-compensation through C_{M1} and C_{M2}, as explained with Fig. 6.22.

With this model the change in input bias current and offset voltage and saturation effects are properly modeled when the common-mode input voltage passes from below the negative rail up even across the positive rail. The noise is also being properly modeled together with the slew rate and frequency characteristic. At the output a proper saturation behavior near the negative and positive rail voltage is modeled.

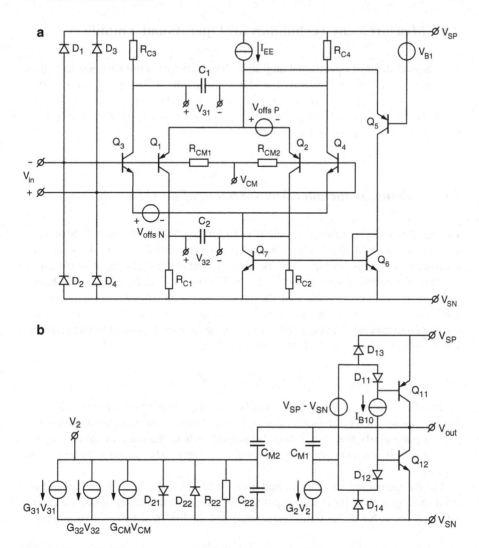

Fig. 2.7 (**a**) A SPICE macromodel of a low-voltage bipolar R-R input stage for a three-stage bipolar OpAmp. (**b**) A SPICE macromodel of a low-voltage-nested-Miller-compensated intermediate and output stage of a three stage bipolar OpAmp with R-R output

2.5.4 Conclusion

In conclusion, in SPICE simulations of large systems where many OpAmps and other components function together the simulation time can be much shortened by using simplified macromodels for the operational amplifiers. These macromodels often use transistor models at the input and at the output, where the largest non-linearities occur, while linear controlled current sources are used in the central part of the model.

2.6   Measurement Techniques for Operational Amplifiers

The measurement of operational amplifier characteristics is not easy because their parameters can seldom be directly found in an open circuit. The gain is so large that any offset and noise will drive the output in complete saturation. Hence the OpAmp has to be placed in a stable feedback measurement setup and the parameters to be measured have to be indirectly measured.

2.6.1   Transconductance Measurement of an OTA

Exempt from the mentioned problem is the measuring of the transconductance $Y_t = G_m$ of an OTA (Sect. 7.1). This value is so low that it can be easily directly measured. A voltage source V_s can be directly connected to the input as shown in the first measurement approach of Fig. 2.8. The top-top value should not be larger than about 20 mV, in order that the input stage is not being overdriven in its nonlinear region. At the output a relatively small resistor, i.e., $R_L = 1 \text{ k}\Omega$ can be connected to ground. The G_m can be found as the ratio of the output voltage and the input voltage divided by the load resistance R_L. So:

$$Y_t = G_m = V_o/(V_i R_L) \tag{2.5}$$

The output load resistor R_L may not be too large. Otherwise the parasitic open output conductance Y_o in parallel with $1/R_L$ will make the measurement inaccurate. This is particularly the case at high frequencies where the parallel output capacitance C_L will attenuate the output voltage strongly. The situation is drawn in Fig. 2.8.

For the measurement of offset and noise of an OTA the same setup can still be used. At a grounded input the output voltage indicates the offset through a calculation by the value of the transconductance and the load resistance according to:

$$V_{offs} = V_o/(G_m R_L) \tag{2.6}$$

However, if we want to measure the open voltage gain A_V, the above setup does not fit anymore. The load resistance R_L has to be chosen larger than the open output resistance $R_o = 1/G_o$ and the offset would fully drive the output to one of the

Fig. 2.8 Measurement of the transconductance $Y_t = G_m$ of an OTA

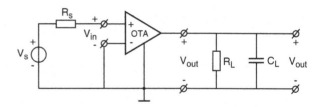

supply rails. Fiddling with a DC input bias voltage would help in some cases, but cannot be a standard measurement procedure. This can only be used in SPICE simulations as we may expect perfectly matched transistors and an inherent low offset.

2.6.2 Voltage Gain Measurement of an OpAmp

The second closest approach is to use feedback for proper biasing in such a way that the negative input terminal is fed back by a resistor R_1 with a value larger than the open output resistance R_o but capacitively grounded. The situation is depicted in Fig. 2.9.

The external loop contains an extra pole at a time constant $\tau_1 = R_1C_1$. This must be larger than the DC gain 2 A_V multiplied by the dominating time constant $\tau_d = R_oC_p$ of the open-loop amplifier, to satisfy the stability of the loop. As the dominating pole frequency $1/2\ \pi\tau_d$ may easily be as low as 1.6 kHz at $R_o = 10$ MΩ and $C_p = 10$ pF, the external pole frequency τ_1 must be lower than 0.008 Hz at a voltage gain A_v of 10^5. When we choose $R_1 = 100$ MΩ, C_1 must have a value larger than 0.2 μF. The voltage gain as a function of the frequency (Bode plot) can now be measured as $A_V = V_o/V_i$ with a network analyzer with active probes ($C_{in} < 3$ pF). If we do not take τ_1 large enough, the phase margin could become too low and the circuit may multivibrate at a low frequency.

The above solution may work well with most CMOS OTAs. But if the open output resistor R_o becomes larger, a problem arises as the value of R_1 is limited.

If the input stage has bipolar transistors, the feedback resistor has to supply the input bias current. This requires a much lower value of R_1 and a higher value of C_1.

A possible solution will then be to insert a single-transistor voltage-follower buffer between the output and the connection to the feedback resistor R_1 in order not to load the open output too much. In SPICE simulations we can easily solve the problem by replacing R_1 by a large inductance L_1, so that there is no DC loss across this element.

With an OpAmp with two or more gain stages the open output impedance is normally so low that a much lower value of R_1 is allowed.

Fig. 2.9 Measuring the open-loop frequency response $A_V = V_o/V_i$ of an operational amplifier

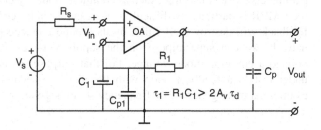

Fig. 2.10 Inverting
feedback configuration for
measuring the open-loop
frequency response
$A_V = V_o/V_i$ of an operational
amplifier

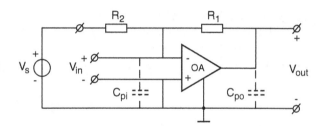

2.6.3   Voltage Gain and Offset Measurements of an OpAmp

A third, simple, and effective solution is to use an inverting amplifier feedback
configuration as shown in Fig. 2.10.

The idea is to apply the amplifier in an inverting configuration with a minus
unity gain at $R_2 = 10\,k\Omega, R_1 = 10\,k\Omega$, or i.g. with a gain of -100 at $R_2 = 1\,k\Omega$,
$R_1 = 100\,k\Omega$ inverting configuration. The voltage gain can be measured by a
network analyzer by simply connecting one active probe to the output and the
other active probe to the input of the amplifier.

The only care we must take is that the probe capacitance C_{pi} at the input does not
cause a pole in the loop at a frequency where the loop gain is still larger than unity.
This should not be a problem at a strong feedback attenuation $B_v = R_2/R_1$. The
frequency characteristic can be measured with this third method up to several
hundred megahertz.

If the open output impedance R_o is high, then R_1 must also be chosen sufficiently
high to avoid degradation of the measured voltage gain. This inverting-amplifier
setup also allows us to simply measure the offset between the input terminals. The
input voltage noise can also be measured at the input in a frequency band where the
feedback is active, or $A_vB_v > 1$.

In conclusion, of these three simple measuring setups, the third measuring circuit
is quite powerful.

2.6.4   General Measurement Setup for an OpAmp

The above methods do not satisfy a general measurement setup in which more
parameters can be measured, such as common-mode rejection. The measurement of
the CMRR is particularly difficult as we cannot apply feedback for biasing, as the
feedback will destroy the CMRR by its ground connection (see Sect. 4.3). Alternatively we can superimpose on the supply voltages and the output reference
voltage a common-mode voltage. For that purpose we need to place the device
under test in a feedback loop with the aid of a second amplifier. As a fourth
example, a more general low-frequency setup is given in Fig. 2.11 [5].

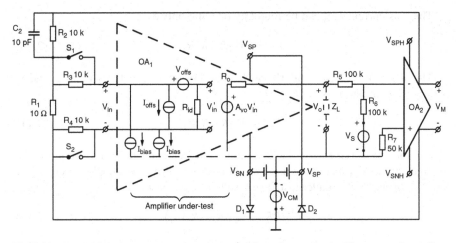

Fig. 2.11 General low-frequency measurement setup for voltage gain A_V, bias current I_{bias}, offset voltage V_{offs}, noise voltage V_n, and common-mode rejection CMRR of an operational amplifier OA_1 under test aided by a helping operational amplifier OA_2

The amplifier under test, OA_1, is supplied with positive and negative supply voltages, respectively V_{SP} and V_{SN}, while a common-mode voltage source V_{CM} moves the supply voltages and the output reference voltage in regard to the input voltages. A broadband helping amplifier OA_2 compares the output voltage V_o of the amplifier under test with a source voltage V_s and amplifies the difference to the output measurement voltage V_M. This voltage is being fed back to the input of the amplifier under test by an attenuation network $B_v = R_1/R_2$ and bias current measurement resistors R_3 and R_4. The capacitor C_2 provides phase lead to stabilize the long loop through two amplifiers. The low-frequency voltage gain can be measured by dividing V_s by $(R_2/R_1)V_M$, so:

$$A_V = V_S R_2 / R_1 V_M \qquad (2.7)$$

This can be done, for instance, by a network analyzer. Active probes need not be used, depending on the impedance levels of the source and output of OA_2. Load conditions can be changed by applying Z_L. The measurement of the frequency characteristic is correct up to about 1 MHz, which is the pole of the feedback network.

For an offset measurement the switches S_1 and S_2 have to be closed. The output voltage V_M represents the input offset voltage amplified by the inverse feedback attenuation.

$$V_{offs} = V_M R_1 / R_2 \qquad (2.8)$$

The spectral input noise can also be read from V_M.

$$V_{nin} = V_{nM} R_1 / R_2 \qquad (2.9)$$

The bias current I_{bias} can be found by opening only S_1.

$$I_{bias} = V_M R_1 / R_2 R_3 \qquad (2.10)$$

The offset current I_{offs} similarly can be found by opening both switches S_1 and S_2 and by using equal resistors $R_3 = R_4$.

$$I_{offs} = V_M R_1 / R_2 R_3 \qquad (2.11)$$

For the measurement of the common-mode rejection ratio a common-mode voltage source V_{CM} has to be used which is able to drive the supply-voltages V_{SP} and V_{SP} of the device under test. The result can be measured as:

$$CMRR = V_{CM} R_2 / (R_1 V_M) \qquad (2.12)$$

Note that the loop gain of the measurement setup is high ($A_{01} A_{02} R_2 / R_1$). The high frequency behavior of the loop is determined by the two dominating time constants of both amplifiers. The stability of the loop is ensured if the help amplifier OA_2 has a high dominating pole frequency, say 1 MHz, and if the feedforward capacitor C_2 corrects the phase lag of OA_2 above that frequency. The measurement setup may function up to frequencies of 1 MHz.

In conclusion, we have seen that it is not easy to measure OpAmp parameters for we have to take care of proper biasing simultaneously. For SPICE simulations, the second method can be used to measure frequency characteristics. The third method is powerful in practical situations. The measurement of the CMRR is particularly difficult, as we cannot apply feedback to properly bias the OpAmp as this destroys the CMRR. A proper method is to alternatively move the supply with a common-mode voltage. This is done in the fourth general measurement setup.

2.7   Problems and Simulation Exercises

2.7.1   Problem 2.1

Figure 2.6 shows a Boyle type macromodel developed for CMOS operational amplifiers. Starting from that picture, calculate the parameters of an operational amplifier with a differential gain $a_{VD} = 100$ dB, common-mode rejection ratio $CMRR = 90$ dB, unity gain bandwidth $f_{0dB} = 10$ MHz with a phase margin $\varphi_m = 76°$. The slew-rate specifications are $S_{rp} = 10$ V/µs for rising output voltage and $S_{rn} = 9$ V/µs for falling output voltage. A short-circuit current $I_{SC} = 7$ mA should be supplied at the output, with an output DC impedance $R_{out} = 1$ kΩ. Supply voltages are $V_{SP} - V_{SN} = 3$ V, and the macromodel should dissipate $P_d = 3$ mW. No capacitive or resistive load is present.

2.7.2 Solution

The gain in the first stage is chosen to be equal to unity:

$$R_{S1} = R_{D1} - 1/g_{m1} \tag{2.13}$$

As R_{D1} usually has values around $2/g_{m1}$ for convenient common mode input range,

$$R_{D1} = 2/g_{m1} \tag{2.14}$$

In order to calculate g_{m1}, weak inversion transistors will be considered for the input stage with an approximate g_{m1} given by

$$g_{m1} = \frac{I_{SS}}{100\text{mV}} \tag{2.15}$$

The tail current I_{SS} results from S_{rp} value and a convenient Miller capacitor:

$$\begin{aligned} C_2 &= 2\text{pF} \\ I_{SS} &= C_2 S_{rp} = 20\mu A \end{aligned} \tag{2.16}$$

Returning to R_{D1} and R_{S1} values:

$$\begin{aligned} R_{D1} &= 2/g_{m1} = 200\text{mV}/I_{SS} = 10\text{k}\Omega \\ R_{S1} &= R_{D1} - 1/g_{m1} = 1/g_{m1} = 5\text{k}\Omega \end{aligned} \tag{2.17}$$

Using these values, the unity gain bandwidth results:

$$f_{0dB} = 1/2\pi R_{D1} C_2 = 8\text{MHz} \tag{2.18}$$

which is close enough to the desired bandwidth. It can be increased by choosing a lower value for R_{D1}. The gain in the virtual intermediate stage of the model is larger than unity, given by its components G_a and R_2:

$$\begin{aligned} G_a &= 1/R_{D1} = 0.1\text{mS} \\ R_2 &= 100\text{k}\Omega \end{aligned} \tag{2.19}$$

The output impedance will be equal to R_{01} for frequencies above:

$$f_c = 1/2\pi R_{02} C_2 (1 + R_2 G_b) = 79\text{Hz} \tag{2.20}$$

Below this corner frequency, the DC output impedance is

$$R_{out} = R_{01} + R_{02} \tag{2.21}$$

Because the short-circuit current will be given by the ratio of a diode voltage and R_{01}, R_{01} should be less than DC output impedance:

$$R_{01} = 100\Omega \tag{2.22}$$

The rest of the gain up to $a_{VD} = 100$ dB is given by the third stage:

$$\begin{aligned} R_{02} &= R_{out} - R_{01} = 900\Omega \\ G_b &= a_{VD}R_{D1}/R_2R_{02} = 11.1S \end{aligned} \tag{2.23}$$

The gain of the R_cG_c voltage repeater should be equal to unity with an R_c value much smaller than R_{01}, so:

$$\begin{aligned} R_c &= 10\Omega \\ G_c &= 0/R_c = 0.1S \end{aligned} \tag{2.24}$$

This voltage repeater puts the difference between the voltages at the output and at the internal output on R_{01}, such as if a difference exists, the maximum current supplied is the short-circuit current:

$$I_{SC} = V_D/R_{01} = 0.6V/100\Omega = 6mA \tag{2.25}$$

To correct the short circuit current value to the desired 7 mA, the diode threshold voltage can be changed to 0.7 V for simulation purposes. The S_{rn} value is controlled by C_S:

$$C_S = (I_{SS}/S_{rn}) - C_2 = 0.22pF \tag{2.26}$$

The common-mode rejection ratio is given by G_{cm}:

$$G_{cm} = 1/R_{D1}CMRR = 3nS \tag{2.27}$$

For the desired phase margin, C_1 introduces a second pole

$$p2 = 1/2R_{D1}C_1 \tag{2.28}$$

which produces

$$C_1 = (C_2 \tan \varphi_m)/2 = 46fF \tag{2.29}$$

The power dissipated by the whole macromodel is modeled by R_p which takes the value:

$$R_P = \frac{V_{SP} + V_{SN}}{P_d - (V_{SP} + V_{SN})I_{SS}} = 1.0k\Omega \tag{2.30}$$

2.7.2   Solution

The gain in the first stage is chosen to be equal to unity:

$$R_{SI} = R_{DI} - 1/g_{m1} \tag{2.13}$$

As R_{DI} usually has values around $2/g_{m1}$ for convenient common mode input range,

$$R_{D1} = 2/g_{m1} \tag{2.14}$$

In order to calculate g_{m1}, weak inversion transistors will be considered for the input stage with an approximate g_{m1} given by

$$g_{ml} = \frac{I_{SS}}{100\text{mV}} \tag{2.15}$$

The tail current I_{SS} results from S_{rp} value and a convenient Miller capacitor:

$$\begin{aligned} C_2 &= 2\text{pF} \\ I_{SS} &= C_2 S_{rp} = 20\mu A \end{aligned} \tag{2.16}$$

Returning to R_{DI} and R_{SI} values:

$$\begin{aligned} R_{DI} &= 2/g_{ml} = 200\text{mV}/I_{SS} = 10\text{k}\Omega \\ R_{SI} &= R_{DI} - 1/g_{ml} = 1/g_{ml} = 5\text{k}\Omega \end{aligned} \tag{2.17}$$

Using these values, the unity gain bandwidth results:

$$f_{0dB} = 1/2\pi R_{DI} C_2 = 8\text{MHz} \tag{2.18}$$

which is close enough to the desired bandwidth. It can be increased by choosing a lower value for R_{DI}. The gain in the virtual intermediate stage of the model is larger than unity, given by its components G_a and R_2:

$$\begin{aligned} G_a &= 1/R_{DI} = 0.1\text{mS} \\ R_2 &= 100\text{k}\Omega \end{aligned} \tag{2.19}$$

The output impedance will be equal to R_{0I} for frequencies above:

$$f_c = 1/2\pi R_{02} C_2 (1 + R_2 G_b) = 79\text{Hz} \tag{2.20}$$

Below this corner frequency, the DC output impedance is

$$R_{out} = R_{01} + R_{02} \tag{2.21}$$

Because the short-circuit current will be given by the ratio of a diode voltage and R_{01}, R_{01} should be less than DC output impedance:

$$R_{01} = 100\Omega \qquad (2.22)$$

The rest of the gain up to $a_{VD} = 100$ dB is given by the third stage:

$$\begin{aligned} R_{02} &= R_{out} - R_{01} = 900\Omega \\ G_b &= a_{VD}R_{D1}/R_2R_{02} = 11.1\text{S} \end{aligned} \qquad (2.23)$$

The gain of the R_cG_c voltage repeater should be equal to unity with an R_c value much smaller than R_{01}, so:

$$\begin{aligned} R_c &= 10\Omega \\ G_c &= 0/R_c = 0.1\text{S} \end{aligned} \qquad (2.24)$$

This voltage repeater puts the difference between the voltages at the output and at the internal output on R_{01}, such as if a difference exists, the maximum current supplied is the short-circuit current:

$$I_{SC} = V_D/R_{01} = 0.6\text{V}/100\Omega = 6\text{mA} \qquad (2.25)$$

To correct the short circuit current value to the desired 7 mA, the diode threshold voltage can be changed to 0.7 V for simulation purposes. The S_{rn} value is controlled by C_S:

$$C_S = (I_{SS}/S_{rn}) - C_2 = 0.22\text{pF} \qquad (2.26)$$

The common-mode rejection ratio is given by G_{cm}:

$$G_{cm} = 1/R_{D1}CMRR = 3\text{nS} \qquad (2.27)$$

For the desired phase margin, C_1 introduces a second pole

$$p2 = 1/2R_{D1}C_1 \qquad (2.28)$$

which produces

$$C_1 = (C_2 \tan \varphi_m)/2 = 46\text{fF} \qquad (2.29)$$

The power dissipated by the whole macromodel is modeled by R_p which takes the value:

$$R_P = \frac{V_{SP} + V_{SN}}{P_d - (V_{SP} + V_{SN})I_{SS}} = 1.0\text{k}\Omega \qquad (2.30)$$

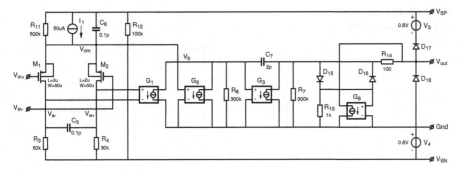

Fig. 2.12 Boyle macromodel for a two-stage operational amplifier

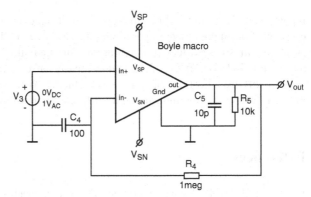

Fig. 2.13 Differential gain simulation for an OpAmp macromodel

2.7.3 Simulation Exercise 2.1

The Boyle macromodel depicted in Fig. 2.12 can be simulated for AC analysis using a simulation setup as shown in Fig. 2.13.

The use of ideal high-valued capacitors and resistors allow the simulator to solve correctly the biasing point for the transistors inside the operational amplifier. At DC, the amplifier is basically a repeater with the positive input connected to ground. What is the low limit frequency for a DC voltage gain simulation, considering the effect of C_4? What can be the additional benefits of replacing R_4 with an inductor?

2.7.4 Simulation Exercise 2.2

The circuit shown in Fig. 2.14 is used to simulate a Boyle macromodel of an operational amplifier for AC solution of common-mode to differential crosstalk.

Using this crosstalk gain and the differential gain, the common-mode rejection ratio can be calculated. At DC, the inductor keeps the operational amplifier in a

Fig. 2.14 Common-mode
to differential crosstalk
simulation

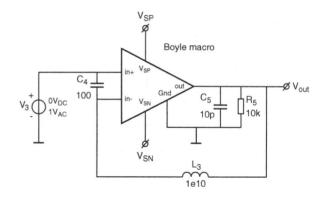

repeater configuration, while above a limit frequency given by the L_3C_4 time constant, both inputs of the operational amplifier are tied together and the circuit is placed in an open-loop configuration. Simulate this circuit using the Boyle macromodel shown in Fig. 2.12 and decrease the common-mode rejection ratio with 20 dB by adjusting the G_5 transconductor. What other circuit elements in Fig. 2.12 affect the common-mode to differential crosstalk?

References

1. J.C. Lin, J.H. Nevin, A modified time domain model for nonlinear analysis of an operational amplifier. IEEE J Solid-State Circuits **SC-21**(3), 478–483 (1986)
2. G.R. Boyle, B.M. Cohn, D.O. Pederson, J.E. Solomon, Macromodelling of integrated circuit operational amplifiers. IEEE J Solid-State Circuits **SC-9**(6), 353–364 (1974)
3. M. Alexander, D.F. Bowers, New Spice compatible OpAmp model boosts AC simulation accuracy. EDN 15 February 1990
4. J.H.A. Feyes, R. Hogervorst, J.H. Huijsing, *Macromodeling Operational Amplifiers*, vol. 5, ISCAS 93 Conference, London, Proceedings, pp. 681–684 (1990)
5. J.A. Connelly, *Analog Integrated Circuits* (John Wiley and Sons, New York, 1975)

Chapter 3
Applications

Abstract This chapter describes a number of general applications suitable for quantifying the requirements of universal active devices or operational amplifiers. The transfer of each example is described by a matrix containing, firstly, one or more nominal values, and secondly, error terms having low values. The nominal values are determined by the circuit configuration and by the gain-setting passive components in the circuit. The error terms are determined by the nonidealities of the active devices as discussed in Chap. 2.

This chapter describes a number of general applications suitable for quantifying the requirements of universal active devices or operational amplifiers. The transfer of each example is described by a matrix containing, firstly, one or more nominal values, and secondly, error terms having low values. The nominal values are determined by the circuit configuration and by the gain-setting passive components in the circuit. The error terms are determined by the nonidealities of the active devices as discussed in Chap. 2. Each of the error terms can be relatively easily expressed as a sum of errors caused by:

Firstly, the non-zero input voltage.
Secondly, the non-zero input current.
Thirdly, if the input port is floating, the CM input current; and
Fourthly, if the output port is floating, the non-zero CM output current.

With the aid of these error terms one can estimate the required specifications of the active devices or operational amplifiers.

In the last section particularly the requirements of OpAmps are evaluated based on the desired dynamic range.

It appears that for an optimum ratio of dynamic range and supply power the output of an operational amplifier should have a current-efficient class-AB biasing and a voltage-efficient rail-to-rail voltage range. In some cases the input should also have a voltage-efficient rail-to-rail common-mode voltage range.

J. Huijsing, *Operational Amplifiers*, DOI 10.1007/978-3-319-28127-8_3

3.1 Operational Inverting Amplifier

The operational inverting amplifier has both the input and output ports grounded which restricts the feedback to parallel connections. This results in applications with a low input and output impedance.

3.1.1 Current-to-Voltage Converter

As we have already discussed in Sect. 3.1.1, the transimpedance amplifier or current-to-voltage converter is the most basic application of the OIA. This amplifier configuration is drawn in Fig. 3.1 together with the source and load circuits.

For a description of the transfer we choose a kind of matrix which clearly presents the nominal transfer and additionally presents all errors in relation to the input quantities V_i and I_i. For a two-port the total matrix is equal to the chain matrix. The matrix of Fig. 3.1 is given in Eq. 3.1:

$$\begin{vmatrix} I_1 \\ V_1 \end{vmatrix} = \begin{vmatrix} 1/Z_{tn} + 1/Z_{te} & 1/A_{ie} \\ 1/A_{ve} & 1/Y_{te} \end{vmatrix} \begin{vmatrix} V_2 \\ -I_2 \end{vmatrix} + \begin{vmatrix} I_{1offs} \\ V_{1offs} \end{vmatrix}; \qquad (3.1)$$

with:

$$1/Z_{tn} = -1/Z_2$$
$$1/Z_{te} \simeq -Y_2(Y_2 + Y_0)/Y_t - Y_i(Y_2 + Y_0)/Y_t$$
$$1/A_{ie} \simeq -Y_2/Y_t - Y_i/Y_t$$
$$1/A_{ve} \simeq -(Y_2 + Y_0)/Y_t$$
$$1/Y_{te} \simeq -1/Y_t$$
$$1_{1offs} \simeq V_{i\,offs}/Z_2 + I_{i\,offs}$$
$$V_{1offs} \simeq V_{i\,offs}$$

and with $Y_2 = 1/Z_2$

The nominal term $1/Z_{tn}$ represents the minus reciprocal value of the nominal impedance $-1/Z_2$. All other terms are error terms caused by nonidealities as described by the error matrix (Eq. 2.1). Each error term can be written as a

Fig. 3.1 Transimpedance amplifier with source and load circuits

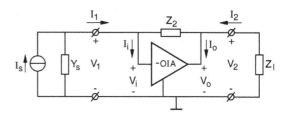

sum of partial errors which originate, firstly, from the non-zero OIA input voltage V_i and secondly, from the non-zero input current I_i. The first error term $1/Z_{te}$ represents an entrance error current V_2/Z_{te} as a function of the exit voltage V_2 at $-I_2 = 0$. This error current is composed of two partial error currents: $V_2Y_2(Y_2 + Y_0)/Y_t$ following through Y_2 as a consequence of the non-zero input voltage $V_i = -V_2$ $(Y_2 + Y_0)/Y_t$ and the non-zero input current $I_1 = -V_2Y_i(Y_2 + Y_0)/Y_t$. The second error term $1/A_{ie}$ represents an entrance error current I_2/A_{ie} as a function of $-I_2$ at $V_2 = 0$. This error current is composed of two partial error currents: I_2Y_2/Y_t flowing through Y_2 as a consequence of the input voltage $V_i = I_2/Y_t$, and the input currents $I_i = I_2Y_i/Y_t$. The other error terms are built up in the same way. The input offset voltage V_{ioffs} and current I_{ioffs} give rise to the entrance offset voltage V_{1offs} and current I_{1offs}, as presented. The spectral input noise voltage V_{in} and current I_{in} can be thought of as superimposed on the relevant offset quantities [1, 2]. An extra entrance noise current, caused by the noise current I_{Z2n} of the impedance Z_2, is present.

3.1.2 Inverting Voltage Amplifier

Another important application of the OIA is the inverting voltage amplifier, a configuration widely used in analog computer circuits. The circuit is shown in Fig. 3.2. The matrix is given in Eq. 3.2:

$$\begin{vmatrix} V_1 \\ I_1 \end{vmatrix} = \begin{vmatrix} 1/A_{vn} + 1/A_{ve} & 1/Y_{te} \\ 1/Z_{te} & 1/A_{ie} \end{vmatrix} \begin{vmatrix} V_2 \\ -I_2 \end{vmatrix} + \begin{vmatrix} V_{1offs} \\ I_{1offs} \end{vmatrix}; \qquad (3.2)$$

with

$$1/A_{vn} \simeq -Z_1/Z_2$$
$$1/A_{ve} \simeq -(Z_1 + Z_2)(Y_2 + Y_0)/Y_tZ_2 - Z_1Y_i(Y_2 + Y_0)/Y_t$$
$$1/Z_{te} \simeq -1/Z_2$$

Fig. 3.2 Inverting voltage amplifier with source and load circuits

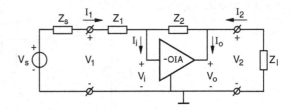

An important disadvantage of the inverting voltage amplifier is its large entrance current error $V_2/Z_{te} \simeq -V_2/Z_2$. The non-inverting voltage amplifier using an OVA, which will be described in the next section, does not have this disadvantage.

3.2 Operational Voltage Amplifier

The operational voltage amplifier (OVA) has a floating input port that provides applications with a high input impedance by a serial feedback through the input port.

3.2.1 Non-inverting Voltage Amplifier

The most basic application of the OVA is the non-inverting voltage amplifier (see Sect. 3.1.2) as shown in Fig. 3.3.

The transfer of the circuit can be described by one nominal value and error terms having low values, as described by matrix (Eq. 3.3):

$$\begin{vmatrix} V_1 \\ I_1 \end{vmatrix} = \begin{vmatrix} 1/A_{vn} + 1/A_{ve} & 1/Y_{te} \\ 1/Z_{te} & 1/A_{ie} \end{vmatrix} \begin{vmatrix} V_2 \\ -I_2 \end{vmatrix} + \begin{vmatrix} V_{1offs} \\ I_{1offs} \end{vmatrix}; \tag{3.3}$$

with:

$$1/A_{vn} = Z_1/(Z_1 + Z_2)$$
$$1/A_{ve} \simeq (Y_0 + Y'_s)/Y_t + Y_{id}(Y_0 + Y'_s)/Y_pY_t + 1/A_{vn}H_i - Y_{ic1}/Y_PA_{vn}$$
$$1/Y_{te} \simeq 1/Y_t + Y_{id}/Y_pY_t$$
$$1/Z_{te} \simeq Y_{id}(Y_0 + Y'_s)/Y_t + Y_{ic2}/A_{vn}$$
$$1/A_{ie} \simeq Y_{id}/Y_t$$
$$V_{ioffs} \simeq V_{ioffs} + I_{1offs}/Y_p + I_{ibias}/Y_p$$
$$I_{1offs} \simeq +I_{ibias}$$

and with $Y_i = 1/Z_1$, $Y_2 = 1/Z_2$, $Y_p = Y_1 + Y_2$, $Y'_s = 1/(Z_1 + Z_2)$

Fig. 3.3 Voltage amplifier
with source and load
circuits

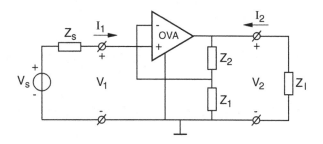

The reciprocal nominal value of the voltage amplification factor is $1/A_{vn} = Z_1/(Z_1 + Z_2)$. All other matrix elements are error terms caused by nonidealities. The first error term $1/A_{ve}$ represents the entrance error voltage V_2/A_{ve} as a function of the exit voltage V_2 at $-I_2 = 0$. The error voltage is the sum of four partial error voltages. The first of these partial error voltages $V_2(Y_0 + Y'_s)/Y_t$ represents the differential input voltage V_{id} needed to generate the exit voltage V_2.

The second partial error voltage $V_2 Y_{id}(Y_0 + Y'_s)/Y_p Y_t$ is the voltage loss across the parallel admittance $Y_p = Y_1 + Y_2$ of the feedback network as a result of the differential input current $I_{id} = V_2 Y_{id}(Y_0 + Y_s)/Y_t$. The third partial error voltage $V_2/A_{vn} H_i$ is the equivalent differential input voltage V_{id} evoked by the crosstalk $1/H_i$ from the CM input voltage $V_{ic} = V_2/A_{vn}$. The fourth partial error voltage $-V_2 Y_{ic1} /Y_p A_{vn}$ is the voltage loss across the parallel admittance $Y_p = Y_1 + Y_2$ of the feedback network as a result of the common-mode input current $I_{ic} = V_2 Y_{ic1}/A_{vn}$. The other error terms are built up in the same way. In addition to the offset and noise quantities of the OIA (see Sect. 3.1), there are the input bias current I_{ibias} and the input bias noise current I_{ibn} and their effects. An extra entrance noise voltage, caused by the noise voltage V_{Zpn} of the parallel impedance $Z_p = 1/Y_p$, is present.

3.2.2 Voltage Follower

If we choose $Z_1 = \infty$ and $Z_2 = 0$, the voltage amplifier becomes a voltage follower with a nominal voltage amplification factor $A_{un} = 1$ in that all error contributions with a Y_p in the denominator disappear.

3.2.3 Bridge Instrumentation Amplifier

A bridge instrumentation amplifier can be built with an OVA and bridge of four impedances Z_{11} through Z_{22} (Fig. 3.4). The circuit can be thought to be composed partly from the inverting voltage amplifier (Fig. 3.4) and the non-inverting voltage amplifier (Fig. 3.3). If the bridge is well balanced, which means that $\Delta Z_b/Z_b = 1 - Z_{21}Z_{12}/Z_{22}Z_{11} \ll 1$, the circuit only amplifies the differential-mode entrance voltage V_{id} while the common-mode entrance voltage V_{ic} is rejected. The circuit of a bridge instrumentation amplifier is given in Fig. 3.4 and the matrix with the error terms in Eq. 3.4

$$
\begin{vmatrix} V_{1d} \\ I_{1d} \\ I_{1c} \end{vmatrix} = \begin{vmatrix} 1/A_{vn} + 1/A_{ve} & 1/Y_{te} & 1/H_e \\ & 1/A_{te} & 1/A_{ie} \\ & \cdot & 1/Z_{1cc} \end{vmatrix} \begin{vmatrix} V_2 \\ -I_2 \\ V_{ic} \end{vmatrix} + \begin{vmatrix} V_{ioffs} \\ I_{ioffs} \\ I_{ibias} \end{vmatrix} ; \tag{3.4}
$$

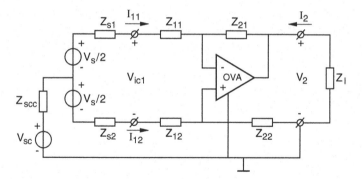

Fig. 3.4 Bridge instrumentation amplifier with source and load circuits

with:

$$1/A_{vn} = -Z_{11}/Z_{21}$$

$$1/A_{ve} \simeq -\Delta Z_b/2Z_b(A_{vn}+1)A_{vn}$$
$$-(Z_{11}+Z_{21})(Y_0+Y_{21})/Y_tZ_{21} - (Z_{11}+Z_{12})Y_{id}(Y_0+Y_{21})/Y_t$$
$$-1/2H_iA_{vn} - Y_{ic1}/2Y_{11}A_{vn} + Y_{ic2}/2Y_{12}A_{vn}$$

$$1/H_e \simeq -\Delta Z_b/Z_b(A_{vn}+1)$$
$$-1/Y_tZ_{21} - (Z_{11}+Z_{12})Y_{id}/Y_t(Z_{11}+Z_{21})$$
$$-1/H_i - Y_{ic}/Y_{11} - Y_{ic}/Y_{12}$$

$$1/Z_{1cc} \simeq 1/(Z_{11}/Z_{21})$$

and with:

$$\Delta Z_b/Z_b = 1 - Z_{21}Z_{12}/Z_{22}Z_{11} \ll 1$$
$$Y_{11} = 1/Z_{11}, Y_{12} = 1/Z_{12}, Y_{21} \simeq 1/Z_{21}, Y_{22} = 1/Z_{22}$$

The main disadvantage of this type of instrumentation amplifier is that the bridge resistors establish a connection between the input port of the OVA and the output and ground. This destroys the CM isolation barrier of the input port of the OVA and thus destroys the CMRR of the application. Therefore, the common-mode crosstalk ratio (CMCR) $1/H_e$ is directly determined by the imbalance $\Delta Z_b/Z_b$ of the bridge and a factor $1/(A_{vn}+1)$ depending on the nominal amplification factor A_{vn}. Another disadvantage is the relatively low CM entrance impedance $1/Z_{1cc} = 1/(Z_{11}+Z_{21})$ which is also caused by the lack of a CM isolation barrier.

The latter disadvantage can be overcome by connecting the input terminals in cascade with either two voltage followers, one for each terminal, or two voltage amplifiers of the type of Fig. 3.3, joined in a balanced configuration by connecting the bottom sides of Z_l to each other, instead of to ground.

A basically better way to build an instrumentation amplifier which does have a CM isolation barrier is to use two OFAs. An example of such an instrumentation will be given in Sect. 3.4.

3.3 Operational Current Amplifier

The operational current amplifier has a floating output port, which allows a high output impedance or current output by a serial feedback through the output port.

3.3.1 Current Amplifier

The most basic application of the OCA is the current amplifier (Sect. 3.4). This configuration is given in Fig. 3.5 together with its source and load circuits.

The nominal amplification factor and error terms with small values can be described by the matrix of Eq. 3.5.

$$\begin{vmatrix} I_1 \\ V_1 \end{vmatrix} = \begin{vmatrix} 1/A_{in} + 1/A_{ie} & 1/Z_{te} \\ & 1/Y_{te} & 1/A_{ve} \end{vmatrix} \begin{vmatrix} -I_2 \\ V_2 \end{vmatrix} + \begin{vmatrix} -I_{1offs} \\ V_{1offs} \end{vmatrix}; \tag{3.5}$$

with:

$$1/A_{in} = Y_1/(Y_1 + Y_2)$$

$$1/A_{ie} \simeq Y'_s/Y_t + Y_i/Y_t + 2H_o/A_{in} - Y_{oc1}/Y_p A_{in}$$

$$1/Z_{te} \simeq Y_{se}Y_{od}/Y_t + Y_i Y_{od}/Y_t + Y_{oc2}/A_{in}$$

$$1/Y_{te} \simeq 1/Y_t$$

$$1/A_{ve} \simeq Y_{od}Y_t$$

$$I_{1offs} \simeq V_{ioffs}Y_t + I_{ioffs} + 2I_{obias}/A_{in}$$

$$V_{1offs} \simeq V_{ioffs}$$

and with: $Y_p = Y_1 + Y_2$, $Y'_s = 1/(1/Y_1 + 1/Y_2)$, neglecting Y_{od} against Y_p

Fig. 3.5 Current amplifier with source and load circuits

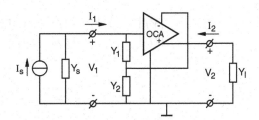

The reciprocal value of the nominal current amplification factor is $1/A = Y_1/(Y_1 + Y_2)$. All other terms are error terms caused by nonidealities. The first error term $1/A_{ie}$ can be written as the sum of four partial error terms.

The first partial error Y'_s/Y_t represents the entrance error current $-I_2Y_s/Y_t$. This results from the current V_iY_s flowing through the series conductance $Y's = 1/(1/Y_1 + 1/Y_2)$ as a consequence of the input voltage $V_i = -I_2/Y_t$ needed to deliver the exit current $-I_2$. The second partial error Y_i/Y_t simply represents the input current $I_i = -I_2Y_i/Y_pY_t$ which is needed to deliver the exit current $-I_2$.

The third partial error $2/H_oA_{in}$ is caused by the crosstalk $1/H_o$ of the DM output current $I_{od} = -I_2$ on the CM output current I_{oc}. Double this value (from both output terminals) must be counted.

The fourth partial error $-Y_{oc1}/Y_pA_{in}$ is a result of the error current I_2Y_{oc1}/Y_p which flows into the CM output impedance Y_{oc1} in parallel with the parallel admittance $Y_p = Y_1 + Y_2$ as a function of $-I_2$. The other error terms are made up in the same way. In addition to the offset quantities mentioned with the OIA, there are the output bias current I_{obias} and output bias noise current I_{obn}. There is an extra entrance noise current caused by the noise current I_{YSN} of the series admittance $Y'_s = 1/(1/Y_1 + 1/Y_2)$.

3.4 Operational Floating Amplifier

The operational floating amplifier is the most versatile OpAmp. It permits series feedback through the input and output ports. This allows for applications with a high input and output impedance.

3.4.1 Voltage-to-Current Converter

The most basic application of the OFA is the transadmittance amplifier or the voltage-to-current converter (Sect. 3.5). The circuit is shown in Fig. 3.6 with the source and load circuit.

The nominal transadmittance and error terms are presented in the matrix of Eq. 3.6.

$$\begin{vmatrix} V_1 \\ I_1 \end{vmatrix} = \begin{vmatrix} 1/Y_{tn} + 1/Y_{te} & 1/A_{ve} \\ & 1/A_{ie} & 1/Z_{te} \end{vmatrix} \begin{vmatrix} -I_2 \\ V_2 \end{vmatrix} + \begin{vmatrix} V_{1offs} \\ I_{1offs} \end{vmatrix} ; \tag{3.6}$$

Fig. 3.6 Voltage-to-current converter or transadmittance amplifier with source and load circuits

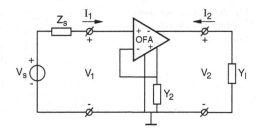

with:

$$1/Y_{tn} = -1/Y_2$$

$$1/Y_{te} \simeq -1/Y_t - Y_{id}/Y_tY_2 - 1/H_iY_2 + Y_{ic2}/Y_2Y_2 - 2/H_oY_2 + Y_{oc2}/Y_2Y_2$$

$$1/A_{ve} \simeq -Y_{od}/Y_t - Y_{id}Y_{od}/Y_tY_2 - 2Y_{od}/Y_tH_oY_2 - Y_{oc1}/Y_2$$

$$1/A_{ie} \simeq -Y_{id}/Y_t - Y_{ic1}/Y_2$$

$$1/Z_{te} \simeq -Y_{id}Y_{od}/Y_t$$

$$V_{1offs} \simeq V_{ioffs} + I_{ioffs}/Y_2 + I_{ibias}/Y_2 + 2I_{obias}/Y_2$$

$$I_{1offs} = I_{ioffs} + I_{obias}$$

The reciprocal value of the nominal transadmittance is $1/Y_{tn} = -1/Y_2$. All other terms are error terms caused by nonidealities. Each error term is the sum of partial errors which can have six origins (see Sect. 3.4): firstly $V_{id} = f(-I_2, V_2)$, secondly $I_{id} = f(-I_2, V_2)$, thirdly $V_{id} = f(V_{ic})$, fourthly $I_{ic} = f(V_{ic})$, fifthly $I_{oc} = f(-I_2)$, and lastly $I_{oc} = f(V_{oc})$. We neglected Y_{od} against Y_2.

The partial errors are placed in six columns according to the above sequence. The partial errors have already been explained in the preceding sections of this chapter. There is an extra entrance noise caused by the noise voltage V_{Y2n} of the admittance Y_2.

3.4.2 Inverting Current Amplifier

An inverting current amplifier or current mirror can be made by adding an entrance admittance Y_1 to the transadmittance amplifier (Fig. 3.6). The entrance admittance Y_1 converts the entrance current I_1 into a voltage V_1 which is in turn converted into the exit current I_2 by the transadmittance amplifier containing Y_2. A special case appears if $Y_1 = Y_2$. In that case the circuit is known as the "current mirror." The circuit is drawn in Fig. 3.7, while the matrix is given by Eq. 3.7.

Fig. 3.7 Inverting current
amplifier with source and
load circuits

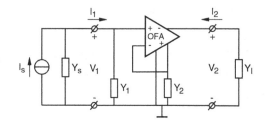

$$\begin{vmatrix} I_1 \\ V_1 \end{vmatrix} = \begin{vmatrix} 1/A_{in} + 1/A_{ie} & 1/Z_{te} \\ 1/Y_{te} & 1/A_{ve} \end{vmatrix} \begin{vmatrix} -I_2 \\ V_2 \end{vmatrix} + \begin{vmatrix} I_{1offs} \\ V_{1offs} \end{vmatrix}; \qquad (3.7)$$

with:

$$1/A_{in} = -Y_1/Y_2$$
$$1/A_{ie} \simeq -Y_1/Y_t - Y_{id}/Y_t$$
$$\quad -1/H_iA_{in} - Y_{ic1}/Y_2 + Y_{ic2}/Y_2A_{in}$$
$$\quad -2/H_oA_{in} + Y_{oc2}/Y_2A_{in}$$
$$1/Y_{te} \simeq -1/Y_2$$

The first term is the reciprocal nominal value of the current amplification factor. The other terms are error terms, as has been explained earlier. A disadvantage of the inverting current amplifier is the relatively high entrance voltage $V_1 = -I_2/Y_{te} = I_2/Y_2$ which is not present in the non-inverting current amplifier of Fig. 3.3.

3.4.3 Differential Voltage-to-Current Converter

The floating character of the input of an OFA together with the current-source character of the output of an OFA having series feedback at its output can be applied to obtain the CM-voltage isolation needed for the entrance circuit of an instrumentation amplifier. This is shown by the instrumentation transadmittance amplifier of Fig. 3.8 [3]. Two OFAs, connected as voltage and current followers (VCFs), firstly, transfer the DM entrance voltage V_{id} at unity gain to the terminals of a conductance Y_2, and secondly, transfer the current $-I_{2d} = -V_1Y_2$ through Y_2 towards the output terminals. The CM input voltage only affects the output current through the nonideal properties of the OFAs, as described by the matrix equation (Eq. 3.8).

Fig. 3.8 Instrumentation
voltage-to-current or
transadmittance amplifier
with source and load
circuits

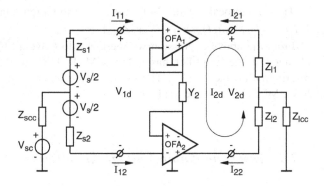

$$
\begin{vmatrix} V_{ld} \\ I_{ld} \\ I_{lc} \\ I_{2c} \end{vmatrix}
=
\begin{vmatrix}
1/Y_{tn}+1/Y_{te} & 1/A_{ve} & 1/H_e & \cdot \\
1/A_{ie} & 1/Z_{te} & \cdot & \cdot \\
\cdot & \cdot & 1/Z_{ice} & \cdot \\
\cdot & \cdot & 1/Z_{tcce} & 1/Z_{oce}
\end{vmatrix}
\begin{vmatrix} -I_{2d} \\ V_{2d} \\ V_{1c} \\ V_{2c} \end{vmatrix}
+
\begin{vmatrix} V_{1offs} \\ I_{1offs} \\ I_{1bias} \\ I_{2bias} \end{vmatrix};
\qquad (3.8)
$$

with:

$$
1/Y_{tn} = -1/Y_2
$$

$$
1/Y_{te} \simeq -2/Y_t - Y_{id}/Y_tY_2 - 1/2H_{i1}Y_2 + 1/2H_{i2}Y_2 + Y_{ic}/2Y_2Y_2
$$
$$
- 1/H_{o1}Y_2 + 1/H_{o2}Y_2 + Y_{oc}/2Y_2Y_2
$$

$$
1/H_e \simeq Y_{od1}/Y_{t1} + Y_{od2}/Y_{t2} + Y_{id1}Y_{od1}/Y_{t1}Y_2 + Y_{id2}Y_{od2}/Y_{t2}Y_2
$$
$$
- 1/H_{i1} + H_{i2} + Y_{ic12}/Y_2 - Y_{ic21}/Y_2 + Y_{oc12}/Y_2 - Y_{oc21}/Y_2
$$

$$
1/Z_{tcce} \simeq +Y_{id}Y_{od}/Y_t - Y_{ic} - Y_{oc}
$$

and   with: $V_{1d} = V_{11} - V_{12}$, $V_{1c} = (V_{11} + V_{12})/2$, $I_{1d} = (I_{11} - I_{12})/2$, $I_{1c} = (I_{11} + I_{12})/$; idem for the exit voltages and currents; all OFA parameters without OFA number are average values.

The nominal value of the reciprocal admittance is $1/Y_{tn} = -1/Y_2$. All other terms are error terms which have been explained in the preceding part of this chapter. The common-mode crosstalk ratio (CMCR), or the reciprocal value of the common-mode rejection ratio (CMRR), is $1/H_i$.

A complete instrumentation voltage amplifier arises if we include the current-to-voltage conversion function of the load impedances Z_{L1} and Z_{L2}. In that case the reciprocal nominal overall voltage gain $1/A_{vn}$ and the overall CMCR $1/H_i$ are given by Eq. 3.9.

$$
1/A_{vn} = -1/Y_2 Z_{ld}
$$
$$
1/H_i \simeq 1/H_e + 1/F_e H_{iL}
\qquad (3.9)
$$

with: $Z_{Ld} = Z_{L1} + Z_{L2}$, $1/H_{iL} = \Delta Z_{L1}/Z_{1d}$, $\Delta Z_{L1} = Z_{L1} - Z_{L2}$.

The reciprocal value of the discrimination factor is defined as:
$1/F_e = (I_{2c}/V_{ic})/I_{2d}/V_{id}) = 1/Z_{tcce}Y_{tn}.$

From the above expression we see that a low overall CMCR $1/H_i$ can only be obtained if the amplifier has a low CMCR $1/H_e$ as well as a low reciprocal discrimination factor $1/F_e$. The effect of the latter term is further reduced by a low CMCR $1/H_{iL}$ of the load circuit.

3.4.4 Instrumentation Voltage Amplifier

An instrumentation voltage amplifier with a voltage-source character between the output terminals arises if we insert a balanced floating impedance amplifier with an OFA and two impedances Z_{l1} and Z_{l2} between the preceding example's instrumentation transadmittance amplifier and its load circuit [3, 4].

The complete instrumentation voltage amplifier circuit is drawn in Fig. 3.9 together with the source and load circuits.

The reciprocal nominal overall differential voltage gain and the CMCR of the complete amplifier are described by Eq. 3.10.

$$1/A_{vn} = -1/Y_2 Z_{ls}$$
$$1/H_i \simeq 1/H_e + 1/F_e H_i$$

(3.10)

with: $Z_{ls} = Z_{l1} + Z_{l2}$, $1/H_l = \Delta Z_l/Z_s$, $\Delta Z_l = Z_{l1} - Z_{l2}$

The instrumentation amplifier has a floating entrance port and an independently floating exit port. It measures the differential entrance voltage at nominal zero entrance currents. The differential exit port has a voltage source character.

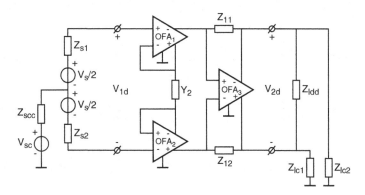

Fig. 3.9 Instrumentation voltage amplifier with source and load circuits

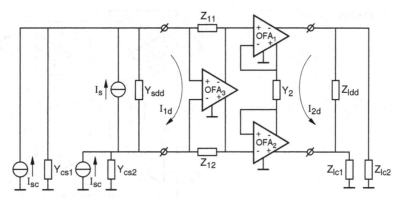

Fig. 3.10 Instrumentation current amplifier

3.4.5 Instrumentation Current Amplifier

If we change the sequence of the transadmittance and transimpedance amplifiers of the preceding example we obtain an instrumentation current amplifier. This circuit is drawn in Fig. 3.10 together with the source and load circuits.

The overall reciprocal nominal current gain is given by Eq. 3.12.

$$1/A_{vn} = -1/Y_2 Z_{ls} \tag{3.11}$$

The instrumentation current amplifier has a floating entrance port and an independently floating exit port. It measures the differential entrance current at nominal zero differential entrance voltage. The exit port has a current source character.

3.4.6 Gyrator Floating

From among the applications of the OFA the gyrator should not be left out. A fully floating gyrator can be composed of two instrumentation transadmittance amplifiers as shown in Fig. 3.11 [5, 6]. The nominal and error terms of the transfer are given in Eq. 3.13.

$$
\begin{vmatrix} I_{1d} \\ I_{1c} \\ I_{2d} \\ I_{2c} \end{vmatrix} =
\begin{vmatrix}
1/Z_{t2ln} + 1/Z_{t2le} & \cdot & 1/Z_{1de} & \cdot \\
\cdot & \cdot & \cdot & 1/Z_{1ce} \\
1/Z_{2de} & \cdot & 1/Z_{t2ln} + 1/Z_{tl2e} & \cdot \\
\cdot & 1/Z_{2ce} & \cdot & \cdot
\end{vmatrix}
\begin{vmatrix} V_{2d} \\ V_{2c} \\ V_{1d} \\ V_{1c} \end{vmatrix} ; \tag{3.12}
$$

Fig. 3.11 Floating gyrator

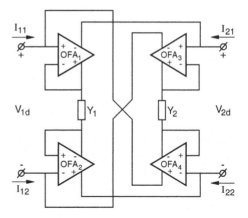

with:

$$1/Z_{t2ln} = -Y_2$$
$$1/Z_{t2ln} = Y_2$$
$$1/Z_{t2le} \simeq 2Y_2 2Y_2/2Y_t + Y_{id}2Y_2/2Y_t$$
$$\quad - Y_{ic}/2 + 1/H_iY_2 - Y_{oc}/2 + 2/H_oY_2$$
$$1/Z_{t2le} \simeq -2Y_2 2Y_1/2Y_t - Y_{id}2Y_1/2Y_t + Y_{ic}/2$$
$$\quad - 1/H_iY_2 + Y_{oc}/2 - 2/H_oY_2$$
$$1/Z_{1de} \simeq 2Y_2 Y_{od}/2Y_t + Y_{id}(Y_{od} + 2Y_1)/2Y_1 + Y_{ic}/2 + Y_{oc}/2$$
$$1/Z_{2de} \simeq 2Y_1 Y_{od}/2Y_t + Y_{id}2Y_2/2Y_t + Y_{ic}/2 + Y_{oc}/2$$

and with average OFA parameters.

The nominal resonant frequency ω_o circuit which is composed of a gyrator with $Y_1 = Y_2 = 1/R$ and loaded on both sides with a capacitance $C_1 = C_2 = C$ is given in Eq. 3.14.

$$\omega_n = 1/RC$$
$$1/Q \simeq 2R/R_d - 2C_t/C \tag{3.13}$$

with:
$1/R_d = real(1/Z_{de})$, $\omega_n C_t = im.\ (1/Z_{te})$, $Z_{de} \simeq Z_{1de} \simeq Z_{2de}$, $Z_{te} \simeq Z_{t2le} \simeq Z_{t2le}$

The second term in the expression of $1/Q$ describes the phase-lag in admittance amplifiers. This phase-lag undamps the circuit at higher resonant frequencies.

3.4.7 Conclusion

The application examples given in this section have shown the relation between the specifications of the active devices and the accuracy of the applications mentioned. This is necessary for determining how far we have to go in improving the specifications of the active devices whose designs will be the subjects of the following chapters. The important overall specification of dynamic range brings about special requirements, as we will see in the next section.

3.5 Dynamic Range

The total amount of information that can be processed in an analog signal-processing step is determined by the product of dynamic range and bandwidth. The dynamic range over power limitations will be evaluated in this section, while the bandwidth over power limitations will be extensively covered in Sect. 3.6.2.

The fundamental specification of low power is in contradiction to the fundamental specification of dynamic range. Therefore, it is important to see how these specifications relate in several OVA applications. An optimum for low power and dynamic range can be found if the output stage and input stage possess a rail-to-rail voltage range. This will be shown in this section, in which the OVA will simply be called operational amplifier (OpAmp).

3.5.1 Dynamic Range over Supply-Power Ratio

The trend towards smaller dimensions in VLSI circuits firstly leads to smaller break-down voltages across isolation barriers. The supply voltages will go down from 5 through 3 to 2 V or even 1 V. Secondly, the high density of circuit cells on a chip limits the power that can be dissipated per circuit cell. Moreover, the increased use of batteries or solar power in wireless applications emphasize the above trend. As a consequence, the dynamic range (DR) of analog signals is squeezed down between a lower supply-voltage ceiling and a higher noise-voltage floor. The latter is a consequence of a lower supply current.

The maximum top value of a single-phase signal voltage is equal to half the supply voltage $V_{sst} = V_{sup}/2$, as is shown in Fig. 3.12a. Its RMS value is $V_{ss} = V_{sup}/2\sqrt{2}$. If this signal is present across a signal-processing resistor R_s, the supply power needed to drive this resistor in class-B mode is $P_{sup} = V_{sup}I_{av} = V_{sup}^2/2\pi R_s$. The thermal noise voltage across this resistor equals: $V_N = (4kTB_eR_s)^{1/2}$, in which k is Boltzmann's constant, T the absolute temperature, and B_e the effective bandwidth. The maximum dynamic range as a function of the supply power can now be calculated as

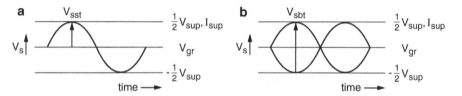

Fig. 3.12 (**a**) and (**b**) Single and balanced rail-to-rail voltage $V_{sst} = V_{sup}/2$ and $V_{sbt} = V_{sup}$, respectively, across a class-B driven signal-processing resistor has a maximum DR $= (\pi/4)P_{sup}/4kTB_e$ ($= 89$ dB at $P_{sup} = 16\,\mu$W, $B_e = 1$ MHz)

$$DR_{max} = \frac{V_{ss}^2}{V_N^2} = \frac{\pi}{4} \cdot \frac{P_{sup}}{4kTB_e} \qquad (3.14)$$

from which we can find the dynamic-range over supply-power ratio as

$$\frac{DR_{max}}{P_{sup}} = \frac{\pi}{4} \cdot \frac{1}{4kTB_e} \qquad (3.15)$$

Exactly the same expressions are found for the balanced case, where the top value of a balanced signal voltage is equal to the full supply voltage $V_{sbt} = V_{sup}$, instead of half the supply voltage in the single case, see Fig. 3.12b. To consume the same power, the value of the balanced resistor R_b must be taken four times that of the single one: $R_b = 4R_s$.

3.5.2 Voltage-to-Current Converter

A simple example is shown in the single and balanced voltage-to-current converter shown in Fig. 3.13a, b with a single resistor of $R_s = 10$ kΩ or balanced resistor of $R_b = R_{b1} + R_{b2} = 40$ kΩ respectively, at a supply voltage $V_{sup} = V_{SP} - V_{SN}$ of 1 V, in a bandwidth of 1 MHz. In this case the supply power P_{sup} is 16 μW at a maximum sinusoidal signal. The result is a maximum dynamic range DR_{max} of 89 dB. This maximum can only be obtained if the signal processing resistors can be driven in class-B and rail-to-rail, and when the amplifier is noise free.

If the output stage is biased in class-A instead of in class-B, the bias current must be equal to the maximum current and the DR/P_{sup} ratio loses minimally a factor of π, or 5 dB from its maximum value. This loss for class-A in regard to class-AB may easily be a factor of 100, or 40 dB, in the many cases where the signals are much lower than their maximum values most of the time. This is the case in audio, telecommunications, hearing aids, etc. If the output voltage range is restricted to one-third of the supply voltage, for instance when a diode voltage V_{BE} is lost at a supply voltage of 1 V, the DR/P_{sup} ratio loses another factor 3, or 5 dB.

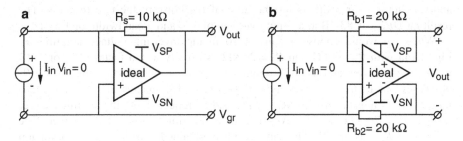

Fig. 3.13 (**a**) and (**b**) Single and balanced voltage-to-current converter with $R_s = 10$ kΩ and $R_{b1} = R_{b2} = 20$ kΩ, respectively, with a $DR = 89$ dB at $V_{sup} = V_{SP} - V_{SN} = 1$ V, $P_{sup} = 16$ μW, R-R class-B output stage, and a bandwidth of 1 MHz

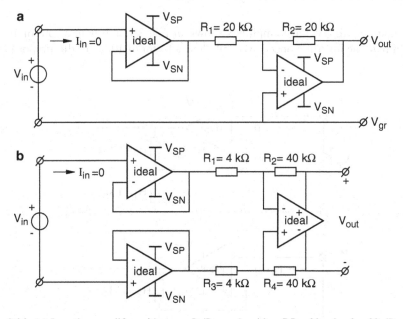

Fig. 3.14 (**a**) Inverting amplifier with $A = -R_2/R_1 = -1$, with a $DR = 89 - 3 - 3 = 83$ dB, at a supply voltage $V_{sup} = V_{SP} - V_{SN}$ of 1 V, $P_{sup} = 16$ μW, R-R, class B, 1 MHz. If $R_1 = 2$ kΩ, $R_2 = 20$ kΩ, we obtain $A = -10$, with a $DR = 89 - 3 - 10 = 76$ dB. (**b**) Balanced inverting amplifier with $A = -(R_2 + R_4)/(R_1 + R_3) = -10$, with a $DR = 89 - 3 - 10 = 76$ dB, at a supply voltage $V_{sup} = V_{SP} - V_{SN}$ of 1 V, $P_{sup} = 16$ μW, R-R, Class-B, and 1 MHz

3.5.3 Inverting Voltage Amplifier

The inverting voltage amplifier of Fig. 3.14a, firstly, loses a factor of 2, or 3 dB in its DR/P_{sup} ratio because an additional input buffer is needed to supply the power in the resistor R_1. Otherwise, this power has to be supplied by the source. Secondly, another factor 2, or 3 dB, is lost because of the noise of the two resistors. The resulting DR_{max} is 83 dB at a supply voltage of 1 V and a bandwidth of 1 MHz.

When we would choose a gain of 10 in the inverting voltage amplifier of Fig. 3.14a with $R_1 = 2$ kΩ and $R_2 = 20$ kΩ, we firstly lose a factor of 2, or 3 dB into the input buffer, and secondly another factor of 10, or 10 dB because resistor R_1 only uses one-tenth of the supply voltage range. This means that the $\sqrt{10}$ times larger current noise of resistor $R_1 = 2$ kΩ will be reflected into the ten times larger resistor $R_2 = 20$ kΩ, which gives rise to a ten times larger noise power. The resulting DR_{max} is 76 dB. The same result is obtained with the balanced version given in Fig. 3.14b.

3.5.4 Non-inverting Voltage Amplifier

The non-inverting voltage amplifiers shown in Fig. 3.15a, b with a gain of 10 do better than the inverting one(s). We only lose a factor of 10, or 10 dB, proportional

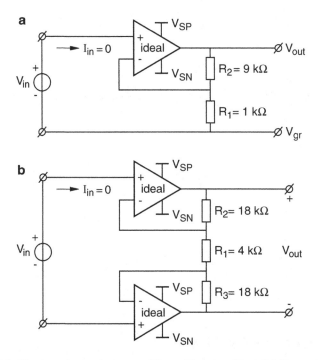

Fig. 3.15 (a) Non-inverting voltage amplifier with $A = (R_2 + R_1)/R_1 = +10$, with a $DR = 89 - 10 = 79$ dB, at a supply voltage $V_{sup} = V_{SP} - V_{SN}$ of 1 V, $P_{sup} = 16$ μW, R − R, Class-B, and 1 MHz. (b) Balanced non-inverting voltage amplifier with $A = (R_3 + R_2 + R_1)/R_1 = +10$, with a $DR = 89 - 10 = 79$ dB, at a supply voltage $V_{sup} = V_{SP} - V_{SN}$ of 1 V, $P_{sup} = 16$ μW, R − R, Class-B, and 1 MHz

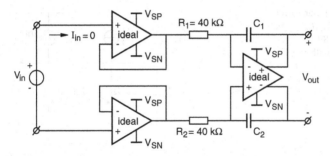

Fig. 3.16 Balanced inverting voltage integrator with $A = -(1/jwC_1 + 1/jwC_2)(R_1 + R_2)$ with a $DR = 89 - 3 = 86$ dB, at $V_{sup} = V_{SP} - V_{SN} = 1$ V, $P_{sup} = 16$ μW, $R - R$, Class-B, and 1 MHz

to the gain because R_1 only uses one-tenth of the supply voltage range. We do not lose the factor 2, or 3 dB, because we do not need an additional input buffer. The DR_{max} is 79 dB in a frequency band of 1 MHz and a supply voltage of 1 V.

3.5.5 Inverting Voltage Integrator

The balanced inverting voltage integrator shown in Fig. 3.16 only loses the factor 2, or 3 dB, because of the use of input buffers. The capacitors do not add to the noise.

Within the effective bandwidth of $B_e = 1/2\pi RC$, with $R = R_1 = R_2 = 40$ kΩ and $C = C_1 = C_2$, at a supply voltage of 1 V, the DR_{max} is 86 dB. The resistor values have been chosen such that the supply power is again 16 μW at a maximum sinusoidal signal [7].

The dynamic range of an inverting voltage integrator is generally large:

$$DR = (\pi/4)P_{sup}/4kTB_e = (\pi^2/2)P_{sup}RC/4kT$$
$$DR = \pi V_{sup}^2 C/4kT \tag{3.16}$$

with: $B_e = 1/(2\pi RC), R = R_1 = R_2, C = C_1 = C_2$

3.5.6 Current Mirror

A very severe loss of the DR is found in current mirrors. The current mirror of Fig. 3.17 firstly, loses a factor π, or 5 dB in the DR_{max} because the circuit operates in class-A and not in class-AB, and, secondly, a factor 40, or 16 dB with bipolar transistors because the signal is compressed in a voltage range of $V_T \approx kT/q = 25$ mV across the gain-setting base-emitter resistors. These resistors are

Fig. 3.17 Bipolar current
mirror with $I_{out}/I_{in} = -n$,
with a $DR = 89 - 5 -$
$16 = 68$ dB, with $n = 1$,
$V_{sup} = V_{SP} - V_{SN} = 1$ V,
$P_{sup} = 16$ μW, class-A,
and 1 MHz

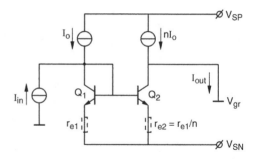

small in regard to V_{sup}/I_{sup} and therefore their noise current is unnecessarily large.
The resulting DR is only 68 dB. This is a factor 30, or 15 dB lower than the DR of
the inverting voltage amplifier. Emitter degeneration resistors will help in decreasing the loss. A CMOS mirror will also do better, because a better use of the supply-
voltage range is made because of the larger intrinsic source resistances.

The dynamic range of a current mirror is generally:

$$DR = \left(n/(n+1)^2\right)\left(V_T/V_{sup}\right)P_{sup}/4kTB_e$$
$$DR = (1/4)(1/40)P_{sup}/4kTB_e = 89 - 5 - 16 = 68 \text{ dB}$$

(3.17)

with: $V_T = kT/q = 1/40, n = 1, V_{sup} = 1$ V, $P_{sup} = 16μW$, $class - A$, and 1MHz

3.5.7 Conclusion Current Mirror

From this equation it follows that in all amplifier realizations, in which the signal
has to pass through a current mirror, the dynamic range over power ratio is quite
low. This is the case in so-called "current-mode" amplifiers. Moreover, in these
current-mode solutions, where the signal passes through one or more internal
current mirrors, the accuracy of the transfer is limited to that of the matching of
non-linear transistor characteristics of the internal current mirrors. Further, current-
mode active network elements do not fit the basic and accurate nullor approach of
Chap. 1. Instead, they have more complicated network models and cannot describe
the overall transfer function more accurately than of the order of 0.5 %.

Another approach, the so-called current-mode feedback does have a better high-
frequency behavior than the normal approach, due to the elimination of a pole in the
feedback network. However, this solution has higher noise and offset due to the
feeding of an emitter or source bias current through the feedback network. And
again it does not fit the basic nullor approach of Chap. 1. For these reasons,
so-called "current-mode" solutions, like current-mode amplifiers, and current-
mode feedback are not discussed in this book.

3.5.8 Nonideal Operational Amplifiers

If we take into account the nonidealities of the amplifiers, then the DR/P_{sup} ratio is further reduced. Important nonidealities are caused by the input noise voltage and a restricted output voltage range.

The problem at the input is that we do not want to spill supply current in the input stage, while on the contrary, we need a large bias current for the active input devices in order to lower the input voltage noise. The input voltage noise can easily be estimated by the equivalent input series noise resistance R_{neqs}. For bipolar transistors $R_{neqs} = r_e/2 = kT/2qI_e = V_T/2I_e$, with $V_T \approx 25$ mV at room temperature. For field-effect transistors we find $R_{neqs} = \gamma/g_m = \gamma/(2\,\mu C_{ox}(W/2)I_D)^{1/2}$, which is of the order of $R_{neqs} = 10\gamma/(I_D)^{1/2}$ for transistors with a W/L ratio of 100, while γ is of the order of 2. The W/L ratio has been chosen as large for analog applications to increase g_m and lower noise and offset input voltages.

An optimal solution would be to choose no separate input transistors, but to use one-stage amplifiers in which the input transistors are used as output transistors as well. This interesting realization will be shortly evaluated here. In Fig. 3.18a, b, a single and balanced current-to-voltage application is shown with a one-stage single or balanced class-A transistor amplifier. The transistors T_1 and T_2 symbolize either bipolar or field-effect transistors.

Because of the class-A operation, we lose at least 5 dB. With bipolar transistors the equivalent input noise resistor will be $R_{neqs} \approx 700\ \Omega$ for the single and 1500 Ω for the balanced version at a total supply current of 16 μA. The resulting extra noise is much lower than 1 dB. With CMOS transistors, the equivalent input noise resistor will be $R_{neqs} = 5000\ \Omega$ for the single and 7000 Ω for the balanced version. The resulting extra noise is of the order of 2 dB. At the output the signal cannot reach the rail within 100 mV. This results in a loss of 1 dB for bipolar as well as CMOS transistors.

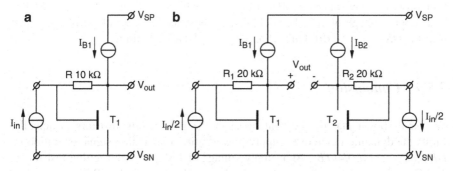

Fig. 3.18 (a) and (b) Single and balanced current-to-voltage converter in a single-stage class-A realization with a $DR = 89 - 5 - 1 = 83$ dB for bipolar transistors, and $DR = 89 - 5 - 2 = 82$ dB for CMOS transistors, R-R, $V_{sup} = 1$ V, $P_{sup} = 16$ μW, $I_B = 16$ μA, and 1 MHz

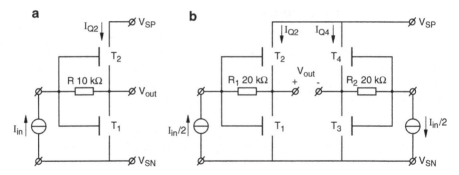

Fig. 3.19 (**a**) and (**b**) Single and balanced current-to-voltage converter in a single-stage class-AB realization with a $DR = 89 - 2 = 87$ dB for bipolar transistors and $DR = 89 - 5 = 84$ dB for CMOS transistors at $V_{sup} \approx 1.2$ V, $P_{sup} = 20$ μW, 1 MHz

We can avoid the −5 dB loss of class-A biasing if we choose a class-AB push–pull configuration as shown in Fig. 3.19a, b.

The push–pull transistors have been connected with the emitters or sources between the rails, while their bases or gates are connected. To ensure proper class-AB biasing the supply rail voltage has to be regulated at two diode voltages at a proper quiescent current I_Q. The circuit may function at roughly 1.2 V. While we have now avoided the −5 dB loss of the class-A circuit with a class-AB approach, the input noise voltage has been increased because the quiescent current has been reduced strongly with no signal. If we choose the quiescent current at one-tenth of the maximum current, the extra noise reduces the DR by 2 dB for bipolar transistors and 5 dB for CMOS transistors. The result is still better than in class-A. However, we have to build an additional supply-voltage regulator which easily takes away another 1 dB from the dynamic range. As an alternative to a fixed supply voltage we can use a supply-current source.

If we also have to take the DC offset into account, the dynamic range becomes even lower. At DC the dynamic range is $DR_{max\,DC} = V^2_{sup}/V^2_{offs}$. With bipolar transistors we may expect an offset of 0.3 mV, which results in $DR_{max\,DC} = 70$ dB, at a supply voltage of 1 V. For field-effect transistors with an offset of 3 mV the result is $DR_{max\,DC} = 50$ dB. Only chopping can elevate this limit.

3.5.9 Conclusion

We have shown that the thermal noise in the gain-setting resistors principally limits the dynamic-range over supply-power ratio of an analog signal operation to $DR_{max}/P_{sup} = (4/\pi)/4kTB_e$. At a supply voltage of 1 V and a gain setting resistor of 10 kΩ in a current-to-voltage converter, the supply power is 16 μW for sinusoidal signals and the dynamic range in a bandwidth of 1 MHz can never be better than $DR_{max} = 89$ dB, even for an ideal operational amplifier.

This maximum can only be obtained if the operational amplifier is able to, firstly, efficiently feed the full supply voltage range from rail-to-rail (R-R) to the load, and secondly, efficiently feed the supply current in a class-AB mode to the load or feedback resistor. In some cases, for instance in a voltage buffer input application, the operational amplifier must also be able, thirdly, to efficiently allow a common-mode signal from rail-to-rail. These three requirements impose the important requirements in the design of the input stages of Chap. 4 and output stages of Chap. 5.

In amplifiers where the above requirements cannot be met, the dynamic range over power ratio is lower. For instance, when a diode prevents the output to go from rail-to-rail, but only allows an output voltage swing of 0.3 V at a supply voltage of 1 V, the dynamic range loses 10 dB. When the biasing is not class-AB, but class-A, we lose at least 5 dB or much more at low signal levels.

When we process analog signals through a bipolar current mirror without emitter degeneration, we lose 16 dB of dynamic-range over power ratio. With degenerated or CMOS current mirrors the loss is still 10 dB or more. Moreover, the accuracy and linearity of the overall transfer function is only as good as the matching of highly nonlinear internal transistor-diode characteristics. For these reasons we will leave out current-mode amplifiers from this book.

3.6 Problems

3.6.1 Problem 3.1

The circuit in Fig. 3.3 shows a non-inverting voltage amplifier built around the operational amplifier model explained with Fig. 2.2. The impedances in the external circuit are $Z_S = 10$ kΩ, $Z_1 = 10$ kΩ, and $Z_2 = 40$ kΩ. Z_l is the load impedance and has a value of 50 kΩ. The parameters for the operational amplifier macromodel are: $V_{ioffs} = 10$ mV, $I_{ioffs} = 10$ nA, $I_{ibias} = 50$ nA, $Y_{id} = 1/20$ MΩ, $Y_{ic1} = Y_{ic2} = 1/100$ MΩ, $H_i = 80$ dB, $Y_t = 100$ S, and $Y_o = 1/1$ kΩ. Calculate the output voltage for an input voltage of $V_S = 10$ mV, 100 mV, 1 V.

3.6.1.1 Solution 3.1

The matrix equation (Eq. 3.3) shows the relations between input/output voltages and currents for the application circuit. Solution starts by calculating the parameters $A_{vn}, A_{ve}, Y_{te}, Z_{te}, A_{ie}$.

These values can be used to calculate the nominal and error terms in Eq. 3.3:

$$\frac{1}{A_{vn}} = \frac{Z_1}{Z_1 + Z_2} = 0.2$$

$$\frac{1}{A_{ve}} = \frac{Y_o + Y'_s}{Y_t} + \frac{Y_{id}(Y_o + Y'_s)}{Y_p Y_t} + \frac{1}{A_{vn}H_i} - \frac{Y_{ic1}}{Y_p A_{vn}} = 13.5 \times 10^{-6}$$

$$\frac{1}{Y_{te}} = \frac{1}{Y_t} + \frac{Y_{id}}{Y_p Y_t} = 0.010004$$

$$\frac{1}{Z_{te}} = \frac{Y_{id}(Y_o + Y'_s)}{Y_t} + \frac{Y_{ic2}}{A_{vn}} = 2.10^{-9}$$ (3.18)

$$\frac{1}{A_{ie}} = \frac{Y_{id}}{Y_t} = 5.10^{-10}$$

$$I_{1offs} = I_{ibias} = 50 \text{ nA}$$

$$V_{1offs} = V_{ioffs} + \frac{I_{ioffs}}{Y_p} + \frac{I_{ibias}}{Y_p} = 10.48 \text{ mV}$$

With these values replaced in the matrix equation (Eq. 3.3), and considering the voltage-current relations at the input and output

$$V_S - V_1 = Z_S I_1$$
$$V_2 = Z_l I_2$$ (3.19)

With these values replaced in the matrix equation (Eq. 3.3), and considering the voltage-current relations at the input and output

$$V_S - V_1 = Z_S I_1$$
$$V_2 = Z_l I_2$$ (3.20)

the following equation is obtained for V_2:

$$V_2 \frac{V_S - Z_S I_{1offs} - V_{1offs}}{\frac{1}{A_{vn}} + \frac{1}{A_{ve}} - \frac{1}{Z_l Y_{te}} + \frac{Z_S}{Z_{te}} - \frac{Z_S}{Z_l A_{ie}}}$$ (3.21)

Replacing the values for circuit parameters and signal source voltage, the three output voltage values result:

$$V_2(10 \text{ mV}) = -4.9 \text{ mV}$$
$$V_2(10 \text{ mV}) = 445 \text{ mV}$$ (3.22)
$$V_2(1 \text{ V}) = 4.94 \text{ V}$$

3.6.2 Problem 3.2

Figure 3.6 shows a voltage-to-current converter built around the OFA macromodel explained with Fig. 2.4. The external components are $Z_S = 10 \text{ k}\Omega, Y_2 = 1/50 \text{ k}\Omega, Y_l = 1/50 \text{ k}\Omega$. The macromodel parameters have the following values: $V_{ioffs} = 10$ mV, $I_{ioffs} = 10$ nA, $I_{ibias} = 100$ nA, $Y_{idd} = 1/10$ MΩ, $Y_{ic1} = Y_{ic2} = 1/100$ MΩ, $H_i = 80$ dB, $Y_t = 0.1$ S, $Y_{odd} = 1/1$ MΩ, $I_{obias} = 100$ nA, $Y_{oc1} = Y_{oc2} = 1/100$ MΩ, $H_o = 60$ dB. Calculate the output current I_2 for a signal voltage $V_S = 100$ mV, 1 V, 5 V.

3.6.2.1 Solution 3.2

The matrix equation (Eq. 3.6) shows the relations between input/output voltages and currents for the application circuit. Solution starts by calculating the parameters $Y_{tn}, Y_{te}, A_{ve}, A_{ie}, Z_{te}$.

$$\frac{1}{Y_{tn}} = -\frac{1}{Y_2} = -50.10^3$$

$$\frac{1}{Y_{te}} = -\frac{Y_2 + Y_{odd}}{Y_2 Y_t} - \frac{Y_{idd}(Y_2 + Y_{odd})}{Y_2^2 Y_t} - \frac{1}{H_i Y_2} +$$

$$\frac{Y_{ic1}}{Y_2^2} - \frac{2}{H_o Y_2} + \frac{Y_{oc2}}{Y_2^2} = -65.55$$

$$\frac{1}{A_{ve}} = -\frac{Y_{odd}}{Y_t} - \frac{Y_{idd}Y_{odd}}{Y_t Y_2} - \frac{2Y_{odd}}{Y_t H_o Y_2} - \frac{Y_{oc1}}{Y_2} = -1.51.10^{-3} \qquad (3.23)$$

$$\frac{1}{A_{ie}} = \frac{Y_{idd}(Y_{odd} + Y_2)}{Y_2 Y_t} - \frac{Y_{ic1}}{Y_2} = -5.01 \cdot 10^{-4}$$

$$\frac{1}{Z_{te}} = -\frac{Y_{idd}Y_{odd}}{Y_t} = -1.10^{-12}$$

$$V_{1offs} = V_{ioffs} + \frac{I_{ioffs}}{Y_2} + \frac{I_{ibias}}{Y_2} + \frac{2I_{obias}}{Y_2} = 0.025 \text{ V}$$

$$I_{1offs} = I_{ioffs} + I_{ibias} = 110 \text{ nA}$$

With these values replaced in the matrix equation (Eq. 3.6) and considering the voltage-current relations at the input and output

$$V_S - Z_S I_1 = V_1$$
$$I_2 = -Y_l V_2 \qquad (3.24)$$

the following equation is obtained for I_2:

$$I_2 = -\frac{V_S - Z_S I_{1offs} - V_{1offs}}{\frac{1}{Y_{in}} + \frac{1}{Y_{te}} + \frac{1}{Y_l A_{ve}} + \frac{Z_S}{A_{ie}} + \frac{Z_S}{Z_{te} Y_l}} \tag{3.25}$$

Replacing the values for circuit parameters and signal source voltage, the three output voltage values result:

$$I_2(10 \text{ mV}) = -1.47 \text{ μA}$$
$$I_2(1 \text{ V}) = -19.4 \text{ μA} \tag{3.26}$$
$$I_2(5 \text{ V}) = -99.18 \text{ μA}$$

3.6.3 Problem 3.3

The instrumentation amplifier shown in Fig. 3.14b relies on a differential input buffer and an inverting differential amplifier. Using the resistor nominal values $R_1 = R_3 = 4$ kΩ, $R_2 = R_4 = 40$ kΩ, a supply voltage $V_{SP} = 3$ V, $V_{SN} = 0$ V and nonideal operational amplifiers with input common mode voltage range $\Delta_{VIN} = (V_{SP} - 1$ V; $V_{SN} - 0.5$ V), output voltage range $\Delta_{VOUT} = (V_{SP} - 0.2$ V; $V_{SN} + 0.2$ V) and $V_{ioffs} = 10$ mV, calculate the maximum input signal which can be amplified without distortion. Using this value, calculate the ratio of dynamic range over dissipated power which can be obtained with this amplifier over a bandwidth $\Delta f = 1$ MHz if each operational amplifier draws $I_d = 100$ μA and has an input noise of $S_f = 10$ nV/√Hz. The resistors have a precision of $\Delta R = 2\%$. Boltzmann's constant is $K = 1.38 \times 10^{-23}$, temperature $T = 300$ K.

3.6.3.1 Solution 3.3

The first limit is introduced at the input by the differential buffer, as both input operational amplifiers are connected as repeaters, thus not allowing all of the input voltage range to be used:

$$V_{imin} = V_{ioffs} + V_{VOUTmin} = V_{SN} + 0.2 \text{ V} + V_{ioffs} = 0.210 \text{ V}$$
$$V_{imax} = V_{VINmax} - V_{ioffs} = V_{SP} - 1.0 \text{ V} - V_{ioffs} = 1.990 \text{ V} \tag{3.27}$$

Another factor of R_2/R_1 is lost in order not to saturate the output, which limits the input signal to

$$V_{imax} - V_{imin} = \frac{(V_{VOUTmax} - V_{VOUTmin})}{10} = 0.26 \text{ V} \tag{3.28}$$

Considering also the maximum gain

$$A_{Vmax} = \frac{R_2(1 + \Delta R)}{R_1(1 - \Delta R)} = 10.4$$

$$V_{imax} - V_{imin} = \frac{(V_{VOUTmax} - V_{VOUTmin})}{A_{Vmax}} = 0.25 \text{ V}$$
(3.29)

this voltage range is reduced furthermore by the unwanted effect of the third operational amplifier input offset voltage:

$$V_{imax} - V_{imin} = \frac{(V_{VOUTmax} - V_{VOUTmin})}{AV_{max}} - 2V_{ioffs} = 0.248 \text{ V}$$
(3.30)

The dynamic range of the circuit is limited by the noise power and the maximum input signal. The input referred noise power is

$$P_n = 3S_f^2\Delta f + 4KTR_1\Delta f + 4KTR_3\Delta f$$
(3.31)

and is composed from buffer noise and equivalent resistor noise (virtually the equivalent resistors are equal with R_1, R_3). The dynamic range can now be calculated

$$DR = 10\log_{10}\frac{(V_{imax} - V_{imin})^2}{P_n} = 81.5 \text{ dB}$$
(3.32)

The ratio of dynamic range over dissipated power is

$$\frac{DR}{P_d} = 112 \text{ dB}$$
(3.33)

References

1. J.G. Graeme et al., *Operational Amplifiers Design and Applications* (McGraw Hill Book Company, New York, 1971)
2. C.D. Motchenbacher, F.C. Fitchen, *Low-Noise Electronic Design* (John Wiley and Sons, New York, 1973)
3. J.H. Huijsing, Instrumentation amplifiers: a comparative study on behalf of monolithic integration. IEEE Trans. Instrum. Meas. **IM-25**, 227–231 (1976)
4. B.J. Dool, J.H. Huijsing, Indirect current feedback instrumentation amplifier with a common-mode input range that includes the negative rail. IEEE J. Solid-State Circuits **28**(7), 743–749 (1993)
5. J.D. Voorman, The gyrator as a monolithic circuit in electronic systems, Thesis, Katholieke Universiteit Nijmegen, Gema B.V., Eindhoven, June 1977
6. K.M. Adams, E.F.A. Deprettere, J.O. Voorman, The gyrator in electronic systems, in *Advances in Electronics and Electron Physics*, vol. 37 (Academic, San Francisco, 1975), pp. 79–179
7. G. Groenewold, Optimal dynamic range integrated continuous-time filters, PhD thesis, Delft University of Technology, Delft, The Netherlands, 1992

Chapter 4
Input Stages

Abstract The input stage of an operational amplifier has the task of sensing the differential input voltage. This process is disturbed by interference signals such as offset, bias, drift, noise, and common-mode crosstalk. The modeling of these signals has been given in Chap. 2. The level of these additive interference signals determines the useful sensitivity of the amplifier. The design of the input stage should aim at low values of these interference signals, while the current consumption should be low, and a large portion of the rail-to-rail range should be available for common-mode signals.

The input stage of an operational amplifier has the task of sensing the differential input voltage. This process is disturbed by interference signals such as: offset, bias, drift, noise, and common-mode crosstalk. The modeling of these signals has been given in Chap. 2. The level of these additive interference signals determines the useful sensitivity of the amplifier. The design of the input stage should aim at low values of these interference signals, while the current consumption should be low, and a large portion of the rail-to-rail range should be available for common-mode signals.

The discussion of input stages will be divided into aspects of: offset, bias, and drift in Sect. 4.1, noise in Sect. 4.2, common-mode crosstalk in Sect. 4.3; and the design of rail-to-rail input stages in Sect. 4.4.

4.1 Offset Bias, and Drift

The quiescent input voltage and current, which are needed to drive the ,active elements at the input of an amplifier into their normal working range, result in equivalent offset and bias quantities at the input of an amplifier (see Chap. 2). Variations in these quantities as a function of time, supply voltage, or ambient temperature are referred to as drift.

The product of differential input voltage and input current is the input sensing power which must be supplied by the external source circuit. It is clear that a high sensitivity requires a low input sensing power in order to distinguish the differential DC signal from offset, bias, and drift.

© Springer International Publishing Switzerland 2017   57
J. Huijsing, *Operational Amplifiers*, DOI 10.1007/978-3-319-28127-8_4

Generally, the lowest input sensing power, or product of input voltage and current, is achieved if the active elements at the input are connected in the general-amplifier (GA) connection (common-emitter or common-source connection). This is the reason why all effective input stages have a GA connection.

One can apply two general techniques to reduce the effect of quiescent input voltages and currents: isolation and balancing. We will review these techniques.

4.1.1 Isolation Techniques

The input offset and bias quantities are the manifest results of the quiescent voltages and currents of the active elements in the input circuit. One of the most successful ways of reducing offset and bias is to apply active elements whose input is electrically isolated from the internal quiescent voltages and/or currents.

Examples of this electrical isolation can be found in the group of parametric amplifiers, such as the magnetic amplifier and the vibrating capacitor electrometer. A monolithically integrable variant of the latter is the varactor electrometer amplifier with varactor diodes as voltage-dependent capacitances. With a varactor amplifier an input offset current as low as 10 fA can be obtained, while the offset voltage is of the order of 1 mV [1].

Another elaboration of the principle of isolation is the separation of useful frequency regions from frequency bands where additive interference signals such as offset, drift, and 1/f-noise can be expected. This idea is realized in the chopper amplifier or in the chopper-stabilized amplifier [2] (see Chap. 10 of this book). The residual offset quantities are mainly caused by capacitive crosstalk of the chopper driving signals. With CMOS chopper switches offset values of 1 μV and 1 nA can be obtained.

The principle of isolation can also be applied to reduce the effects of variations in the environmental conditions. Some examples are: stabilizing the supply voltages, isolating the chip from changes in the ambient temperature or stabilizing the chip temperature, isolating the chip from mechanical vibrations, and avoiding the influence of chemical reactions by the use of stable materials or an effective shielding.

A design which aims at a low value of the offset and drift must generally have an input stage with a sufficient amount of gain to shield the influence of offset and drift of stages behind the input stage.

Basically, isolation, shielding, or stabilization can be applied to any degree of perfection by increasing the "isolation barrier." This is in contrast to balancing where the result is limited by the accuracy with which components can be matched, which is in turn dependent on the precision of the integration process. However, the abovementioned isolation techniques cannot always be applied. Then, we must rely on balancing techniques, which is the subject of the following section.

4.1.2 Balancing Techniques

Balancing techniques reduce the input offset voltage and current of a single device into those of a differential transistor pair. This will be expressed for bipolar transistors in Eqs. 4.1, 4.2, and 4.3 for the balanced stage of Fig. 4.1 and for CMOS transistors in the Eqs. 4.4a, 4.4b, 4.5a, 4.5b, and 4.6a, 4.6b for the balanced stage of Fig. 4.2 respectively.

For the single bipolar transistor [3] the input bias voltage and current is:

$$V_{BE} = \frac{kT}{q} \ln \frac{I_C W_b}{A_e K} + V_G,$$

$$I_B = I_C / \beta_F; \tag{4.1}$$

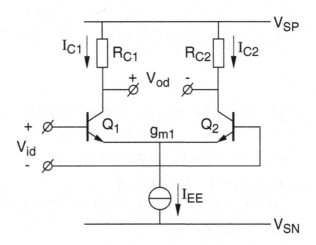

Fig. 4.1 Balanced bipolar-transistor input stage

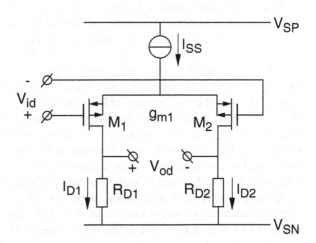

Fig. 4.2 Balanced CMOS-transistor input stage

in which: k is the Boltzmann constant, T the absolute junction temperature, q the absolute value of the charge of an electron, I_C the collector current, W_b the base width, A_e the effective emitter area, K a constant proportional to T^{-4} and dependent on the doping profile, β_F the large signal current gain, and V_G the bandgap voltage which is about 1.2 V in silicon [4].

For the balanced configuration of Fig. 4.1 the input offset voltage, offset current and bias current are:

$$V_{IOFFS} = V_{BE1} - V_{BE2} = \frac{kT_1}{q} \ln \frac{I_{C1} W_{b1}}{A_{e1} K_1} - \frac{kT_2}{q} \ln \frac{I_{C2} W_{b2}}{A_{e2} K_2} + V_{G1} - V_{G2}$$

$$\simeq \frac{kT}{q} \left(\frac{\Delta R_C}{R_C} + \frac{\Delta W_b}{W_b} - \frac{\Delta A_e}{A_e} - \frac{\Delta K}{K} \right) + \Delta T \frac{k}{q} \ln \frac{I_C W_b}{A_e K} + \Delta V_G$$

$$\tag{4.2}$$

$$I_{IOFFS} = (I_{B1} - I_{B2})/2 \simeq \frac{I_C}{2\beta_F} \left(\frac{-\Delta R_C}{R_C} - \frac{\Delta \beta_F}{\beta_F} \right)$$

$$I_{IBIAS} = (I_{B1} + I_{B2})/2 \simeq \frac{I_C}{\beta_F}$$

in which

$$I_{C1} = (V_{SP} - V_{REF})/R_{C1}, I_{C2} = (V_{SP} - V_{REF})/R_{C2},$$

$$I_C = (I_{C1} + I_{C2})/2, \Delta R_C = R_{C1} - R_{C2}, R_C = (R_{C1} + R_{C2})/2,$$

$$\Delta W_b = W_{b1} - W_{b2}, W_b = (W_{b1} + W_{b2})/2, \Delta K = K_1 - K_2, K = (K_1 + K_2)/2,$$

$$\Delta A_e = A_{e1} - A_{e2}, A_e = (A_{e1} + A_{e2})/2,$$

$$\Delta T = T_1 - T_2, T = (T_1 + T_2)/2, \Delta V_G = V_{G1} - V_{G2}, \Delta \beta_F = \beta_{F1} - \beta_{F2},$$

$$\beta_F = (\beta_{F1} + \beta_{F2})2$$

Balancing greatly reduces the offset voltage to a value of the order of 0.2 mV. The balancing owes its success to the well-determined voltage and current relations of bipolar transistors. The offset voltage of the balanced stage consists of three terms. The first term $(kT/q)(-\Delta R_C/R_C + \Delta W_b/W_b - \Delta A_e/A_e - \Delta K/K)$ represents layout mismatches and doping inequalities of the two transistors Q_1 and Q_2. This term is the dominating one. It has a value of the order of 0.2 mV for the present state of the art. The second term $(\Delta Tk/q) \ln (I_C W_b/A_e K)$ can be rated lower than 20μV if the difference between the junction temperatures is supposed to be lower than 0.01 K [5]. The third term ΔV_G involves differences between the bandgap voltages, which can arise from differences in the mechanical strains of the two junctions. For two transistors situated close together in the center of the chip the voltage difference ΔV_G can be estimated lower than 20μV [3].

A way to reduce the bias currents for bipolar transistors is to compensate these currents with internal current sources, see Fig. 6.6b. Here the matching accuracy is limited to a value of the order of the offset current. This method raises the input noise current minimally with a factor $\sqrt{2}$. To further lower the offset and bias current super-β transistors can be applied [6], although, this requires an additional step in the integration process.

The temperature drift of the balanced bipolar circuit of Fig. 4.1 can be expressed as (for definitions see Eqs. 4.1 and 4.2):

$$
\begin{aligned}
\frac{\partial V_{IOFFS}}{\partial T} &\simeq \frac{k}{q} \ln \frac{R_{C2} A_{e2} K_2 W_{b1}}{R_{C1} A_{e1} K_1 W_{b2}} \simeq \frac{V_{IOFFS}}{T}, \\
\frac{\partial I_{IOFFS}}{\partial T} &\simeq \frac{\partial \beta_F}{\partial T} \frac{I_{IOFFS}}{\beta_F}.
\end{aligned}
\tag{4.3}
$$

Note that if the offset of a bipolar pair is trimmed to zero, also the offset drift is zero. With CMOS transistors the offset voltage and drift behave more complicatedly. There are two basic operation ranges: Weak inversion at low current densities, with an exponential V_{GS}–I_D function like that of bipolar transistors, and strong inversion at high current densities, with a quadratic V_{GS}–I_D function. For most input stages operation in moderate inversion close to weak inversion is chosen, as that results in the highest transconductance G_M at a given tail current [7], while the parasitic capacitors are still relative small. But sometimes, for a better linearity or better high frequency behavior, even smaller transistors with small input capacitors are chosen to operate in strong inversion. In contrast to bipolar transistors, where the NPN is often better than the PNP transistor, in CMOS the P-channel transistor is often better for an input stage at low frequencies than the N-channel transistor. The P-channel one has lower 1/f noise and in most processes the back gate can be connected to the source, which improves the CMRR of the input stage (see Sect. 4.3).

In weak inversion the V_{GS}–I_D function can be expressed as:

$$
V_{GS} = V_{TH} + \frac{nkT}{q} \ln \frac{I_D I}{I_{DO} W}
\tag{4.4a}
$$

in which V_{TH} is the threshold voltage, n is a factor of approximately 1.6 in weak inversion, I_{DO} is a leakage current, and W/L is the width over length ratio of the CMOS transistor. The resulting V_{GS} is roughly 60 mV above the threshold voltage V_{TH} in moderate inversion close to weak inversion. This value represents a reasonable compromise between a small transistor size $W \times L$ and a large as possible G_M.

The offset voltage can be expressed like that of the bipolar transistor as:

$$
V_{IOFFS} \simeq \Delta V_{TH} + \frac{nkT}{q} \left(\frac{\Delta R_D}{R_D} - \frac{\Delta I_{DO}}{I_{DO}} + \frac{\Delta L}{L} - \frac{\Delta W}{W} \right) + \Delta T \frac{nk}{q} \ln \frac{I_D L}{I_{DO} W}
\tag{4.5a}
$$

Unlike the offset voltage of the bipolar transistor pair, the offset voltage of a CMOS transistor pair has a large threshold voltage offset term ΔV_{TH} in addition to the geometric offset voltage term [8]. The difference in threshold voltage ΔV_{TH} is most difficult to control. This voltage difference depends on irregularities of the channel doping and charge inclusion in the gate oxide of CMOS transistors. Therefore, in weak inversion the threshold offset voltage is dominating and can be of the order of 2 mV, which is ten times larger than the offset of bipolar transistors.

The drift over temperature in weak inversion is:

$$\frac{\partial V_{IOFFS}}{\partial T} \simeq \frac{\partial V_{TH}}{\partial T} + \frac{nk}{q} \ln \frac{R_{D2}I_{DO2}W_2L_1}{R_{D1}I_{DO1}W_1L_2} \simeq \frac{\partial V_{TH}}{\partial T} + \frac{V_{IOFFS}}{T} \qquad (4.6a)$$

Unlike the temperature drift of a bipolar transistor pair, the temperature drift of a CMOS transistor pair in weak inversion is not zero if the offset is zero, but the additional drift term due to the threshold voltage is present.

In strong inversion the gate-source voltage V_{GS} of a single CMOS transistor stage is a square-root function of the current:

$$V_{GS} = V_{TH} + \sqrt{2I_D/K} \qquad (4.4b)$$

in which V_{TH} is the threshold voltage; I_D is the drain current; $K = \mu C_{ox}W/L$ a main CMOS transistor parameter; W/L is the width over length ratio of the channel; C_{ox} is the normalized gate-oxide capacitance; μ is the mobility of the charge carriers in the channel, which is about a factor 3 more for N doped channels than for P doped ones.

For the balanced input stage of Fig. 4.2 in strong inversion the offset is:

$$V_{IOFFS} = V_{GS1} - V_{GS2} = \ \Delta V_{TH} + \frac{1}{2}\left(\frac{-\Delta R_D}{R_D} + \frac{\Delta K}{K}\right)\sqrt{2I_D/K}$$

$$V_{IOFFS} = \Delta V_{TH} + \frac{1}{2}\left(\frac{-\Delta R_D}{R_D} + \frac{\Delta K}{K}\right)(V_{GS} - V_{TH}) \qquad (4.5b)$$

in which: $\Delta V_{TH} = V_{TH1} - V_{TH2}$; $\Delta R_D = R_{D1} - R_{D2}$; $\Delta K = K_1 - K_2$; and $\sqrt{(2I_D/K)} = V_{GS} - V_{TH}$, and average values I_D, K, V_{GS}, V_{TH}.

The offset voltage clearly consists of two terms. The difference in threshold voltage ΔV_{TH} in the left-hand term is nearly a constant voltage, while the two components $\Delta R_D/R_D$ and $\Delta K/K$ in the right-hand term are proportional to the square root of the drain current I_D, or directly proportional to $(V_{GS}-V_{TH})$.

In strong inversion, at relative high currents, the right-hand term may dominate the offset.

The bias and offset current is very low. With junction FETs the bias current is equal to that of the saturation current I_S of a bipolar diode, of the order of nA. With CMOS the bias current is equal to the leakage current of the gate oxide, of the order of pA to fA.

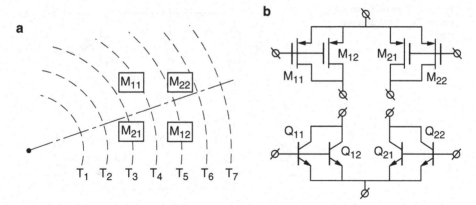

Fig. 4.3 (**a**) Four transistors in a cross-coupled layout subjected to a temperature gradient. (**b**) Balanced input circuit with four transistors in the cross-coupled quad layout of Fig. 4.1 and 4.2

The temperature drift of the offset voltage in strong inversion is

$$\frac{\delta V_{I\,OFFS}}{\delta T} = \frac{\delta\,\Delta V_{TH}}{\delta T} - \frac{1}{2}\left(\frac{\Delta R_D}{R_D} + \frac{\Delta K}{K}\right)\left(\frac{\delta I_D}{I_D\delta T} - \frac{\delta K}{K\delta T}\right)\sqrt{\frac{2I_D}{K_9}}$$

$$= \Delta\alpha - \frac{31}{4T}\left(\frac{\Delta R_D}{R_D} + \frac{\Delta K}{K}\right)(V_{GS} - V_{TH})$$

(4.6b)

with: $V_{TH} = V_{TH,THo} - \alpha(T - T_o)$, $\Delta\alpha = \alpha_1 - \alpha_2$, $\mu = k_\mu T^{-3/2} I_D =$ constant, and $K = \mu C_{ox} W/L$

The effect of gradients in doping, temperature and strain can be canceled in first-order approximation by using double-balanced transistors in a cross-coupled quad layout [6]. This method is visualized in Fig. 4.3a, b. For linear gradients the internal pair $M_{12}M_{21}$ or $Q_{12}Q_{21}$ and external pair $M_{11}M_{22}$ or $Q_{11}Q_{22}$ cancel each other's effects.

An additional consideration related to doping profiles is to give all transistors the same orientation. For equal stress the aluminum and oxide profiles should be equal around all transistors.

4.1.3 Offset Trimming

For bipolar transistors (Eq. 4.2), we can draw the conclusion that if we trim the ratio R_{C2}/R_{C1} such that $R_{C2}/R_{C1} = A_{e1}K_1W_{b2}/A_{e2}K_2W_{b1}$, then both the main term of the offset voltage $V_{ioffs} \approx (kT/q)\ln(R_{C2}A_{e2}K_2W_{b1}/R_{C1}A_{e1}K_1W_{b2})$ (see Eq. 4.2) and the temperature drift of the offset voltage will be zero. We can regard the two transistors of the balanced stage and their collector resistors as a kind of geometric bridge circuit. Once the bridge balance is set to zero, its temperature drift will also

Fig. 4.4 A basic trimming
circuit for a bipolar input
stage

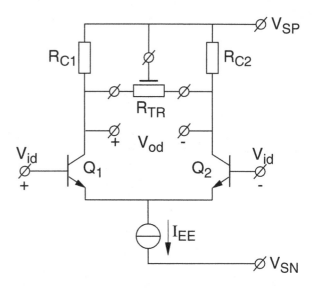

Fig. 4.4 A basic trimming circuit for a bipolar input stage

be set to zero. A basic trim circuit for a bipolar input stage is drawn in Fig. 4.4 with a potentiometer. The potentiometer can be trimmed, i.e., with laser trimming or Zener zapping, or by a multiplying DAC. The stage behind should not disturb this balance.

For trimming a CMOS stage we encounter the difficulty that the temperature coefficient of the left-hand threshold term differs from the temperature coefficient of the right-hand geometric term in (Eqs. 4.5a and 4.5b). So two trim actions are needed. The left-hand threshold offset term can be trimmed by inserting an adjustable voltage in series with the gate-source circuit, as shown in Fig. 4.5. Or alternatively, by adding a parallel input stage with same current density but lower bias current and with a trim voltage at its input. The right-hand geometric term can be resistively trimmed in the same way as the bipolar circuit such that $\Delta R_D/R_D$ compensates $\Delta K/K$ over temperature voltage. In weak inversion at low bias current the voltage trim is dominant, while in strong inversion at high bias current the resistive trim is dominant.

A well-trimmed bipolar input stage may have a temperature coefficient of the offset voltage lower than 1 μV/K [9], while a junction-FET or CMOS stage has a coefficient lower than 10 μV/K [10, 11].

The supply voltage dependence of the input offset voltage is equivalent to the terms of the common-mode voltage crosstalk if these terms are separated for parasites connected to the positive and negative supply voltages. This will be treated in Sect. 4.3.

With a balanced input stage one can only profit from a low offset voltage V_{IOFFS} and offset current I_{IOFFS} if the source circuit is also balanced to cancel the bias current I_{IBIAS}. This is particularly important for a bipolar input stage. Figure 4.6 shows a balanced source circuit connected to the input of an OpAmp, whose offset and bias quantities are shown. Both bias currents of the OpAmp must flow back into

Fig. 4.5 A basic trimming
circuit for a CMOS input
stage

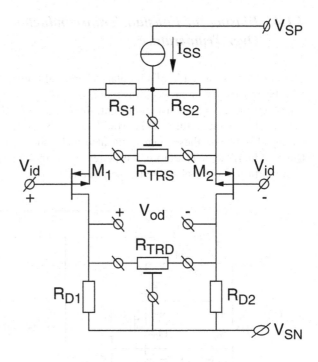

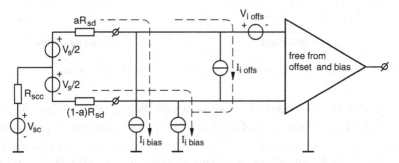

Fig. 4.6 A balanced source circuit connected with the input of an OpAmp whose offset and bias
quantities are shown separately

ground via the source circuit. This results in two error voltages $-aR_{sd}I_{I\ BIAS}$ and $+$
$(1 - a)R_{sd}I_{I\ BIAS}$ of opposite polarities in series with the input terminals of the
OpAmp. These error voltages cancel each other if the two parts aR_{sd} and $(1-a)R_{sd}$
of the differential source resistance R_{sd} are equal, i.e., if a $= 1/2$. Hence, when a
balanced source circuit is used only the offset current $I_{I\ OFFS}$ results in an error
voltage of $R_{sd}I_{I\ OFFS}$. But when an asymmetric source circuit is used with a $= 1$, we
get the much larger error voltage of $R_{sd}I_{I\ BIAS}$. We can expect the offset current to be
a factor of ten or more lower than the bias current.

4.1.4   Biasing for Constant Transconductance G_m Over Temperature

The transconductance G_m of a bipolar differential transistor pair is I_{CQ}/kT, and of a CMOS transistor pair in weak inversion is I_{DQ}/nkT, with about $n = 2.4$. If these stages are biased by a constant current, their G_m would decrease proportionally to the absolute temperature. This would not only affect the gain of an OpAmp equipped with one these stages, but also make the frequency compensation (see Chap. 6) inefficient. Therefore, it is better to bias these input stages by a current that is proportional to the absolute temperature (PTAT).

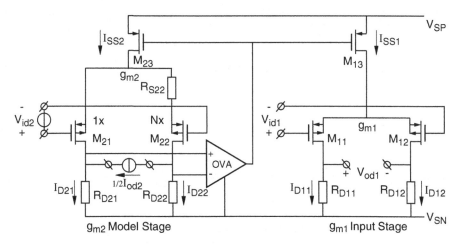

Fig. 4.7 Input stage with constant G_m over temperature by model bias generator

For a CMOS pair biased in strong inversion the G_m is $2I_D/(V_{GS} - V_{TH}) = \mu Cox(W/L)(V_{GS} - V_{TH}) = \sqrt{\{\mu Cox(W/L)(2I_D)\}}$. This function is proportional to the root of mobility times drain current $\sqrt{\mu I_D}$. The sensitivity of μ to the absolute temperature is less than inverse proportional. If we want a constant G_m over the temperature range, we have to bias the stage by a current I_D that is proportional to the inverse mobility (PTIM).

The question is of course: how to build a biasing circuit that can handle the different requirements?

The idea is to just force a constant G_m on a model pair by feedback, and use the same current generated in the model pair for the input stage. The situation is shown in Fig. 4.7.

Firstly, we look at the model stage with an artificial input voltage source V_{id2} and an artificial output current source $1/2I_{od2}$, while the value of the asymmetrical source resistor R_{S22} is zero and M_{22} is equal to M_{21}, so Nx is 1x. The OVA regulates the tail current in such a way that the boundary conditions of input voltage and output current are met. This forces the G_m of the model to be:

$$G_m = I_{od2}/V_{id2} \qquad (4.7)$$

for relative small signals of V_{id2} and I_{od2}. The transconductance G_m has a nonlinear slanted S-shaped function of the input voltage. But it is quite linear in the middle (see Sect. 6.4). If we do not take too large signals, then the value of G_m is approximately valid for input voltages with a maximum of 50 mV for bipolar and 100 mV for CMOS transistors.

When we apply the same tail current of the model stage for the input stage we can expect the same G_m, under the condition of equal type of transistors and same current densities as those of the model. If the input voltage V_{id2} is derived from a voltage across a resistor R_m through which a current is flowing equal to the output current I_{od2}, the transconductance G_m becomes $G_m = 1/R_m$. This result is approximately independent of the process and even independent of the type of transistor. Measures have to be taken to ensure start-up of the bias circuit.

Secondly, we try to avoid the use of an artificial input voltage V_{id2} and output current I_{od2}. So we give them a zero value. This means a shortcircuit input and an open output. Instead, we take the transistor M_{22} N times larger (wider) than M_{21} and insert a resistor R_{S22} in series with the source of M_{22}. The result is that the output currents I_{D21} and I_{D22} are regulated to be equal, supposedly R_{D21} and R_{D22} are equal. This means that an N times wider M_{22} in regard to M_{21} has to catch up for the voltage loss across R_{S22}. Derived we get for the transconductance of the model:

$$G_m = \ln N/R_{S22}. \qquad (4.8)$$

This result is again approximately independent of the process and even independent of the type of transistor. If we bias the input stage with the same tail current as that of the model the input stage will have the same transconductance G_m as that of the model, under the condition of equal type of transistors and same current densities as those of the model [12].

The model circuit can be simplified [12] to that of Fig. 4.8. The model contains four transistors M_{21} through M_{24} in a nearly unity gain positive feedback loop. The transistor M_{22} is N times larger than M_{21}. The resistor R_{S22} determines the current I_{S22}. If the current is smaller than the nominal one, the transistor M_{22} dominates by its multiple N and the current increases. If the current is larger than the nominal one, the source resistor R_{S22} reduces the current of transistor M_{22} and the current decreases. If the tail current I_{SS1} of the input stage is biased by two times the current $I_{ang1024\ S22}$ in the model, so $I_{SS1} = 2I_{S22}$, the G_m of the input stage is described by (Eq. 4.8). This is under the condition of equal type of transistors and same current densities as those of the model.

A disadvantage of the simple bias circuit of Fig. 4.8 is that the progressive Early voltages of the transistors in the loop deteriorate its accuracy and supply-voltage rejection. A measure we can take is to aid the transistors of the bias circuit and the

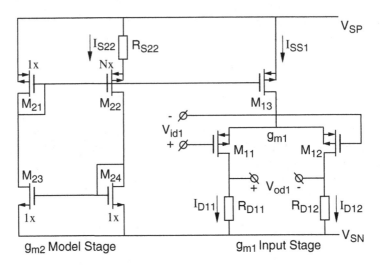

Fig. 4.8 Simple bias circuit for an input stage with constant G_m over temperature

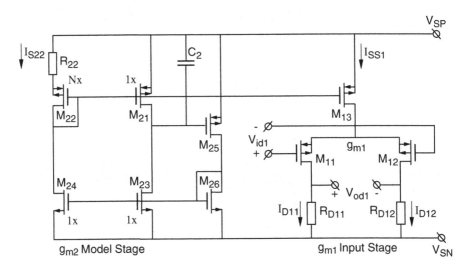

Fig. 4.9 Low voltage bias circuit for an input stage with constant G_m over temperature

tail current of the input stage with cascode transistors. A disadvantage of this measure is that the minimum supply voltage is lowered.

A circuit that can work at very low supply voltage is drawn in Fig. 4.9.

The circuit has two loops: Firstly, a fast but not dominating loop through transistors M_{25}, M_{26}, and M_{23} with positive feedback. This can become oscillative. Therefore a small capacitor C_2 is placed across the gate and source of M_{25}. Secondly, a dominating loop through transistors M_{25}, M_{26}, M_{24}, M_{22}, and M_{21}

with negative feedback. The advantage of this circuit is that all current-determining transistors have equal Early voltages for all supply voltages [13]. The minimum supply voltage is equal to one V_{GS} and one saturation voltage V_{SAT}, which can be totally of the order of 1 V. If the tail current I_{SS1} of the input stage is biased by two times the current I_{S22} in the model, so $I_{SS1} = 2I_{S22}$, the G_m of the input stage is described by (Eq. 4.8). This is under the condition of equal type of transistors and same current densities as those of the model.

The bias circuits for input stages with constant G_m over temperature have to be given some support circuitry to start up reliably [14].

4.2   Noise

Noise can be regarded as a fluctuation of the input bias and offset quantities. Therefore, the same techniques can be applied as with bias, viz., isolation and balancing. Balancing will help less with noise because of the random character of noise.

4.2.1   Isolation Techniques

An equivalent series spectral noise voltage source V_n and an equivalent parallel spectral noise current source I_n at the input port of an amplifier represent all noise sources of a linear amplifier, as shown in Fig. 4.10. Often it gives us more feeling for the amount of noise, if we compare the noise with that of a resistor. Therefore, we translate the noise voltage into that of an equivalent series noise resistance $R_{eqs} = V_n^2/4kT$ and the noise current into that of an equivalent parallel noise resistance $R_{eqp} = 4kT/I_n^2$.

The first step for low-noise design is to make the noise voltage and current as low as possible by themselves. This means R_{eqs} low and R_{eqp} high.

With bipolar transistors the values of the equivalent noise resistors are $r_{eqs} \simeq r_e/2 + r_{bb}$ and $r_{eqs} \simeq 2\beta r_e$ in which r_e is the small-signal emitter resistance, r_{bb} the ohmic base resistance and β the small-signal current amplification factor.

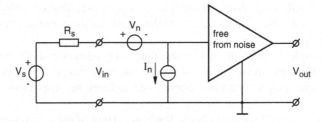

Fig. 4.10 Noise sources in a single input stage

A practical value of r_{eqs} is between 1 and 10,000 Ω or higher, and of r_{eqp} between 0.1 and 1000 kΩ or higher depending on the quiescent current. With junction FETs we find an equivalent series noise resistance $Z_{eqs} = \delta/g_m$, with $\delta \approx 2/3$. The equivalent parallel noise resistor represents the isolation of the gate by the gate oxide. This resistance is very high. A practical value of r_{eqs} is of the order of 500 Ω and of r_{eqp} is of 100 MΩ at low frequencies. In addition to the given noise values, we must regard l/f-noise which dominates with bipolar transistors in the frequency range roughly below 100 Hz [9].

With CMOS transistors the value of the 1/f-noise below 10 kHz is so high that these transistors are not used in the input stage of low-noise amplifiers, unless a chopper amplifier configuration is used to convert the low-frequency signals into a high-frequency band, as explained in Chap. 10. The noise voltage of CMOS transistors is inverse proportional to the root of their gate area.

The second step in low-noise design is to optimally adapt the ratio of V_n and I_n to the signal-source resistance R_s.

The overall noise behavior may be described by the noise figure F [9]. This figure F can be calculated as the total available noise power of the source P_{ns} increased by the equivalent noise power $P_n = V_n I_n$ of the amplifier divided by the noise power of the source P_{ns}.

$$F = \frac{P_{ns} + P_n}{P_{ns}} = \frac{\left(V_{ns}^2 + V_n^2 + I_n^2 R_s^2\right)/4R_s}{V_{ns}^2/4R_s}$$

$$F = 1 + \left(V_n^2/4R_s + I_n^2 R_s/4\right)/kT = 1 + R_{eqs}/R_s + R_s R_{eqp} \qquad (4.9a)$$

$$F_{min} = 1 + V_n I_n/2kT = 1 + 2\sqrt{R_{eqs} R_{eqp'}}$$

$$\text{at}: R_{sopt} = V_n/I_n = \sqrt{R_{eqs} R_{eqp}}$$

The minimum value of F can be found by choosing an optimum value for the source resistance

$$R_{sopt} = V_n/I_n = \sqrt{R_{eqs} R_{eqp}} \qquad (4.9b)$$

with an equivalent series and parallel noise resistance, respectively $R_{eqs} = V_{n2}/4kT$ and $R_{eqp} = 4kT/I_n^2$. This only makes sense under the condition that the source power $P_s = V_s^2/R_s$ itself does not decrease. So optimalization by adding a series resistance or applying a parallel resistor to the source only deteriorates the available signal power of the source.

As an example in which the source resistance may be enlarged or reduced while maintaining the same available signal power, one may choose a pick-up coil of a microphone with a larger or a lower number of windings, but with the same volume of copper.

As another example we may choose the bias current of the active input devices larger or smaller in such a way that a better optimization can be obtained.

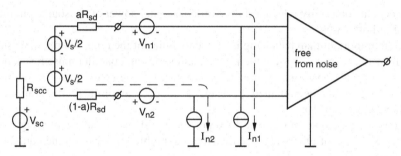

Fig. 4.11 Noise sources of the balanced input stage

4.2.2 Balancing Techniques

The noise figure of a balanced input stage of Fig. 4.11 is described by Eq. 4.10 in the case of a symmetrical source circuit (a = 1/2).

$$F = 1 + \left(V_n^2/2R_{sd} + I_n^2 R_{sd}/8\right)/kT = 1 + 2R_{eqs}/R_{sd} + R_{sd}/2R_{eqp'}$$

$$F_{min} = 1 + V_n I_n/2kT = 1 + 2\sqrt{R_{eqs} R_{eqp'}} \quad (4.10)$$

$$\text{at}: R_{sd\,opt} = 2V_n/I_n = 2\sqrt{R_{eqs}R_{eqp'}}$$

For an asymmetrical source (a = 1) with a balanced input circuit we find:

$$F = 1 + \left(V_n^2/2R_{sd} + I_n^2 R_{sd}/4\right)/kT = 1 + 2R_{eqs}/R_{sd} + R_{sd}/2R_{eqp'}$$

$$F_{min} = 1 + V_n I_n/2kT = 1 + 2\sqrt{2R_{eqs}R_{eqp'}} \quad (4.11)$$

$$\text{at}: R_{sd\,opt} = \sqrt{2}V_n/I_n = \sqrt{2R_{eqs}R_{eqp'}}$$

In the asymmetrical case (a = 1) the noise contribution of one noise current source is multiplied by the full source resistance. While in the symmetrical case the noise currents of both sources is multiplied by half the source resistance and then added as the root of the squares. This results in a factor √2 lower noise.

Low-noise operation with F < 2 is possible at values of R_{sd} between the equivalent series noise resistance R_{eqs} and the equivalent parallel noise resistance R_{eqp}.

4.2.3 Conclusion

From Eqs. 4.9a–4.11 we can draw the conclusions that the minimum noise figure F_{min} of the balanced input stage with a symmetrical source circuit equals that of a single transistor stage, while the use of an asymmetrical source circuit together with a balanced input stage leads to a √2 larger value of the minimum noise figure.

This conclusion is only relevant if the available signal power of the source circuit is equal in both cases.

More important than balancing is a proper choice of the type of input transistors and their bias currents, so that their equivalent series and parallel noise resistances R_{eqs} and R_{eqp} are positioned geometrically around the source resistance R_s. Finally, low 1/f noise can be obtained by using a chopper amplifier configuration, as explained in Chap. 10.

Low-noise design implies that the input stage will have a sufficiently high gain so that the noise of other stages will have a negligible influence on the equivalent input noise sources.

4.3   Common-Mode Rejection

The common-mode rejection ratio (CMRR) H (used here without index i) or the supply-voltage rejection ratio (SVRR) describe the influence of a common-mode (CM) input voltage or supply-voltage variation on the differential-mode (DM) driving of the amplifier. In Chap. 2 the common-mode crosstalk ratio (CMCR) was defined as the inverse ratio 1/H. This is the ratio of an equivalent DM input voltage and the CM input voltage, which brings it about.

Chapter 3 showed that a low value of the CMCR 1/H is needed in all those applications in which an accurate equation is needed between the two input voltages. A low CMCR can be obtained by two methods: by isolation techniques and by balancing techniques. These two methods are symbolized in Figs. 4.12 and 4.13 respectively, and will be discussed in this section.

4.3.1   Isolation Techniques

Electrical isolation of the input circuit is a basic method for obtaining a low common-mode crosstalk ratio (CMCR). Electrical isolation naturally adapts to the requirement of a floating input port, because if we isolate the input circuit, a

Fig. 4.12 Common-mode rejection by balancing techniques

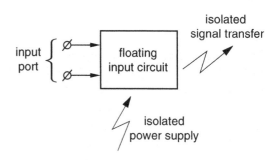

Fig. 4.13 Common-mode
rejection by isolation
techniques

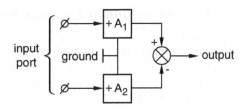

common-mode voltage cannot influence the current distribution in the circuit. Imperfections of the isolation determine the value of the CMCR. Isolation can principally be realized to any large degree of perfection.

Isolated signal transfer and power supply (Fig. 4.12) can be obtained by all kinds of energy carriers. One can envision coupling by optical energy, magnetic energy, by the magnetic field energy of a transformer, or flow of voltage-isolated electrical charge. The last method is the most obvious to cope with both signal transfer and power supply on a chip. There are two different ways in which this method can be applied.

Firstly, floating capacitors can be switched to and from the isolated input circuit to the other parts of the amplifier. The capacitors can carry energy and signal values. Secondly, electronic current sources can continuously carry power and signal without making a voltage-dependent connection. The latter way of isolation by electronic current sources will be discussed in more detail in combination with balancing.

4.3.2   Balancing Techniques

Balancing of the input circuit is another method for obtaining a low common-mode crosstalk ratio (CMCR). Figure 4.13 shows a basic configuration of grounded amplifiers with the gain factors of A_1 and A_2. The CMCR $1/H$ of the balanced circuit equals the relative inaccuracy of the matching of the amplification factors:

$$1/H = A_c/A_d = \Delta A/A \qquad (4.12)$$

with: $A_c\ A_1 - A_2 = \Delta A, A_d = (A_1 + A_2)/2 = A$

If only balancing techniques were used in a monolithically integrated amplifier without trimming, the CMCR could not be guaranteed to be lower than of the order of 1/1000 because the integrated resistors cannot be matched better than of the order of 0.1 %. This is in contrast to isolating techniques which can basically be applied to any grade of perfection at low frequencies. At high frequencies parasitic coupling capacitors limit the isolation quality.

4.3.3   Combination of Isolation and Balancing

We conclude that a low common-mode crosstalk ratio (CMCR) which is inverse to
the CMRR can best be achieved by using both techniques together. Current source
isolation provides the main step on the way to a low CMCR. Balancing further
reduces the CMCR.

The combination of electronic current-source isolation and balancing is depicted
in the long-tailed bipolar transistor pair of Fig. 4.14a. The simplified equivalent
circuit of Fig. 4.14b clearly shows that if the tail contains an electronic current
source I_{EE}, and if the input transistors have a current-source output, the complete
input circuit is only connected by current sources with the surrounding parts. This
means that the common-mode input voltage has no influence on the distribution of
the current in the input circuit. With the ideal current sources of Fig. 4.14b it does
not even matter whether the circuit is balanced or not.

Fig. 4.14 (a) Bipolar long-
tailed-pair input stage.
(b) Simplified equivalent
circuit of the bipolar
long-tailed-pair input
stage which shows the
current-source isolation
of the CM input voltage

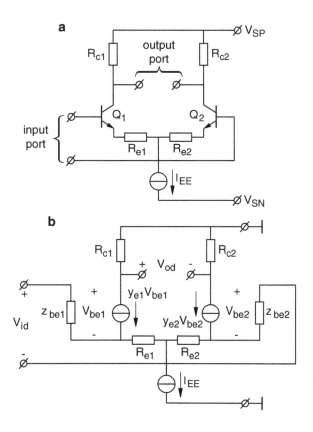

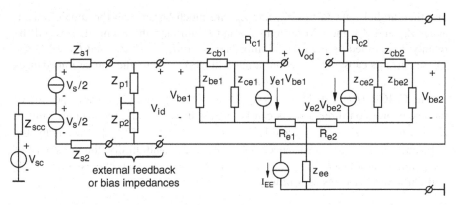

Fig. 4.15 Small-signal equivalent circuit of a bipolar long-tailed transition pair with parasitic impedances across the current sources

4.3.4   Common-Mode Cross-Talk Ratios

In practice, the equivalent circuit must be made complete by adding parasitic impedances in parallel with the current sources. This is shown in Fig. 4.15 where each transistor is represented by its hybrid-π equivalent circuit. The hybrid-π model for bipolar transistors has been chosen because it can easily be adapted to represent the HF behavior or to symbolize JFETs and CMOS transistors. The tail current source has also been provided with its parallel impedance Z_{ee}. At the input side any externally grounded feedback or bias impedances or stray capacitances have been represented by Z_{p1} and Z_{p2}.

The values of the parasitic impedances determine the degree of isolation, and the mutual inequalities of the parasitic impedances determine the degree of balancing. Together they determine the common-mode crosstalk ratio (CMCR).

The effect of the parasitic impedances on the CMCR will be discussed separately for each of the parasitic types. A new calculation method will be presented which distinguishes the aspect of isolation from that of balancing. With this method a better estimation of the crosstalk ratio as a function of a certain design aspect can be made than with preceding calculation methods.

4.3.5   Parallel Input Impedance

The effect of the parallel input impedances Z_{p1} and Z_{p2} on the CMCR will be discussed in detail, while other effects are dealt with only briefly. The common-mode crosstalk ratio (CMCR) is defined as the ratio of the equivalent DM input voltage which is evoked by the CM input voltage. Hence, let us supply a CM input voltage V_{ic} and see what is the effect on the DM input voltage V_{id}. If we suppose that the parallel impedances Z_{p1} and Z_{p2} are much larger than the source impedances Z_{s1} and Z_{s2}, the CM currents I_{s1} and I_{s2} through these impedances will be mainly determined by the parallel impedances, so: $I_{s1} \approx V_{ic}/Z_{p1}$ and $I_{s2} \approx V_{ic}/Z_{p2}$. These currents cause a differential input voltage V_{id} across the source impedances of: $V_{id} = -Z_{s1}I_{s1} + Z_{s2}I_{s2} \approx V_{ic}\left(-Z_{s1}/Z_{p1} + Z_{s2}/Z_{p2}\right)$.

The CMCR $1/H_1$ is $1/H_1 = V_{id}/V_{ic} = -Z_{s1}/Z_{p1} + Z_{s2}/Z_{p2}$. On inspection of the last expression three aspects can be distinguished: firstly, the average ratio of Z_s/Z_p with $Z_s = (Z_{s1} + Z_{s2})/2$ and $1/Z_p = Y_p = \left(1/Z_{p1} + 1/Z_{p2}\right)/2$, secondly, the imbalance of the source resistances $\Delta Z_s = Z_{s1} - Z_{s2}$, and thirdly, the imbalance of the parallel admittances $\Delta Y_p = 1/Z_{p1} - 1/Z_{p2}$. We can now express the CMCR with these three aspects as follows:

$$1/H_1 \approx \left(Z_s/Z_p\right)\left(-\Delta Z_s/Z_s - \Delta Y_p/Y_p\right) \tag{4.13}$$

The first factor of the CMCR (Z_s/Z_p) will be called the "isolation factor." It expresses the relative isolation of the source when loaded by the parallel impedances. The second factor $(\Delta Z_s/Z_s - \Delta Y_p/Y_p)$ will be called the "balancing factor." It expresses the relative imbalance of the source and parallel impedances. These impedances are connected in the form of a bridge circuit. Overall, the expression for the CMCR has a clear structure and can be easily understood. In the extreme situation of a fully unbalanced source or parallel load circuit, when one of the source impedances equals zero or one of the parallel impedances infinity, the balancing factor equals plus or minus two. In that situation the suppression of the crosstalk fully relies on the isolation factor (Z_s/Z_p).

To depict the frequency dependency, the CMRR H_1 is drawn in Fig. 4.16 as a function of the frequency in the case in which both source impedances amount to 1 kΩ and each of both parallel or grounded impedances consists of only one capacitance whose values differ by 1 pF. The example gives an idea about which values are attainable as a function of the frequency.

The effects of the other parasitic impedances will be treated in the same way as has just been seen. However, the explanation will be shortened because of their similarity, the definitions for average values and unbalanced values will not be repeated.

4.3.6   Collector or Drain Impedance

All effects are supposed ideal except for the collector-emitter impedance.

The collector-emitter impedances z_{ce1} and z_{ce2} carry the full input CM voltage V_{ic}. This results in equivalent base-emitter voltages $V_{be1} = -V_{ic}z_{e1}/z_{ce1}$ and $V_{be2} = -V_{ic}z_{e2}/z_{ce2}$. The equivalent differential input voltage $V_{id} = V_{be1} - V_{be2} = V_{ie}(-z_{e1}/z_{ce1} + z_{e2}/z_{ce2})$ represents a CMCR:

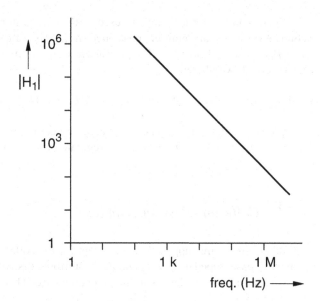

Fig. 4.16 CMRR of a balanced source circuit with two series resistors of 1 kΩ and an imbalance of 1 pF in the grounded load capacitors

$$1/H_2^* = (z_e/z_{ce})(-\Delta z_e/z_e - \Delta y_{ce}/y_{ce}) \qquad (4.14)$$

The reciprocal gain ratio $(1/\mu) = (z_e/z_{ce})$ of transistors of the same type tends to match better than their z_e and z_{ce} values separately. This holds for bipolar transistors, where z_e and z_{ce} are both inversely proportional to the quiescent emitter current, as well as for JFETs and CMOS transistors, where the $1/g_m$ and z_{ds} are both likewise dependent on the quiescent source current and on the length-to-width ratio. For this reason, it is realistic to replace (Eq. 4.15) for (Eq. 4.16):

$$1/H_2 = (1/\mu)(-\Delta\mu/\mu) \qquad (4.15)$$

A lower CMCR can only be obtained by cascoding the input stage while the reference base or gate voltage of the cascode stage is bootstrapped by the CM input voltage.

4.3.7 Tail Impedance

The tail impedance Z_{ee} also carries the full CM input voltage V_{ic}. This causes a tail current $I_{ee} = V_{ic}/Z_{ee}$. This current brings about a crosstalk if the emitter circuit or the collector circuit is unbalanced:

$$1/H_3^* = \{(R_e + z_e)/2Z_{ee}\}\{-\Delta(R_e + z_e)/(R_e + Z_e) + \Delta R_c/R_c\} \qquad (4.16)$$

When a tail current source is used with a transistor Q_3 in a common base connection with an emitter resistance R_{e3}, the value of $1/Z_{ee}$ is $1/Z_{ee} = 1/\mu(R_{e3} + z_{e3}) + 1/\beta_3 r_{ce3} + 1/r_{cb3}$. Taking only the first term of this expression into account, we obtain:

$$1/H_3 = \{(R_e + z_e)/2\mu(R_{e3} + z_{e3})\}\{-\Delta(R_e + z_e)/(R_e + z_e) + \Delta R_c/R_c\} \quad (4.17)$$

A lower CMCR can only be obtained by cascoding the tail-current source, or if the tail-current source can be actively regulated for a constant output current.

4.3.8   Collector–Base Impedance

All other effects are supposed ideal, except for the collector–base impedance. The collector–base impedances Z_{cb1} and Z_{cb2} contribute firstly to the crosstalk ratio like the parallel input impedances in the preceding case. This contribution is:

$$1/H_{41} = (Z_s/z_{cb})(-\Delta Z_s/Z_s - \Delta y_{cb}/y_{cb}) \quad (4.18)$$

Secondly, the collector–base impedances cause a direct signal transfer from the input towards the output. The collector–base impedances and the collector load resistances R_{c1} and R_{c2} make up a bridge circuit. The CMCR can be calculated by dividing the transfer of that bridge by the differential-to-differential voltage gain factor $A_{dd} \approx R_c/(R_e + z_e)$. This results in:

$$1/H_{42} = \{(R_e + z_e)/z_{cb}\}(-\Delta R_c/R_c - \Delta y_{cb}/y_{cb}) \quad (4.19)$$

This CMCR component is absent in CMOS at low frequencies.

4.3.9   Base Impedance

The base-emitter impedances z_{be1} and z_{be2} connect the input terminals with the emitter circuit. Via these connections the base currents can reach the input terminals. The common-mode base currents are a factor of $1/\beta = z_e/z_{be}$ lower than the CM tail current $I_{ee} = V_{ic}/Z_{ee}$ and the collector current $I_c = V_{ic}/z_{ce}$. These base currents load the source circuit and a CMCR arises:

$$1/H_5 = (Z_s/\beta z_{ce} + Zs/\beta 2Z_{ee})\{-\Delta Z_s/Z_s - \Delta(1/\beta)/(1/\beta)\} \quad (4.20)$$

This CMCR component is absent if the tail-current source is ideal.

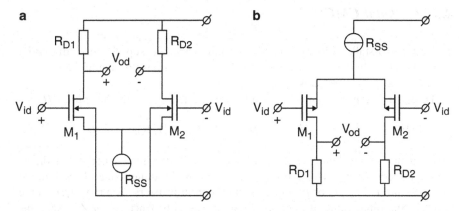

Fig. 4.17 (a) N-channel CMOS input stage with substrate-connected back-gate bias. (b) P-channel CMOS input stage with source-connected back-gate bias

4.3.10   Back-Gate Influence

N-channel CMOS input stages have a strong disadvantage regarding CMCR. Figure 4.17a shows the situation. Besides the normal gate, there is the back-gate, which is connected to the substrate in most CMOS processes for the N-channel transistor. The back-gate has a transconductance g_{mb} of the order of ten times smaller than the normal transconductance g_m. This limits the isolation factor g_{mb}/g_m to about 10^{-1}.

If we estimate differences in back-gate transconductances of the order of $\Delta g_{mb}/g_{mb} = 1\%$, then an additional partial crosstalk ratio will be added

$$1/H_{N\ CMOS} = g_{mb}/g_m \cdot \Delta g_{mb}/g_{mb} \tag{4.21}$$

which is of the order of 10^{-3} or -60 dB.

This severely limits the minimum CMCR of N-channel CMOS stages. With P-channel CMOS stages of Fig. 4.17b the situation is better, because we have access to the back-gate and are able to connect this to the source. The remaining CMCR of P-channel CMOS input stages in strong inversion is maximally 10^{-4} or -80 dB, which is still a factor of 10 worse than that of bipolar transistors because the internal gain factor μ of FETs is lower than that of bipolar transistors. However, the lower the bias current is taken, or the larger the W/L ratio, the more the FET approaches the gain factor of bipolar transistors, because the g_m is proportional to the root of the current and W/L ratio. In weak inversion the g_m is only roughly a factor of about 2 lower than that of the bipolar transistor. Again, cascoding of the input stage with bootstrapping of the cascode gate voltage by the CM input voltage can strongly improve the CM crosstalk.

4.3.11   Total CMCR

The total crosstalk ratio $1/H_t$ can be found when we superimpose all partial crosstalk ratios. In the worst case, the total crosstalk ratio is the sum of the absolute values of the partial crosstalk ratios:
The most dominating partial crosstalk ratios are $1/H_2$ and $1/H_3$. The ratio $1/H_2$ is a

$$|1/H_t| \le |1/H_1| + |1/H_2| + |1/H_3| + |1/H_{41}| + |1/H_{42}| + |1/H_5| \qquad (4.22)$$

consequence of the finite value of the collector-emitter impedance z_{ce} or drain-source impedance Z_{ds}. This ratio is determined by the finite value of the internal voltage gain $\mu = r_{ce}/r_e$ or $\mu = r_{ds}g_m$ of the input transistors. The ratio $1/H_3$ is the result of the finite value of the tail current-source impedance Z_{ee} or Z_{ss}. Also this ratio is mainly determined by the finite value of the internal voltage gain μ_3 of the current source transistor Q_3 or M_3 and its degeneration emitter or source resistor R_3.

With the bipolar circuit of Fig. 4.13 isolation factors of 2.10^{-4} and balancing factors of 5.10^{-2} can be obtained, leading to a CMCR of 10^{-5}. With CMOS we obtain a CMCR of 10^{-4}. At low frequencies, a practically unlimited improvement of the isolation can be obtained by using composite transistors for the input pair as well as for the tail-current source. Cascoding is one of the basic options.

To obtain a low overall CMCR $1/H$ of a complete OpAmp the input or first stage should have a low total CMCR $1/H_1$ (the index number refers to the stage number here) as well as a low reciprocal discrimination factor $1/F_1$. This reciprocal factor is defined as the ratio of the CM-to-CM and the DM-to-DM voltage gains. The reciprocal discrimination factor for the input circuit of Fig. 4.15 is:

$$1/F_1 = (R_e + z_e)(-1/2Z_{ee} + 1/\beta r_{ce} + 1/r_{cb}) \qquad (4.23)$$

This expression shows that isolation is the only method to obtain a low reciprocal discrimination factor. A low reciprocal discrimination factor reduces the contribution of the CMCR $1/H_2$ of the second amplifier stage to an overall value of $1/F_1H_2$. It further reduces the extra crosstalk which arises if one of the stages is used single-ended, as is required in an OVA. If the second stage with a reciprocal discrimination factor $1/F_2$ is used single-ended, the contribution to the crosstalk is $1/F_1H_2$. The overall CMCR is in that case:

$$|1/H_o| \le |1/H_1| + |1/F_1H_2| + |1/F_1F_2|. \qquad (4.24)$$

4.3.12   Conclusion

From the preceding discussion we can conclude that the common-mode crosstalk ratio (CMCR) of an input stage can be made low by isolation and balancing. Partial crosstalk ratios, as a result of each kind of parasitic current-source impedances, can be expressed as a product of an isolation factor and a balancing factor. Isolation can be performed practically to any degree of perfection at low frequencies by using composite transistor combinations. Balancing depends on the matching inaccuracies of integrated components. The minimum dominating crosstalk is limited by the internal gain of the transistors and is $1/H_2 = (1/\mu)(\Delta\mu/\mu)$.

For bipolar transistors this is of the order of 10^{-5}, for P-channel CMOS transistors with source-connected back-gate 10^{-4}, and for N-channel CMOS transistors 10^{-3} because of the influence of the back-gate. Cascoding of the input transistors and the tail current source can drastically improve the CMRR. The CMRR further decreases as a function of the frequency because of the influence of parasitic parallel capacitors.

4.4 Rail-to-Rail Input Stages

The trend in lower voltages, going from 30, 12, 5, 3, 1.8, and 1.2 V or 0.9 V, forces us to design input stages which maximally utilize the voltage range between the negative and positive supply rails. For instance, a large input range is required for a voltage follower buffer application with a high input impedance, as shown in Sects. 3.2 and 3.4.

Looking at the usable input range of input stages of Fig. 4.18 we find that the common-mode input voltage range of a P-channel input pair, M_3–M_4, is limited between:

$$-V_{GS} + V_{Dsat} + V_{R3,4} + V_{SS} < V_{CM} < V_{DD} - V_{GS} - V_{Dsat} \qquad (4.25)$$

where V_{CM} is the common-mode input voltage, V_{GS} is the gate-source voltage, V_{Dsat} is the saturation voltage across a current-source, V_{DD} and V_{SS} are the positive and the negative supply-rail respectively.

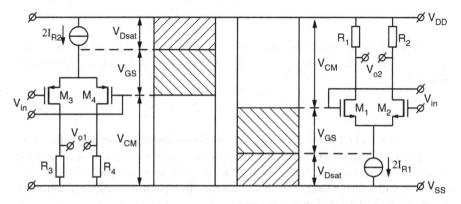

Fig. 4.18 Common-mode input voltage range of a P-channel and an N-channel input stage

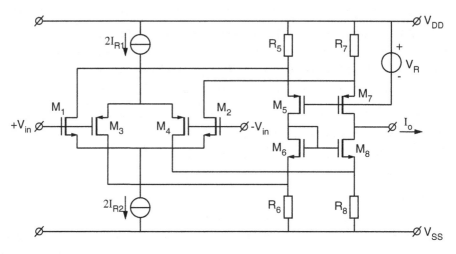

Fig. 4.19 Rail-to-rail CMOS complementary input stage

The common-mode input voltage range of an N-channel input pair, M_1–M_2, is given by:

$$V_{SS} + V_{GS} + V_{Dsat} < V_{CM} < V_{DD} + V_{GS} - V_{sat} - V_{R1,2} \qquad (4.26)$$

The CM range of the P pair may downwards exceeds the negative rail by $-V_{GS} + V_{sat} + V_{R3,4}$. Similarly, the CM range of the N pair may upwards exceeds the positive rail by $V_{GS}-V_{sat}-V_{R1,2}$. This will not be the case if the load resistances are replaced by diodes for current mirrors. For this reason current mirrors cannot be used behind the input stage, but rather folded cascodes.

If the N-channel and P-channel input pairs are placed in parallel, as is shown in Fig. 4.19, the common-mode input voltage range becomes:

$$-V_{GS} + V_{Dsat} + V_{R3,4} + V_{SS} < V_{CM} < V_{DD} + V_{GS} - V_{Dsat} - V_{R1,2} \qquad (4.27)$$

To avoid a forbidden voltage range in the middle of the rail-to-rail voltage range of the complementary input stage [1] the supply-voltage should have a minimum value of:

$$V_{sup,min} = 2V_{GS} + 2V_{Dsat} \qquad (4.28)$$

Using standard CMOS technology, the minimum supply voltage for full rail-to-rail operation is approximately 1.8 V, depending on the bias-current level and the threshold voltage of the transistors. In many CMOS processes transistors are available with much smaller threshold voltages. In bipolar technology a minimum supply voltage of 1.6 V can be obtained.

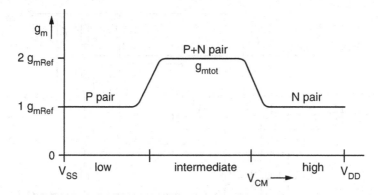

Fig. 4.20 The transconductance g_{mtot} versus the common-mode input voltage for a rail-to-rail complementary input stage

The folded cascode transistors M_5–M_8, together with the current-mirror connection of M_6 and M_8 add the output currents of the complementary input pairs.

A drawback of the simple complementary input stage of Fig. 4.19 is that the transconductance g_{mtot} varies a factor two over the whole common-mode input range depending on the saturating of tail current sources, as is shown in Fig. 4.20 supposedly that the g_m of each of the N en P channel pairs is equal to g_{mRef}. This impedes an optimal frequency compensation of the amplifier [15], as we will see in Chap. 6.

4.4.1 Constant g_m by Constant Sum of Tail-Currents

In bipolar technology and in weak-inversion CMOS, the g_m of a transistor is proportional to the collector or drain current. Therefore, a constant g_{mtot} can be obtained by keeping the sum of the tail-currents of the complementary input stages constant. A realization is shown in Fig. 4.21. Depending on the common-mode input voltage, the current switch Q_5 directs the tail-current $2I_{B1}$ to either one of the input stages [16]. The result is a constant g_{mtot} over the whole common-mode input range, as is shown in Fig. 4.22.

This result for bipolar and weak-inversion CMOS transistors is clear from:

$$g_{mtot} = g_{mN} + g_{mP} = \frac{I_{B1}}{V_T};$$
(4.29)

with: $V_T = kT/q$ for bipolar and about $V_T = 60$ mV for weak-inversion CMOS.

There is an undesirable property of a complementary rail-to-rail input stage. The input offset voltage changes between that of the NPN and PNP pair when the CM input voltage crosses the turnover range. The turnover range is stretching out

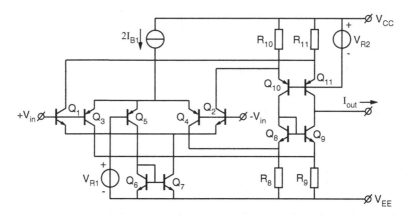

Fig. 4.21 Rail-to-rail complementary bipolar or weak-inversion CMOS input stage with switch Q_5 and 1:1 current mirror Q_6, Q_7 to keep the sum of the tail currents constant, and hence the g_{mtot} constant

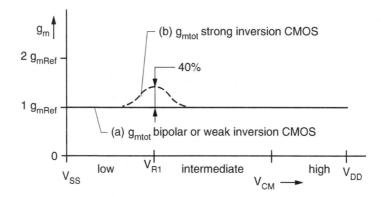

Fig. 4.22 The transconductance g_{mtot} versus common-mode input voltage V_{cm} for the rail-to-rail complementary input stage with a 1:1 current mirror for bipolar technology, and CMOS technology in weak and strong inversion

for about 100 mV for bipolar transistors, centered around the reference voltage V_{R1}. Proper circuit layout with a cross-coupled quad for each input pair can typically keep the untrimmed offset voltage change in the order of 0.1 mV for bipolar transistors.

This means that the CMRR in the turnover CM range is:

$$1/H = 0.1/100 = 1/1200, \text{ or } 60 \text{ dB} \qquad (4.30)$$

The CMCR can be improved by a factor of 10 or 20 dB by inserting a resistor in series with the emitter of Q_5 because this spreads the turn-over range. However, this somewhat increases the minimum supply voltage for rail-to-rail operation.

In CMOS technology, the g_m of a transistor is also proportional to the drain-current if the transistor operates in weak-inversion. This biasing has the advantage that it gives CMOS transistors the highest transconductance of about $g_m = I_{B1}/60$ mV $(4.4.5)$ for a given bias current. Another advantage is that the g_{mtot} can be kept constant in the rail-to-rail input circuit of Fig. 4.21 as in bipolar technology. It is assumed that the P-channel transistors have a W/L that is about three times larger than that of the N-channel transistors, to compensate a three-times smaller mobility μ of the P-channel transistors.

In strong-inversion however, this scheme leads to a g_{mtot} which increases approximately 40 % in the middle of the common-mode range in regard to that in the side ranges, as shown in Fig. 4.22. This is because we have two times the g_m at half the current, while the g_m of a MOS transistor in strong-inversion is proportional to the square-root of the drain-current, according to:

$$g_m = K(V_{GS} - V_{TH}) = \sqrt{2KI_D} \qquad (4.31)$$

with: $K = \mu C_{ox}W/L$, μ is the mobility of the charge carriers, C_{ox} is the normalized gate-oxide capacitance, and W/L is the width over length ratio of the channel.

A P-N complementary circuit is shown in the rail-to-rail input stage of Fig. 4.23. If both input pairs operate equally in the middle of the common-mode (CM) range, the tail-currents of both input pairs are equal and have a value of $2I_{Ref}$. If only one input pair operates, the current-switches M_5 and M_8, and the 1:1 current mirrors M_6–M_7 and M_9–M_{10}, increase the tail-current of the active input pair by a factor 2 by adding the tail current of the inactive pair. Thus, the tail-current of the actual active input pair has a value of $4I_{Ref}$. The result is a constant $g_{mtot} = 2I_{B1}/V_T$ for bipolar transistors, and about $g_{mtot} = 2I_{B1}/60$ mV for CMOS in weak inversion, see Eq. 4.30.

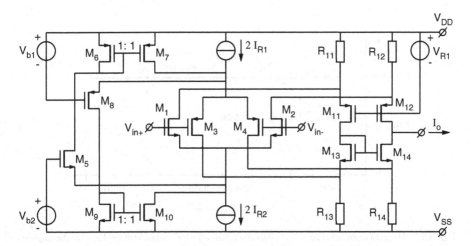

Fig. 4.23 Rail-to-rail CMOS input stage with 1:1 current mirrors to keep the g_{mtot} constant in weak inversion

The R-R input circuit of Fig. 4.23 has two transition ranges in the CM voltage range centered around V_{B2} and V_{DD}–V_{B1}. We estimate a total offset change of 1 mV for CMOS with a carefully designed circuit layout. Then, the CMRR in both the transition ranges of 200 mV for CMOS is:

$$1/H = 1/(2 \times 200), \text{ or } 53\text{dB} \tag{4.32}$$

Outside the transition ranges the CMRR is much higher, depending on the CMRR of the single P and N channel pairs.

4.4.2   Constant g_m by Multiple Input Stages in Strong-Inversion CMOS

If the CMOS input transistors have to operate in strong inversion for high speed or another reason, the proportionality of g_m to the root of the bias current (Eq. 4.32) will cause a 30 % lower total transconductance g_{mtot} outside the switching voltage levels V_{B2} and V_{DD}–V_{B1} of the input CM range than the transconductance of $2g_{mRef}$ in the middle, see Fig. 4.24.

This reduction in g_{mtot} at the extremes can be changed into two hills of 14 % at the switching voltages if the mirrored tail currents are not added to the tail of the active transistor pair, but instead given to one of the tails of an additional parallel N and P CMOS pair. In this way not only the current doubles in the active pairs, but also the total transistor width doubles. Thus the total transconductance comes back at $2g_{mRef}$ at the high and low ends of the transient CM range [11, 17, 18].

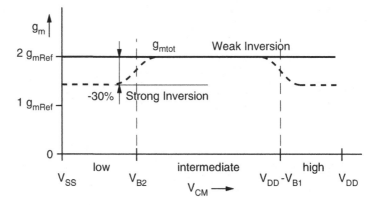

Fig. 4.24 The transconductance g_{mtot} versus the common-mode input voltage V_{CM} for the rail-to-rail complementary input stage with 1:1 current mirrors of Fig. 4.23 for CMOS in weak inversion

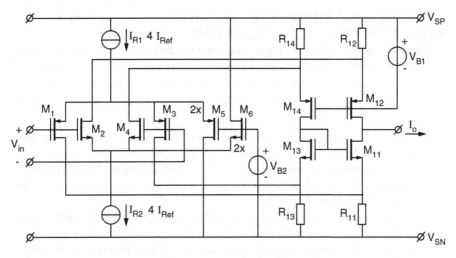

Fig. 4.25 Constant-g_{mtot} rail-to-rail input stage with current spillover control

4.4.3 Constant g_m by Current Spillover Control

Another simple and robust approach to an R-R complementary input stage with a reasonable constant g_{mtot} by means of "current spillover control" [19] is shown in Fig. 4.25. The control transistors M_5 and M_6 spill the current which is not needed by the input pairs M_1M_3 and M_2M_4 into the supply rails.

The transistors M_5 and M_6 together with the input pairs M_1M_3 and M_2M_4 shape a voltage translinear loop:

$$V_{GS1,3} + V_{GS2,4} = V_{GS5} + V_{GS6} \tag{4.33}$$

We choose the W/L ratio of M_5 and M_6 two times that of M_1 through M_4 for biasing in weak inversion. And we choose all P-channel transistors roughly three times wider than the N-channel transistors for compensating the g_ms for the mobility differences between the P and N complementary pairs.

$$
\begin{aligned}
I_{S1,3} \cdot I_{S2,4} &= I_{S5} \cdot I_{S6} \\
I_{S1,3} \cdot I_{S2,4} &= \left(2I_{Ref} - I_{S1,3}\right) \cdot \left(2I_{Ref} - I_{S2,4}\right) \\
I_{S1,3} + I_{S2,4} &= 2I_{Ref} \\
g_{mP} + g_{mN} &= 2g_{mRef}
\end{aligned}
\tag{4.34}
$$

In the middle position of the CM voltage at $V_{CM} = V_{B2}$ the currents through both pairs are equal $I_{S1,3} = I_{S2,4} = I_{Ref}$, and $I_{S5} = I_{S6} = 2I_{Ref}$. The transconductance of the two pairs in parallel equals $g_{mP} + g_{mN} = 2g_{mRef}$.

At the end of the CM range we have $I_{S1,3} = 0$ and $I_{S2,4} = 2I_{Ref}$, or $I_{S1,2} = 2I_{Ref}$ and $I_{S2,4} = 0$ resulting in $g_{mtot} = 2g_{mRef}$ according to (Eq. 4.35).

When we choose to bias the input transistors in strong inversion, the sum of the voltages is proportional to the sum of the roots of the currents in strong inversion, with $g_{mRef} = \sqrt{2KI_{Ref}}$, $K = \mu C_{ox}W/L$. Then the total transconductance of the two pairs in parallel equals $g_{mtot} = 2\sqrt{2}g_{mRef}$ in the middle. This is 40 % larger than the $2g_{mRef}$ of the left and right hand side position.

The input stage with current spillover control is simpler than that with the mirror current control. Moreover, the circuit can switch faster because there are no mirrors. The circuit with current spillover control is very robust and can be adapted to obtain other features [19]. A disadvantage is that it takes two times more bias current than without spillover control (Fig. 4.26).

One of the disadvantages of most R-R input stages is that the CM output drain currents of each pair varies from 0 to $4I_{Ref}$ as a function of the CM input voltage. The summing circuit has to cope with this variation and may produce a CM voltage dependent offset. It is also difficult to optimize the summing circuit for noise and offset in combination with a rail-to-rail CM input range, as the voltage across R_{11}–R_{14} cannot become larger than some hundreds of mVolt. To help solve this problem, the circuit with current spillover control can be adapted to produce a constant CM output current. Therefore, we split up the switching transistor M_5 and M_6 into two transistors each M_5, M_7 and M_6, M_8 and connect the drains with the outputs of the input pairs, as shown in Fig. 4.27.

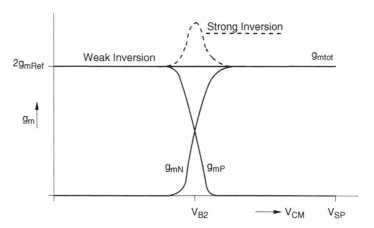

Fig. 4.26 The transconductance g_{mtot} of the rail-to-rail input stage with current spillover control of Fig. 4.25

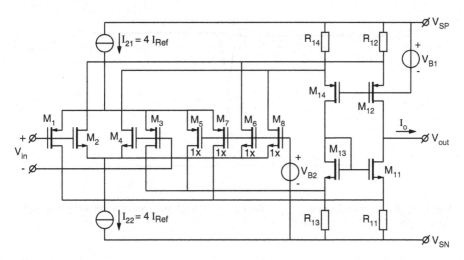

Fig. 4.27 Current spillover switches keep the CM output current of the input pairs constant as well as the g_{mtot} in weak inversion

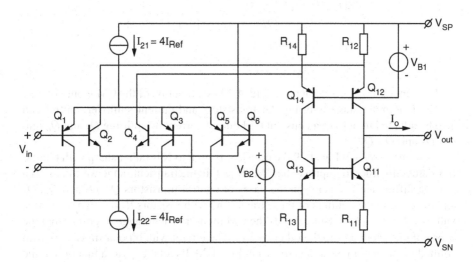

Fig. 4.28 Bipolar rail-to-rail input stage with current spillover control

Current spillover control can also be used to the advantage of a bipolar input stage as shown in Fig. 4.28. The translinear loop keeps the total g_{mtot} theoretically constant as calculated by (Eq. 4.36) and (Eq. 4.37) [19].

$$V_{BE1,3} + V_{BE2,4} = V_{BE5} + V_{BE6} \tag{4.35}$$

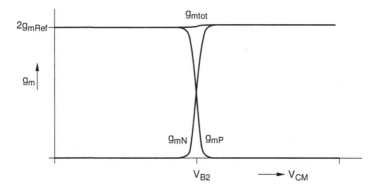

Fig. 4.29 The transconductance g_{mtot} of the bipolar or weak inversion CMOS input stage with current spillover control of Fig. 4.28

When we suppose emitter areas for Q_5 and Q_6 two times larger than that of Q_1, Q_3, and Q_2, Q_4, respectively, we find:

$$
\begin{aligned}
I_{E1,3} \cdot I_{E2,4} &= I_{E5} \cdot I_{E6} \\
I_{E1,3} \cdot I_{E2,4} &= \left(2I_{Ref} - I_{E1,3}\right)\left(2I_{Ref} - i_{E2,4}\right) \\
I_{E1,3} \cdot I_{E2,4} &= 2I_{Ref} \\
g_{mP} + g_{mN} &= 2g_{mRef}
\end{aligned}
\tag{4.36}
$$

The sum of the g_ms is depicted in Fig. 4.29 as a function of the CM input voltage.

The R-R input stage with bipolar transistors and current spillover control can also be adapted to deliver a constant CM output current by splitting Q_5 and Q_6 into Q_5, Q_7 and Q_6, Q_8.

As we have seen in Fig. 4.29 the transient from P to N is rather sharp with bipolar transistors. In order to improve the CMCR [19] in the transient range we can insert a voltage difference V_{LS} between the bases of control transistors Q_5, Q_7 and Q_6, Q_8 and degenerate the emitters of these transistors with resistors R_5 through R_8, with a value of $R = V_{LS}/I_{Ref}$. This extends the CM input transient range of going from the offset of one input pair to the other by V_{LS}. The circuit with bipolar transistors with splitted switching transistors is shown in Fig. 4.30. Its total g_{mtot} as a function of the CM input voltage is shown in Fig. 4.31. This method can, of course, also be used for CMOS transistors in weak inversion. The absence of current mirrors makes this circuit fast in a response on a large CM swing.

4.4.4   Constant g_m in CMOS by Saturation Control

An alternative approach is to use the saturation of CMOS current-source transistors to control the tail currents of the complementary transistors. The circuit is given in Fig. 4.32 [20].

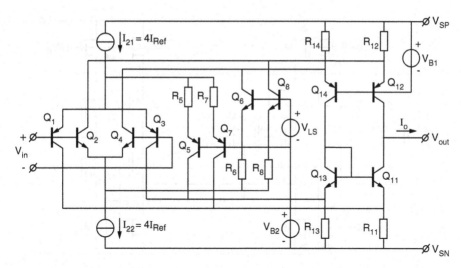

Fig. 4.30 Rail-to-rail input stage with bipolar transistors with splitted and degenerated switching transistors for spillover control having a constant CM output current

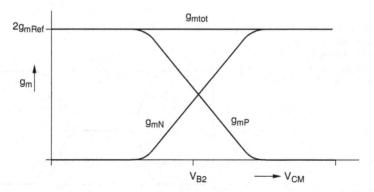

Fig. 4.31 The transconductance g_{mtot} as a function of the input CM voltage of the R-R input stage with degenerated bipolar transistors of Fig. 4.30

The main tail current sources have a value of $4I_{Ref}$. However, a compensating current source from the other side with a value of $2I_{Ref}$ brings the total tail current back to $2I_{Ref}$, which results in one g_{mRef} at one side.

When the CM range is in the middle both tails get $4I_{Ref} - 2I_{Ref} = 2I_{Ref}$. This is I_{Ref} per transistor. This leads to a total $g_{mtot} = g_{mN} + g_{mP} = 2g_{mRef}$ for CMOS in weak inversion.

At the end of each range both current sources at one side are cut off. This results in one active pair with a tail current of $4I_{Ref}$. This results in $g_{mtot} = 2g_{mRef}$, which is equal to the value in the middle situation.

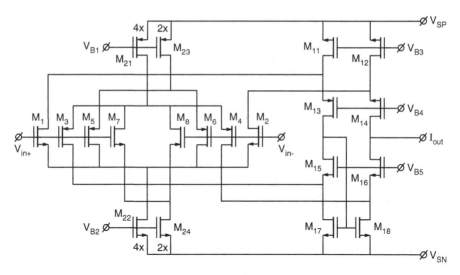

Fig. 4.32 Rail-to-rail input stage with controlled g_{mtot} by the use of saturation CMOS transistors

Fig. 4.33 The transconductance g_{mtot} of the R-R input stage of Fig. 4.32 with saturation control

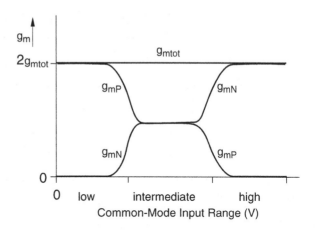

In the case where the current sources at one top or bottom partly saturates, the main tail current source with a nominal value of $4I_{Ref}$ will partly saturate because it is kept at the value of $2I_{Ref}$ of the compensation source from the other side. The compensation current source at that side simultaneously partly saturates from its nominal value of $2I_{Ref}$. The result is that the complementary input pairs keep the sum of their $g_m s$ constant as shown in Fig. 4.33 [20].

4.4.5 Constant g_m in Strong-Inversion CMOS by Constant Sum of V_{GS}

Probably the most essential way to achieve a constant g_{mtot} over the rail-to-rail CM voltage range with a CMOS circuit in strong inversion is to keep the sum of the gate-source voltages constant, as this keeps the sum of the $g_m s$ constant [17]:

$$V_{GSN} + V_{GSP} = V_{THN} + V_{THP} + 2\sqrt{2I_{Ref}/K} = V_C \qquad (4.37)$$

$$g_{mN} + g_{mP} = 2\sqrt{2KI_{Ref}} = g_{mtot} \qquad (4.38)$$

with: V_{GSN} is the gate-source voltage of the N-channel pair, V_{GSP} that of the P-channel pair, V_{THN} the threshold voltage of the N-channel transistors, V_{THP} that of the P-channel transistors, g_{mN} the transconductance of the N-channel pair, g_{mP} that of the P-channel pair $K = \mu C_{ox}W/L$, and supposed is that we have compensated 1 difference in μ_P and μ_N by their W/L ratios, and $g_m = K(V_{GS} - V_{TH}) = \sqrt{(2KI_D)}$.

The implementation of a constant sum of the gate-source voltages of the N-channel and P-channel pair can be obtained by an electronic Zener diode Z, as shown in Fig. 4.34 [17].

A realization of the electronic Zener diode is presented in Fig. 4.35. The diode-connected transistors M_{21} and M_{25} form a reference chain. The current through this chain is set by the W over L ratios of the diode-connected transistors and the voltage across the chain. M_{22} and M_{23} form the two transistor gain stages. Transistor M_{22} is biased by the constant current source, M_{27}, which has a value of I_{Ref}. The current

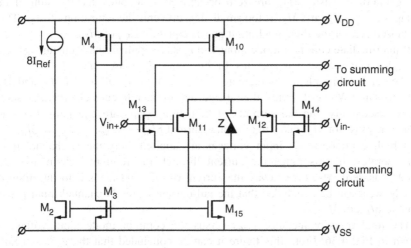

Fig. 4.34 Rail-to-rail input stage. The Zener diode Z makes the g_{mtot} of the input pairs constant

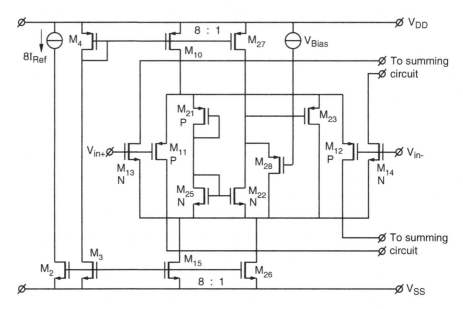

Fig. 4.35 Constant-g_m rail-to-rail input stage with an electronic Zener diode, N and P are the W over L ratios of an N-channel and a P-channel input transistor, respectively [17]

source, M_{26}, drains the extra current of M_{27}. The transistor stage M_{23} controls the tail-currents, and therefore the gate-source voltages of the input pairs.

The principle of the circuit can be best understood by dividing the common-mode input range into three parts.

If low common-mode input voltages are applied, only the P-channel input pair operates. In this range the currents through the reference chain, M_{21} and M_{25}, and the regulator transistor M_{23}, are zero because M_{15} saturates and the sum of the voltages across M_{21} and M_{25} is too small. M_{28} prevents the saturation of M_{27}. Thus the tail-current of the P-channel input pair is equal to $8I_{Ref}$ of M_{10}.

If intermediate common-mode input voltages are applied, the P-channel as well as the N-channel input pair operates. In this range M_{27} biases M_{22}.

The current through M_{22} is regulated to be equal to I_{Ref} of M_{27}. If M_{22} and M_{25} have the same W over L ratios, then the current in the reference chain is also set to I_{Ref}. Consequently, the tail-currents of both input pairs are equal to $2I_{Ref}$. The regulator transistor M_{23} takes the residual current $8I_{Ref} - 2I_{Ref} - 1I_{Ref} = 5I_{Ref}$.

If high common-mode input voltages are applied, only the N-channel input pair operates. In this range the current through the reference chain and M_{22} and M_{23} is zero. M_{28} takes away the current of M_{27} and feeds it to the drain of M_{26}. Now, it can be concluded that the tail-current of the N-channel input pair is equal to $8I_{Ref}$ of M_{15}.

The total transconductance g_{mtot} versus the common-mode input voltage is shown in Fig. 4.36. From this figure it can be concluded that the g_{mtot} is nearly

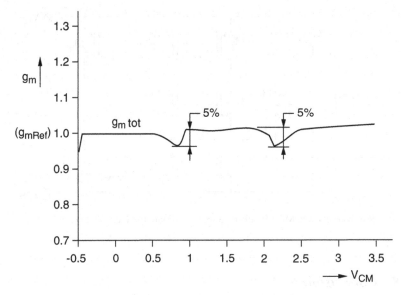

Fig. 4.36 Total transconductance g_{mtot} versus the common-mode input voltage for the constant-g_{mtot} input stage with the electronic Zener circuit of Fig. 4.35

constant over the common-mode input range. In the transition regions the current through transistor M_{23} gradually changes from zero to $5I_{Ref}$, or vice versa. The result is a 5 % variation of the g_m, due to the fact that the voltage across the electronic Zener is slightly current dependent.

4.4.6   Rail-to-Rail in CMOS by Back-Gate Driving

An alternative approach to a 1-V rail-to-rail input stage is depicted in Fig. 4.37 [21].

The back gates of the input transistors are used as input terminals. The g_{mtot} of this stage is roughly a factor 10 lower than the g_{mtot} of a normal differential pair. Therefore, the offset and noise will be reflected a factor of 10 larger than that of a normal driven pair.

4.4.7   Extension of the Common-Mode Input Range

Besides techniques to provide input stages with a rail-to-rail common-mode range, other techniques have been developed to provide input stages with common-mode ranges far below the negative rail voltage or far above the positive rail voltage [22–24].

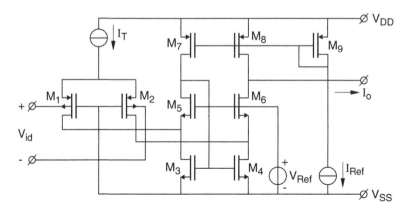

Fig. 4.37 Rail-to-rail 1-V CMOS input stage with back-gate input connection

4.4.8 Conclusion

It can be concluded that low-voltage complementary rail-to-rail input stages are feasible in bipolar as well as in CMOS technology. In bipolar technology and in weak-inversion CMOS the g_{mtot} can be kept constant by mainly keeping the sum of the tail currents constant, while in strong-inversion CMOS the g_{mtot} can be kept constant by mainly keeping the sum of the gate-source voltages constant. In the transition regions the common-mode rejection ratio is degraded because of the transition of the offset of one pair to that of the other pair. Many variations on these themes have been successfully realized. In Chap. 10 it is shown how R-R input stages can be given a very high CMRR by chopping. Also, CM input ranges far beyond the supply rail voltages can be achieved by chopper capacitive coupling at the input.

4.5 Problems and Simulation Exercises

4.5.1 Problem 4.1

The circuits in Fig. 4.18 show two input stages which use NMOS and PMOS devices, respectively. Considering the tail current in both circuits $I_{tail} = 40\,\mu A$, load resistors $R_1 = R_2 = R_3 = R_4 = 10\,k\Omega$ and supply voltage $V_{DD} - V_{SS} = 3\,V$, calculate the common-mode input range for both circuits. Using these ranges, determine the minimum supply voltage so there is no gap between the common-mode input range of the PMOS and NMOS input pairs. At the minimum supply voltage, how far can the supply rails be exceeded by the common-mode input range? The parameters of MOS devices are: $V_{THN} = 0.7$ V, $V_{THP} = -0.8$ V, $K_P = 22$ μA/V^2, $K_N = 75$μA/V^2. All devices have $W/L = 100\mu/1\mu$. The saturation voltage of I_{tail} current source is $V_{Dsat} = 0.2$ V.

4.5.1.1 Solution 4.1

As the circuits are balanced, the DC current through all transistors will be

$$I_D = \frac{I_{tail}}{2} = 20\mu A \tag{4.39}$$

This in turn produces the gain-source voltage for MOS transistors to be

$$V_{GSN} = V_{THN} + \sqrt{\frac{2I_D}{K_N \frac{W}{L}}} = 0.773\text{V}$$

$$V_{GSP} = V_{THP} + \sqrt{\frac{2I_D}{K_P \frac{W}{L}}} = 0.934\text{V} \tag{4.40}$$

The common mode input range for the NMOS pair is limited by ITAL

$$V_{iminN} = V_{SS} + V_{GSN} + V_{Dsat} = V_{SS} + 0.973\text{V}$$
$$V_{imaxN} = V_{DD} - R_1 I_{\hat{D}} - (V_{GSN} - V_{THN}) + V_{GS_N} = V_{DD} + 0.5\text{V} \tag{4.41}$$
$$V_{imaxN} - V_{iminN} = V_{DD} - V_{SS} R_I I_D - (V_{GSN} - V_{THN}) = 2.527\text{V}$$

and for the PMOS pair:

$$V_{iminN} = V_{SS} + R_1 I_D - (V_{GSP} - V_{THP}) + V_{GSP} = V_{SS} - 0.6\text{V}$$
$$V_{imaxP} = V_{DD} - V_{Dsat} + V_{GSP} = V_{DD} - 1.134\text{V} \tag{4.42}$$
$$V_{imaxP} - V_{iminP} = V_{DD} - V_{SS} - R_1 I_D + (V_{GSP} - V_{THP}) = 2.466\text{V}$$

Note that the range for PMOS devices is smaller than the corresponding input range for NMOS devices, due to operating identical sized transistors of both types at the same current. The minimum supply voltage can be calculated by forcing equal values for V_{iminN} (Eq. 4.42) and V_{imaxP} (Eq. 4.43)

$$V_{SS} + 0.973\text{V} = V_{DD} - 1.134\text{V} \tag{4.43}$$

which produces the minimum supply voltage

$$V_{DD} - V_{SS} = 2.1\text{V} \tag{4.44}$$

Independently of supply voltage value, the supply rails are exceeded by common-mode input voltage range with

$$\begin{aligned} V_{imaxN} - V_{DD} &= 0.5\text{V} \\ V_{SS} - V_{iminP} &= 0.6\text{V} \end{aligned} \tag{4.45}$$

4.5.2   Problem 4.2

For the input stage shown in Fig. 4.19, design the output transistors M_5-M_8 so they work with $I_D = 20\mu\text{A}$ at $V_{GSeff} = V_{sat} = 0.2$ V and the biasing source V_R for maximal output range. Then calculate the equivalent G_m of this input stage at common-mode input voltage for which only one of the input pairs or both of them are working, for input devices sized $W/L_{NMOS} = 50\mu/2\mu$, $W/L_{PMOS} = 175\mu/2\mu$ and operated with tail currents $I_{R1} = I_{R2} = 40\mu\text{A}$ from sources with the saturation voltage $V_{Dsat} = 0.2$ V. Estimate the output impedance of this transconductance amplifier. Transistors have parameters: $V_{THN} = 0.5$ V, $V_{THP} = -0.6$ V, $K_P = 16\mu\text{A}/\text{V}^2$, $K_N = 56\mu\text{A}/\text{V}^2$, $\lambda_P = \lambda_N = 0.1$ V^{-1}, λ parameters representing the channel-length modulation of the transistors. The resistors are all equal to 10 kΩ, and the supply voltages are $V_{DD} = -V_{SS} = 2V$.

4.5.2.1   Solution 4.2

The complete expression of $I_D(V_{GS}, V_{DS})$ for a MOS transistor working in saturation is

$$I_D = \frac{K}{2}\frac{W}{L}(V_{GS} - V_{tb})^2(1 + \lambda V_{DS}) \tag{4.46}$$

The equation above shows the role of channel length modulation by drain-to-source voltage and can be used to derive the drain-to-source conductance

$$g_{ds} = \frac{dI_D}{dV_{DS}} = \lambda I_D \tag{4.47}$$

Sizing of transistors M_5-M_8 can be made knowing their drain current and gate overdrive voltage

$$\begin{aligned} \frac{W}{L_{M6,M8}} &= \frac{2I_D}{K_N V_{GSeff}^2} = 18 \\ \frac{W}{L_{M5,M7}} &= \frac{2I_D}{K_P V_{GSeff}^2} = 62 \end{aligned} \tag{4.48}$$

This provides the current through the resistors and the V_{GS} for cascode transistors, so V_R should be at least

$$V_R = I_7 R_7 + V_{GS7} = \left(\frac{I_{R2}}{2} + I_{D7}\right) R_7 - V_{THP} + V_{GSeff} = 1.2V \qquad (4.49)$$

Calculating the equivalent G_m can be reduced to calculating the input pair g_m and the common mode input range domains where each of the input pairs is active. Using Eqs. 4.26 and 4.27, these domains can be calculated based on

$$V_{GS\,M_1,M_2} = V_{THN} - \sqrt{\frac{2I_D}{K_N \dfrac{W}{L}}} = -0.619V$$

$$V_{GS\,M_3,M_4} = V_{THP} - \sqrt{\frac{2I_D}{K_P \dfrac{W}{L}}} = -0.719V \qquad (4.50)$$

$$V_{satN.P} = 0.119V$$

$$V_R = R_5 I_5 = 0.4V$$

$$V_{Dsat} = 0.2V$$

Using these values, with $V_{GS\,M3,M4}$ taken in absolute value, the CM input range limits for the PMOS and NMOS pairs become

$$V_{iminP} = V_{GS\,M_3,M_4} + V_{satN,P} + V_R + V_{SS} = -2.2V$$
$$V_{imaxP} = V_{DD} - V_{GS\,M_3,M_4} - V_{Dsat} = 1.08V$$
$$V_{iminN} = V_{SS} - V_{GS\,M_1,M_2} - V_{Dsat} = 1.18V \qquad (4.51)$$
$$V_{imaxN} = V_{DD} - V_{GS\,M_1,M_2} - V_{satN,P} - V_R = 2.1V$$

Due to transistor sizing, which is proportional with K_N/K_P, g_m of input pairs are equal

$$g_{mN,P} = \frac{2I_D}{V_{satN,P}} = 336\mu S \qquad (4.52)$$

The equivalent G_m is given by

$$G_m(V_{iminP} < V_{in} < V_{iminN}) = g_{mN,P} = 336\mu S$$
$$G_m(V_{iminP} < V_{in} < V_{imaxP}) = g_{mN,P} = 672\mu S \qquad (4.53)$$
$$G_m(V_{imaxP} < V_{in} < V_{imaxN}) = g_{mN,P} = 336\mu S$$

The impedance of the cascoded output stage is given by the voltage gain of cascode transistors and biasing resistors

$$\mu_7 = \frac{g_{m7}}{g_{ds7}} = \frac{1}{\lambda V_{GSeff}} = 50$$

$$Z_o = \left(\mu_7 R_7 + \frac{1}{g_{ds7}} \right) \| \left(\mu_8 R_8 + \frac{1}{g_{ds8}} \right) = 250\text{k}\Omega \qquad (4.54)$$

4.5.3   Problem 4.3

Design the circuit in Fig. 4.25 for a transconductance gain $Y = 400$ µA/V. The transistor parameters, transistor sizes, resistor values and output transistors drain currents are the same as in Problem 4.2 of Sect. 4.5.2.

4.5.3.1   Solution 4.3

The transconductance gain is given by the input g_m

$$Y = 2g_{m1,2} = 2\sqrt{2K_N \frac{W}{L_2} I_{D2}} \qquad (4.55)$$

which in turn makes the drain current of the input devices to be

$$I_{D2} = \frac{\left(\frac{Y}{2}\right)^2}{2K_N \frac{W}{L_2}} = 14\text{µA} \qquad (4.56)$$

The transistors M_5 and M_6 are sized to be six times larger than the input devices

$$\frac{W}{L_5} = 6\frac{W}{L_{1,3}} = \frac{1050\mu}{2\mu}$$

$$\frac{W}{L_6} = 6\frac{W}{L_{2,4}} = \frac{300\mu}{2\mu} \qquad (4.57)$$

and the tail current sources must supply eight times more current than needed for each input device

$$I_{21} - I_{22} = 8I_{D2} = 112\text{µA} \qquad (4.58)$$

V_{B1} differs from the one calculated in Solution 4.2 of Sect. 4.5.2.1 because the current through resistor R_{12} has changed

$$V_{B1} = I_{12}R_{12} + V_{GS12} = (I_{D2} + I_{D12})R_{12} - V_{THP} + V_{GSeff} = 1.14\text{V} \qquad (4.59)$$

V_{B2} can be calculated knowing the V_{GS6} is equal with V_{GS2} as for transistors operated at currents proportional to their size

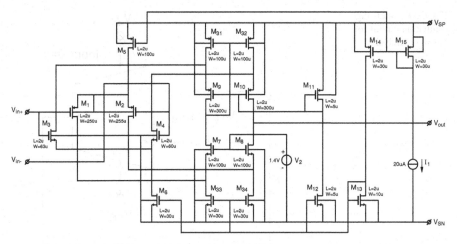

Fig. 4.38 Rail-to-rail input stage with summing circuit

$$V_{B2} = V_{GS6} + V_{Dsat} = 0.819\text{V} \tag{4.60}$$

As a matter of fact, V_{B2} can have any value in the region of common mode input range where both input pairs are active. The other limit for V_{B2} can be easily calculated considering the results of Sect. 4.5.3.

4.5.4   Simulation Exercise 4.1

The input stage shown in Fig. 4.38 is a rail-to-rail input stage which can also be considered a single-stage operational amplifier. Using a circuit as the one shown in Fig. 2.9, simulate the AC behavior of this input stage. If using the same RC load, intended for an output stage, the gain and frequency behavior of the input stage will be strongly degraded. What is the usual load driven by a CMOS input stage?

4.5.5   Simulation Exercise 4.2

A transient analysis circuit for the rail-to-rail input stage is shown in Fig. 4.39. This circuit can be used to plot the $g_m(V_{CM})$ curve which can affect the stability of an entire amplifier if not known in advance. The 10 mV DC input signal is used to obtain a current at the output of the input stage while the V_3 0 V DC voltage source keeps the output of the stage at constant voltage for appropriate biasing. Simulate the $g_m(V_{CM})$ curve for the circuit in Fig. 4.38 and calculate the percentage of g_m change over the whole V_{CM} range. How large can the input voltage V_2 be and what limits this value?

Fig. 4.39 Input stage
simulation circuit for
transient analysis

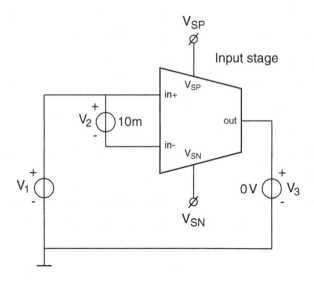

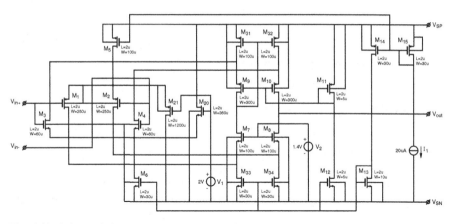

Fig. 4.40 Rail-to-rail input stage with G_m regulation

4.5.6   Simulation Exercise 4.3

For the rail-to-rail input stage in Fig. 4.40 use the circuit depicted in Fig. 4.39 to
simulate the $g_m(V_{CM})$ behavior. A 0 V DC voltage source inserted near the source or
drain of M_{20} and M_{21} can be used to plot the drain current of these transistors as a
function of V_{CM}. What is the ratio of these currents to the drain current of M_6 and M_5
respectively?

References

1. Data sheet, Varactor-bridge operational amplifiers, Model 310 and 311 Analog Devices, 1974
2. E.A. Goldberg, Stabilization of wideband amplifiers for zero and gain, RCA Revue, June 1950, pp. 298
3. R.J. Veen, Piezojunction effect on a planar n-p-n transistor for transducer aims. Electron. Lett. 15(12), 333–334 (1979)
4. R.M. Warner Jr., J.N. Fordemwalt, *Integrated Circuits, Design Principles and Fabrication*. Motorola Series Solid-State Electronics (Mc Graw Hill, New York, 1965)
5. J.E. Solomon, The monolithic opamp: a tutorial study. IEEE J. Solid-State Circuits SC-9, 314–332 (1974)
6. M.A. Maidigne, A high precision monolithic super beta operational amplifier. IEEE J. Solid-State Circuits SC-7, 482–483 (1972)
7. D.J. Comer, D.T. Comer, Using the weak inversion region to optimize input stage design of CMOS OpAmps. IEEE Trans. Circuits Syst. II 51(1), 8–14 (2004)
8. M.J.M. Pelgrom, H.P. Tuinhout, M. Vertregt, Transistor matching in analog CMOS applications, in *IEDM* (1998), pp. 98–915
9. C.D. Motchenbacher, F.C. Fitchen, *Low-Noise Electronic Design* (Wiley, New York, 1973)
10. G.R. Wilson, A monolithic junction FET-NPN operation amplifier. IEEE J. Solid-State Circuits SC-3, 341–348 (1968)
11. R. Hogervorst, J.H. Huijsing, J.P. Tero, Rail-to-rail input stages with g_m-control by multiple input pairs, US Patent 5,561,396, Oct 1996
12. R.F. Wassenaar, Analysis of analog CMOS circuits, Fig. 5.9, PhD Thesis, Twente University, 31 Oct 1996
13. H.C. Nauta, E.H. Nordholt, New class of high-performance PTAT current sources. Electron. Lett. 21, 384–386 (1985)
14. K.J. de Langen, J.H. Huijsing, Compact low-voltage PTAT-current source and bandgap-reference circuits, in *Solid-State Circuits Conference, ESSCIRC '98, Proceedings of the 24th European*, 22–14 Sept 1998, pp. 109–111
15. R. Blauschild, Differential amplifier circuit with rail-to-rail capability, US Patent 4,532,479, 30 July 1985
16. J.H. Huijsing, D. Linebarger, Low-voltage operational amplifier with rail-to-rail input and output ranges. IEEE J. Solid-State Circuits SC-20(6), 1144–1150 (1985)
17. R. Hogervorst, J.P. Tero, R.G.H. Eschauzier, J.H. Huijsing, A compact power-efficient 3 V CMOS rail-to-rail input/output operational amplifier for VCSI cell libraries. IEEE J. Solid-St. Circ. 29(12) (1994)
18. W. Redman-White, A high bandwidth constant g_m and slew-rate rail-to-rail CMOS input circuit and its application to analog cells for low-voltage VLSI systems. IEEE J. Solid-State Circuits 32(5), 701–712 (1997)
19. K.J. de Langen, R. Hogervorst, J.H. Huijsing, Translinear circuits in low-voltage operational amplifiers, in *Analog Circuit Design* (Kluwer, Boston, 1996), pp. 357–385
20. D.L. Knee, C.E. Moore, General-purpose 3 V CMOS operational amplifier with a new constant-transconductance input stage. Hewlett-Packard J. **Aug**, 114–120 (1997)
21. B.J. Blalock, P.E. Allen, G.A. Rincon-Mora, Designing 1-V OpAmps using standard digital CMOS technology. IEEE Trans. Circuits Syst. II 45(07), 769–781 (1998)
22. G. van der Horn, J.H. Huijsing, Extension of the common-mode range of bipolar input stages beyond the supply rails of operational amplifiers and comparators. IEEE J. Solid-State Circuits 28(7), 750–757 (1993)
23. J.F. Witte, K.A.A. Makinwa, J.H. Huijsing, A current feedback instrumentation amplifier with 5 uV offset for bidirectional high-side current sensing, *IEEE Solid-State Circuits Conference 2008*, San Francisco, Session 3.5, 4–6 Feb 2008
24. J.F. Witte, K.A.A. Makinwa, J.H. Huijsing, *Dynamic Offset Compensated CMOS Amplifiers* (Springer, New York, 2009). 250pp. ISBN 978-1-4020-8163-7

Chapter 5
Output Stages

Abstract The output stage of an operational amplifier has to provide the load impedance Z_L with the desired output voltage V_O and current I_O, resulting in an output power $P_O = V_O I_O$. The main requirements of the output stage are: the ability to deliver negative and positive output currents at a high current efficiency, an output voltage range that efficiently utilizes the range between the negative supply rail voltage and the positive one, a high power efficiency, a low distortion, and good high-frequency (HF) performance.

The output stage of an operational amplifier has to provide the load impedance Z_L with the desired output voltage V_O and current I_O, resulting in an output power $P_O = V_O I_O$. The main requirements of the output stage are: the ability to deliver negative and positive output currents at a high current efficiency, an output voltage range that efficiently utilizes the range between the negative supply rail voltage and the positive one, a high power efficiency, a low distortion, and good high-frequency (HF) performance.

Section 5.1 explains some issues of power efficiency. Section 5.2 presents a systematic classification of potential class-AB biased output circuits with a good compromise of power efficiency and distortion. Such a classification enables us to choose one which best matches the requirements of the output stage in combination with the restrictions of a given integration process. The design of output stages is divided into Sect. 5.3 for feedforward-biased class-AB output stages, and Sect. 5.4 for feedback-biased class-AB output stages. Finally, Sect. 5.5 evaluates several current and saturation limiters for bipolar output stages.

5.1 Power Efficiency of Output Stages

Obtaining a high power efficiency is one of the main objects of designing a general-purpose output stage. Power efficiency can be defined as the ratio of output power P_o delivered to the load and the supply power P_s taken from the supply rails. The difference between P_s and P_o is dissipated by the output stage. Power dissipation in the output transistors causes the chip's temperature to rise, while the heat that flows from the dissipating sources towards the cooling surface brings about temperature

© Springer International Publishing Switzerland 2017   105
J. Huijsing, *Operational Amplifiers*, DOI 10.1007/978-3-319-28127-8_5

differences on the chip. Both the temperature rise and differences may deteriorate the amplifier characteristics, as will be discussed shortly.

Above a temperature of 450 K, too many thermally generated electrons appear in the conducting band of extrinsic silicon. This causes excessive diode leakage currents, which double at each 6 K temperature increase.

In order to keep the temperature of the chip T_c below the maximum value T_{cmax} of about 150 °C, the product of the dissipated heat on the chip P_d and the thermal resistance R_{TH} between the chip and the ambience must be lower than the temperature difference ΔT between T_{cmax} and the ambient temperature T_a. So

$$P_d R_{TH} \leq (T_{cmax} - T_a). \tag{5.1}$$

This cooling problem is present not only in large-power amplifiers but also in micro-power amplifiers. In the latter case, the dissipated heat per amplifier determines how many OpAmps can be placed on a single VLSI chip and which measures must be taken to avoid thermal crosstalk to sensitive points on the chip.

The thermal coupling on the chip between the output and input stages causes internal feedback which limits the useful low-frequency gain of the OpAmp [1]. This effect can be divided into two parts. Firstly, there are changes in the temperature of the input transistors which modulate the offset quantities of the amplifier via the temperature drift of these quantities (Sect. 4.1). Secondly, there are temperature gradients which can introduce additional variations in the offset quantities. The first effect can be minimized by placing the output transistors as far away from the critical input transistors as possible. The second effect can be reduced by placing the output transistors on the line of symmetry of the critical transistors, and by precisely balancing or cross-coupling the critical transistors of the input stage as shown in Fig. 5.1.

An output stage for a universal operational amplifier should be able to provide an output voltage and current of both polarities. This leads to the general supply configuration of Fig. 5.2 for an operational amplifier. The ground connection has been shown between two external supply voltages V_{sp} and V_{sn} of opposite polarity. This is according to the general description in Chaps. 1–3. However, with a single supply source, the ground connection is mostly placed at the negative supply rail. In that case, the right-hand side of the load impedance Z_L needs an artificial supply connection in between the supply rails V_{sp} and V_{sn}.

Fig. 5.1 Layout which balances out thermal feedback

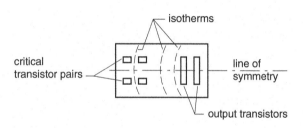

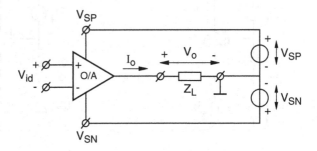

Fig. 5.2 General supply configuration of an operational amplifier

We will now explore several types of possible output circuits with regard to the power efficiency and handling of both output polarities at a low distortion. The efficiency is strongly dependent on the biasing mechanism of the output transistors. Several possibilities of biasing are exemplified by five voltage-follower circuits of Fig. 5.3. These circuits are depicted with CMOS transistors but may also be equipped with bipolar transistors. For this reason we have named these transistors T. The diodes are of the same kind as the transistors.

The single-sided circuits of Fig. 5.3a, b handle their power very uneconomically. The first circuit is not even capable of driving the output down to the negative supply voltage. For a symmetrical sine-wave output of maximally half the supply voltage, its power efficiency is 20 % with a resistor $R_2 = R_L$.

This circuit is much more fit for asymmetrically driving a load resistor connected to the negative supply rail. Like an open collector or drain output, such an open emitter or source output then has a power efficiency of nearly 100 % for a rail-to-rail output block wave, disregarding the voltage loss of the base-emitter or gate-source diode.

The second circuit has a power efficiency of 35 % for a rail-to-rail output sine-wave driving a symmetrical load resistor using a current source $I_2 = \frac{1}{2}V_S/R_L$ connected to the negative rail.

The push–pull circuits of Fig. 5.3c–e handle their power much more efficiently. Neither of these circuits dissipate power if the output is driven against the positive or negative supply voltage, apart from power loss in the base-emitter or gate-source diodes of the output transistors and the auxiliary components.

The class-A circuit of Fig. 5.3c has a maximum sine-wave efficiency of 50 %. The sum of the push and pull currents is constant:

$$I_1 + I_2 = \frac{1}{2}\frac{V_S}{R_L} \tag{5.2}$$

with $V_S = V_{SP} - V_{SN}$. It has a maximum power loss at zero output current. In this point the class-C circuit of Fig. 5.3e is better with a zero power loss at zero output current, as the product of the push and pull currents is zero:

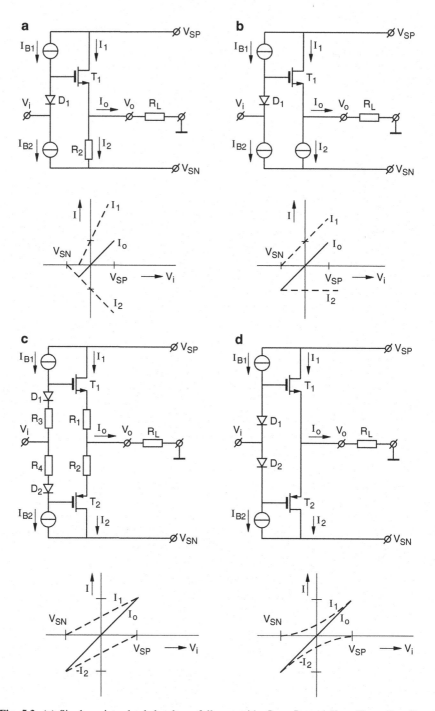

Fig. 5.3 (**a**) Single resistor-loaded voltage follower with: $R_2 = R_L$ and $V_S = V_{SP} - V_{SN}$ Sine-wave efficiency 20 %. (**b**) Single current-source-loaded voltage follower with: $I_2 = \frac{1}{2}V_S/R_L$ and $V_S = V_{SP} - V_{SN}$ Sine-wave efficiency 35 %. (**c**) Push–pull class-A voltage follower: $I_1 + I_2 = \frac{1}{2}V_S/R_L$ Sine-wave efficiency 50 %. (**d**) Push–pull class-AB voltage follower with: $I_B = I_{B1} = I_{B2}$ Sine-wave efficiency nearly 78 %. (**e**) Push–pull class-C voltage follower with a dead current zone around $V_i = 0$, Sine-wave efficiency 78 %

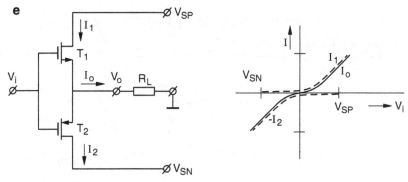

Fig. 5.3 (continued)

$$I_1 \times I_2 = O \tag{5.3}$$

A disadvantage of the class-C stage is its dead band in the transfer function. This can theoretically be reduced to zero by overall feedback with a large loop gain. In that case, the circuit would behave as an ideal class-B amplifier with zero quiescent current in the output transistors and a max. sine-wave efficiency of 78 %. However, the limited slew rate of the preceding driver stage causes the class-C output stage to have a distorted dynamic crossover response. For that reason the compromise of a class-AB output stage with a smooth crossover response is commonly required in analog systems. The class-AB circuit of Fig. 5.3d has push and pull currents whose product is relatively small. It has a smooth crossover behavior. Its sine-wave efficiency is a little below 78 %.

The maximum square-wave response efficiency of the three circuits of Fig. 5.3c–e is nearly 100 %, except that the output voltage cannot completely reach the supply rail voltage due to the diode-voltage loss of the voltage-follower and the saturation voltage of the bias current source. It would be better to connect the output transistors in a common-emitter or common-source configuration, as we will see later on. Then only one saturation voltage would be lost.

In conclusion: The maximum sine-wave efficiency of a pure class-A push–pull stage is 50 %, while that of a pure class-B or class-C stage is 78 % [2]. The sine-wave efficiency of a class-AB stage lies a little below 78 %.

Figure 5.4 gives an overview of the power dissipation as a function of the output voltage of the three push–pull circuits.

A higher efficiency can principally be reached when the supply voltages could be dynamically adapted to the output waveform, i.e., with switchable taps on the supply sources. Another potential way of obtaining a higher efficiency is a class-D amplifier in which the output is switched between zero and both supply voltages and whose sampled output waveform is smoothed by a low-loss filter. These possibilities are not described in this book.

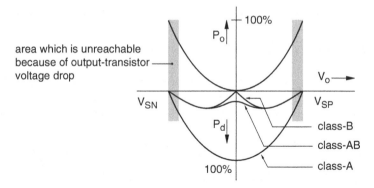

Fig. 5.4 Power dissipation P_d and output power P_o as a function of the output voltage V_o at various biasing classes, disregarding bias currents

5.2   Classification of Output Stages

A systematic classification of potential output stages will be given in this section. This provides the designer with the ability to choose that which best suits the requirements and process parameters.

Output transistors can generally be connected in three different ways: firstly, in a general-amplifier (GA) connection (common-emitter or common-source), secondly, in a voltage-follower (VF) connection (common-collector or common-drain), and thirdly, in a current-follower (CF) connection (common-base or common-gate).

A CF-connected transistor may be inserted as a cascode transistor in the collector or drain lead of a GA-connected output transistor, for instance, to increase the output impedance of the stage. This transistor combination does not change the basic GA configuration of the output stage. A CF-connected transistor may also be inserted in the emitter or source lead of a VF-connected output transistor. This results in a transistor combination which has properties similar to a normal GA-connected transistor. The transistor combination will be classified as a GA-connected circuit. For this reason we may leave the CF-connection out of the classification of output stages, leaving two possible connections: the VF and the GA-connection.

All possible combinations of VF and GA-connected output transistors give rise to three main push–pull configurations: the VF/VF stage or fully VF stage, the compound VF/GA stage or GA/VF stage, and the GA/GA stage or fully GA stage. The VF/GA stage will not be distinguished from a GA/VF stage (Table 5.1).

The three main types are shown in Fig. 5.5a–c.

Figure 5.6 shows the desired class-AB output-current relations. For a large efficiency, the positive and negative maximum output currents should be much larger than the quiescent currents $I_{MP,MN}$. While for a good HF behavior, the minimum positive and negative currents should be as large as possible, $I_{max,min} \approx< I_{quies}$.

Table 5.1 Classification of all possible push–pull output stages

Classification number	Connection of upper output transistor	Connection of lower output transistor
1	VF	VF
2	VF	GA
3	GA	GA

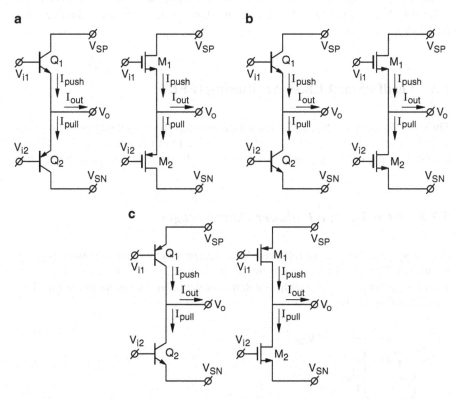

Fig. 5.5 (**a**) Fully voltage-follower (VF) stage in bipolar and CMOS. (**b**) Compound (VF/GA) stage in bipolar and CMOS. (**c**) Fully general-amplifier (GA) stage in bipolar and CMOS

Fig. 5.6 Desired class-AB output-current relations

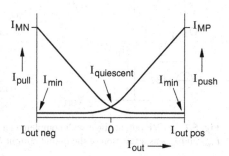

Each basic circuit has a particular feature which makes it appealing. The VF stage can be biased most easily in class-AB. The VF/GA stage can be provided with the best type of transistor which can be made in a certain fabrication process for both push and pull sides. This is important for high output currents or high-frequency capabilities. The GA stage has the largest power gain.

Its output can almost reach both supply rails. This is very important for low-voltage rail-to-rail applications.

In Sects. 5.3 and 5.4, we will see how these three basic output stages can be efficiently biased in class-AB. This can be done by feedforward techniques and feedback techniques.

5.3 Feedforward Class-AB Biasing (FFB)

The term "feedforward biasing" is used if the biasing is fixed by components in series or in parallel with the signal path. This, in distinction to feedback biasing. Here a feedback loop is used to fix the class-AB biasing, which will be discussed in Sect. 5.4.

5.3.1 FFB Voltage Follower Output Stages

For a long time, the voltage follower (VF) configuration has been the most popular feedforward class-AB biasing scheme for its simplicity and robustness. The basic circuit is shown in Fig. 5.7a with complementary bipolar transistors and in Fig. 5.8a with CMOS transistors.

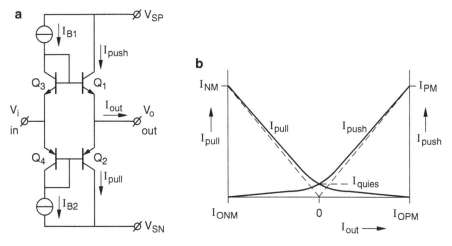

Fig. 5.7 (**a**) VF output stage with complementary bipolar transistors. (**b**) Push and pull currents I_{push} and I_{pull} as a function of the output current I_{out}

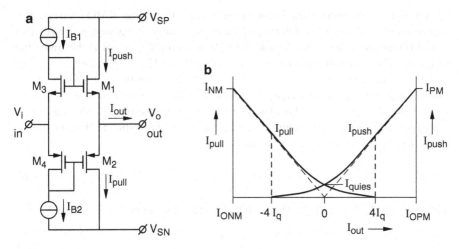

Fig. 5.8 (a) VF output stage with CMOS transistors. (b) Push and pull current I_{push} and I_{pull} as a function of the output current I_{out}

For the bipolar circuit of Fig. 5.7a the product of the push and pull currents I_{push} and I_{pull} is approximately constant. This is because of the logarithmic relation between base-emitter voltage and collector current of bipolar transistors, and the constant sum voltage across the base-emitter voltages, which is set by the diode connected transistors Q_3 and Q_4 in a translinear loop.

So we have:

$$V_{BE1} + V_{BE2} = V_{BE3} + V_{BE4}$$
$$I_{push} \times I_{pull} = I^2_{quies} \tag{5.4}$$

with $I_{quies} = I_{B1} = I_{B2}$, and $V_{BE} = V_T ln\,(I_c/I_{sat}) + V_G$, in which $V_T = kT/q$,

I_{sat} is the saturation current, V_G is the band gap voltage of silicon, and equal parameters for all transistors are supposed.

The CMOS circuit of Fig. 5.8a has a square root characteristic between drain current and gate-source voltage of the output transistors in strong inversion at high current. This, and the constant voltage across the diodes M_3 and M_4 cause the sum of the square roots of the push and pull currents I_{push} and I_{pull} to be constant in a voltage translinear loop. So:

$$V_{GS1} + V_{GS2} = V_{GS3} + V_{GS4}$$
$$\sqrt{I_{push}} + \sqrt{I_{pull}} = 2\sqrt{I_{quies}} \tag{5.5}$$

with $I_{quies} = I_{B1} = I_{B2}$, and $V_{GS} = V_{TH} + \sqrt{2I_0/K}$ and equal parameters for all transistors are supposed. For low output currents the stage behaves in weak inversion like a bipolar stage.

Both VF stages with bipolar or CMOS transistors show a smooth crossover behavior. In the bipolar circuit the maximum output current is $I_{OPM} = \beta_1 I_{B1}$, and

$I_{ONM} = \beta_2 I_{B2}$. No output transistor is ever cut off. In the CMOS circuit the maximum output current is determined by the allowable input voltage drive. The output transistor with the lowest current will be smoothly cut off if the transistor with the highest current has a current larger than $I_O = 4I_B = 4I_{quies}$.

The low-impedance of the two-diode coupling between the input nodes of the transistors in the VF stages makes the relation between the push and pull currents strongly fixed in class-AB. This class-AB biasing is insensitive to wide changes in supply voltages, temperatures and process parameters. The bipolar VF stage has an output impedance in quiescent situation

$$z_o = \frac{1}{2}z_e, \tag{5.6}$$

with $z_e = V_T/I_e$, $V_T = kT/q$. The CMOS VF stage has an output impedance

$$z_o = \frac{1}{2g_m}, \tag{5.7}$$

with $g_m = K(V_{GS} - V_{TH}(index))$, and $K = \mu C_{ox} W/L$ in strong inversion, and $g_m = I_D/(V_{DS} - V_{TH})$ with $(V_{DS} - V_{TH}) = 60$mV in weak inversion.

The successful way of biasing VF stages in class-AB is the basis of a number of further developments, which will be discussed in the next part.

The coupling between the input nodes of the upper and lower output transistors is so firm that the circuit can also be driven asymmetrically. This has been practiced in the bipolar circuits of Figs. 5.9 and 5.10.

The circuit of Fig. 5.9 is the most popular circuit. It can be found in the classic operational amplifier µ A741 of Fairchild Semiconductor [3], and in a number of other OpAmp realizations. In this circuit the VF output stage is asymmetrically driven with a GA-connected NPN transistor. The driver and upper-output transistor are of the NPN type. This type of transistor has the best qualities in terms of current

Fig. 5.9 VF complementary FFB output stage with an asymmetric GA-connected NPN driver transistor, (µA741)

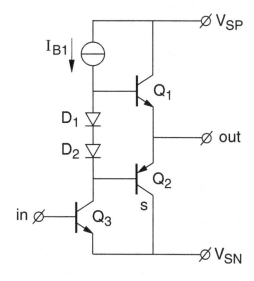

Fig. 5.10 VF
complementary FFB output
stage with an asymmetric
VF-connected NPN driver
transistor

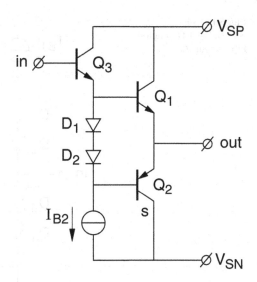

gain, HF-response, and current capability. The lower-output transistor is a substrate PNP, which is the next best transistor with regard to the qualities of an NPN transistor. The maximum positive output current is $I_{OPM} = \beta_1 I_{B1}$.

The circuit of Fig. 5.9 has not often been implemented in the general purpose OpAmp. In this case the output stage is asymmetrically driven by a VF-connected NPN transistor. Its negative output current is limited to $I_{ONM} = \beta_2 I_{B2}$. This is lower than the maximum positive output current of the preceding stage because $\beta_2 < \beta_1$. The output voltage cannot reach the positive rail voltage within two diode voltages and a saturation voltage. The linearity of the current transfer of both previous configurations suffers from the difference in current gains β_1 and β_2 of the two output transistors.

Next, we will look at the group of "Darlington" bipolar variants of the VF output stage in which an emitter follower is inserted in front of the output transistor. The different ways in which a combination of two transistors can be connected allows for more freedom in the circuit configuration than the single VF output stage.

The conventional-Darlington VF output stage of Fig. 5.11 applies a Darlington NPN in the upper half and a Darlington substrate PNP in the lower half. This has the disadvantage that the lower half has a significantly lower current gain (order: $30 \times 30 = 900$) than the upper half (order $100 \times 100 = 10,000$). Moreover, the output cannot reach the supply voltages within two diode voltages and a saturation voltage of the bias current sources. An advantage is the large ratio of maximum output current and bias current. This stage has only sporadically been implemented.

The folded-Darlington variant of Fig. 5.12 uses an NPN and a substrate PNP in both halves, which results in nearly equal current gains for both halves. This results in a more linear current gain than the preceding circuit of Fig. 5.11. It is also simpler and it can reach the supply within only one diode voltage. For these reasons the later stage has frequently been used, for instance in the general-purpose wide-band OpAmp described in Ref. [4]. A disadvantage is the larger bias current I_{B2} needed to drive the lateral PNP Q_2 with a low β.

Fig. 5.11 VF
conventional-Darlington
FFB output stage

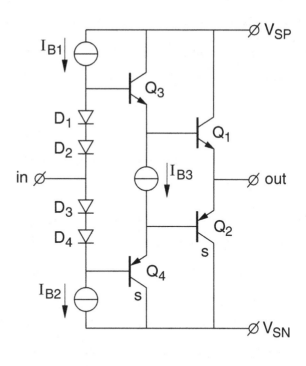

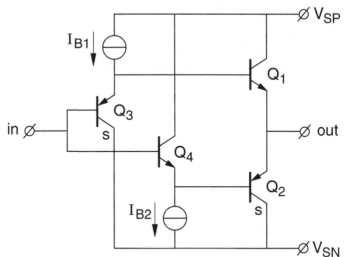

Fig. 5.12 VF folded-Darlington FFB output stage

Another way of increasing the current gain of a bipolar VF output stage, is to add
a GA output booster, as is shown in Fig. 5.13. This stage uses lateral PNP transistors
for Q_1 and Q_4 which have poorer qualities for bandwidth and current gain than those
of the substrate PNPs. Moreover, their current gain β rapidly goes down at higher

Fig. 5.13 GA boosted VF
FFB output stage (LH0021)

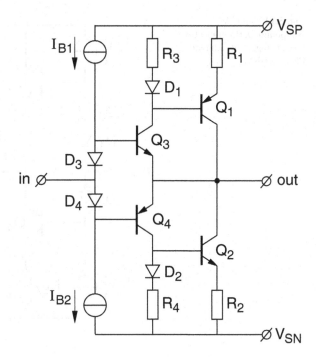

output currents. Consequently, a practical implementation with discrete output transistors can be considered, as in the hybrid GA power OpAmp LH0021 of National Semiconductor [5].

A disadvantage of this stage, in comparison to the previous Darlington stages, is that the GA output transistors Q_1 and Q_2 are driven by current sources. This means that the biasing of the output transistors is dependent on their current gain β. To reduce this disadvantage, additional parallel-base diode chains D_1, D_2 with resistors R_3, R_4 and series-emitter resistors R_1, R_2 have to be inserted. Such resistors reduce the current gain.

5.3.2 FFB Compound Output Stages

In particular, the combination of the GA boosted VF pair in the lower half and the conventional or folded-Darlington pair in the upper half brings about a group of successful compound (VF/GA) stages. This success is due to the presence of an NPN output transistor in the upper and lower halves. Members of this group will be discussed next.

A realization of this compound Darlington output stage is shown in Fig. 5.14 and has been described in Ref. [6] for application in a monolithic power OpAmp which can deliver an output current of the order of 1A. This is possible because the push and

Fig. 5.14 VF/GA
compound Darlington FFB
output stage with a lateral
PNP transistor

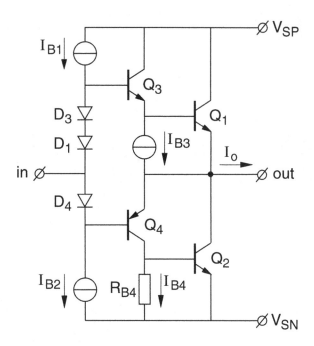

pull output transistors are both of the NPN type. The use of a lateral PNP (Q_4) limits the HF response to some MHz. The biasing stability of this stage is much better than that of the preceding stage in Fig. 5.13. The reason is that the upper half controls its own current due to the inclusion of the V_{BE} of the output transistor Q_1 in its translinear bias loop. And the overall feedback makes that the upper half also controls the biasing of the lower output transistor Q_2. Hence, no low impedance emitter resistor and diode chain has to be included in the lower half, as was the case in the preceding circuit, except for a relatively large bias current-source resistor R_{B4}.

Another realization of the compound Darlington output stage is shown in Fig. 5.15 and has been described in Ref. [7] for application in the (50 MHz) OpAmp CA 3100 of RCA. The wideband application could be obtained by replacing the lateral PNPs of the preceding circuit with P MOS FETs. Again the biasing is controlled by the upper half.

The GA boosted VF pair in the lower half combined with the folded Darlington pair in the upper half results in the relatively simple compound output stage of Fig. 5.16. It has been realized in the general-purpose OpAmp LM101 of National Semiconductor and has been described in Ref. [8]. This stage also has a good output-current capability, although its frequency response is limited to some MHz by the lateral PNP transistor Q_4.

A resistor voltage-level shifter has been used in the lower half of Fig. 5.17a to obtain the compound stage with two equal output N-MOS FETs. The circuit is described in Ref. [9]. The VF-connected N-MOS FETs M_4 and M_5 function as two

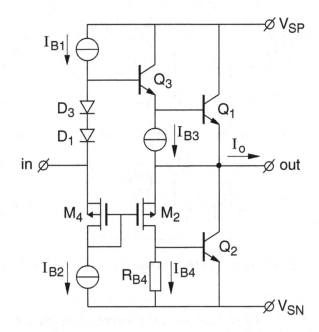

Fig. 5.15 VF/GA compound Darlington output stage with P MOS FETs (CA3100)

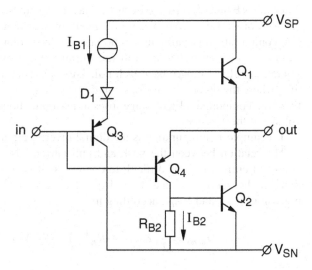

Fig. 5.16 VF/GA compound folded Darlington FFB output stage (LM101)

equal and large source resistances $1/g_4$ and $1/g_5$. These resistors, together with the current mirror M_6M_7, function as a voltage-level shifter. Because the voltage transfer of the level shifter has a negative polarity, the lower output functions as if it were its complementary counterpart. The equivalent VF circuit of the stage is shown in Fig. 5.17b.

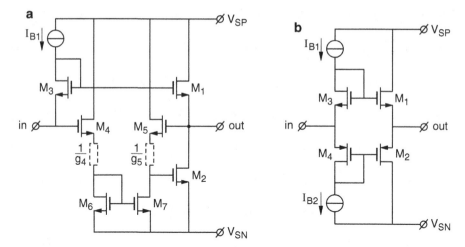

Fig. 5.17 (a) VF/GA all N MOS compound output stage. (b) Equivalent circuit of Fig. 5.17a

5.3.3 FFB Rail-to-Rail General-Amplifier Output Stages

After the VF and VF/GA feedforward biasing schemes, we now have to examine the feedforward class-AB biasing of inverting amplifier (GA) output stages. The GA configuration is particular important for low-voltage operation because the output voltage can utilize the full rail-to-rail voltage range except for two small saturation voltage ranges near each rail. This is in contrast with the VF configuration, where the output range loses two diode voltages plus two saturation voltages. Even the compound VF/GA stage loses at least one diode voltage and a saturation voltage at the VF side.

The simplest configuration is the digital inverter circuit of Fig. 5.18a.

The relation between the push and pull currents is shown in Fig. 5.18b. The quiescent current I_{quies} is difficult to control, as it is strongly dependent on variations of the supply voltage $V_S = V_{SP} - V_{SN}$ and the threshold voltages of the transistors V_{THP} and V_{THN}, according to:

$$I_{quies} = \frac{1}{2}\left(K_P^{-1/2} + K_N^{-1/2}\right)^{-2}(V_S - V_{THP} - V_{THN})^2 \qquad (5.8)$$

To obtain a low sensitivity to supply-voltage variations and to process-dependent variations of the transistor characteristics, these characteristics must have a shallow curve. For this reason the biasing is chosen more A than AB by choosing a low W/L ratio in this configuration. Bipolar transistors have such steep characteristics that they cannot generally be used in this configuration, unless the supply voltage $V_{SP} - V_{SN} = V_S$ is regulated at two diode voltages. The main parameter of the CMOS GA stage is the voltage gain $A_u \simeq g_m z_{ds}$, which is of the

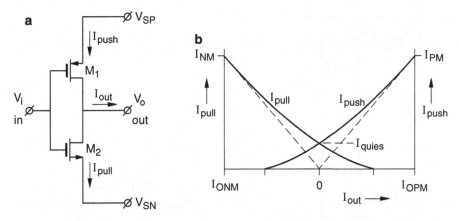

Fig. 5.18 (a) GA FFB R-R class-AB CMOS output stage. (b) Push and pull currents I_{push} and I_{pull} as a function of the output current I_{out}

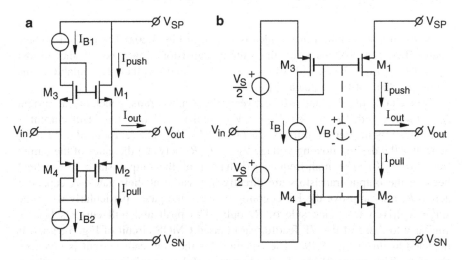

Fig. 5.19 (a) VF feedforward biased class-AB CMOS output stage. (b) GA feedforward biased class-AB CMOS output stage

order of 100. The maximum output current is restricted because of the required shallow driving characteristic. So the digital inverter stage is a rather poor analog GA output stage.

What we really would like to obtain is a behavior that is independent of the supply-voltage and transistor-parameter variations, like the behavior of the VF configuration. To develop this idea, we start with the conventional VF circuit of Fig. 5.19a. Next, we cut this circuit in an upper and lower half, and place these halves reversely on each other in Fig. 5.19b. To this end we have to insert a floating supply voltage source V_S, which is connected at its half to the input. This restores the voltage translinear loop $V_{GS1} + V_{GS2} = V_{GS3} + V_{GS4}$.

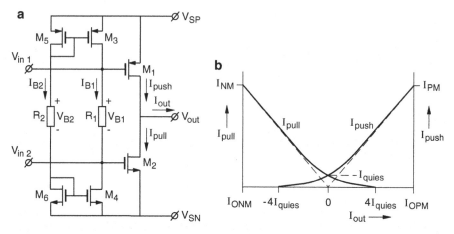

Fig. 5.20 (a) GA FFB R-R class-AB CMOS output stage with common resistive coupling. (b) Push and pull currents I_{push} and I_{pull} as a function of the output current I_{out}

A first implementation of the idea is shown in Fig. 5.20a. The desired voltage source between the gates of the output transistors must be of the format $V_B = V_S - V_{GS1} - V_{GS2}$, with $V_S = V_{SP} - V_{SN}$. Such a voltage is available in the reference chain of Fig. 5.20a as V_{B2}.

This chain M_5, R_2 and M_6 between the supply rails generates a current $I_{B2} = V_{B2}/R_2$, with $V_{B2} = V_S - V_{GS5} - V_{GS6}$ and $V_S = V_{SP} - V_{SN}$. This current is reproduced by the current mirrors M_5, M_3 and M_6, M_4. The reproduced current I_{B1} generates the desired floating voltage $V_{B1} = I_{B1}R_1$ between the gates of the output transistors [10]. The high-frequency behavior of the output stage is excellent because the output transistors are directly accessible at their gates. A capacitor across R_1 may improve the HF coupling between the gates, particularly when the stage is driven from one side of R_1 only. The push and pull currents behave similarly to those of the VF feedforward biased CMOS circuit of Fig. 5.8a and is depicted again in Fig. 5.19a. The equation for the quiescent current is given by (Eq. 5.5). The circuit of Fig. 5.20a is quite useful. The minimum supply voltage equals that of two gate-source or base-emitter diode voltages $2\,V_D$.

For a supply voltage lower than that of two diodes, the resistive level shift voltage $V_{B1} = I_{B1}R_1$ can be prebiased with a negative voltage by cross-coupled bias current sources I_{B3} through I_{B6}. This is shown in Fig. 5.21.

When the bias currents I_{B3} through I_{B6} are equal to I_B, and R_1 and R_2 are equal to R, the value of the bias current can be chosen equal to $I_B = (V_{GS} - V_{sat})/R$, to obtain a minimum supply voltage $V_{SMin} = V_{SP} - V_{SN}$ equal to $V_{SMin} = V_{GS} - V_{sat}$, with V_{GS} equal to the gate source voltage of the output transistors, and V_{sat} equal to the saturation voltage of the current-source transistors.

A disadvantage of the two previous stages is that the quiescent current of the output transistors I_{quies} is proportional to the supply voltages minus two diode voltages according to: $V_{B2} = V_S - V_{GS5} - V_{GS6}$.

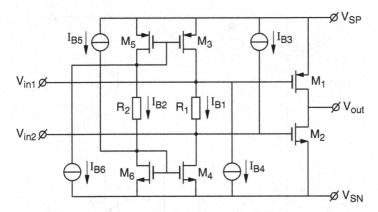

Fig. 5.21 Low-voltage version of the GA feedforward biased rail-to-rail class-AB CMOS output stage with common resistive coupling with a minimum supply voltage of $V_{SMin} = V_{GS} + V_{sat}$

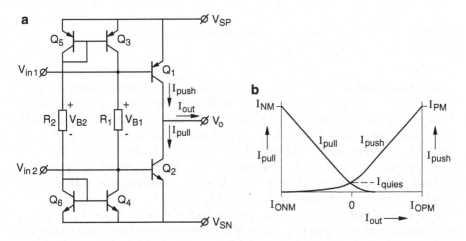

Fig. 5.22 (a) GA FFB R-R class-AB bipolar output stage with CM resistive coupling. (b) Push and pull currents I_{push} and I_{pull} as a function of the output current I_{out}

This can be overcome by decoupling the diode connection of the drain of M_5 and driving a model output transistor M_8 like M_1, by the drain of M_5, and comparing the drain current I_{D8} with a reference, current I_{ref}, and controlling the common gate of M_5 and M_3 so that $I_{D8} = I_{ref}$. Now the quiescent current is fixed at I_{ref} independent at the supply voltage.

The same circuit can be implemented in bipolar technology. The circuit for supply voltages of two diodes or higher is shown in Fig. 5.22a.

The result is similar to that of the circuit of Fig. 5.7a. The product of the push and pull currents should be constant according to Eq. 5.4, except for an asymmetric behavior of the currents below the quiescent value I_{quies} and that one output

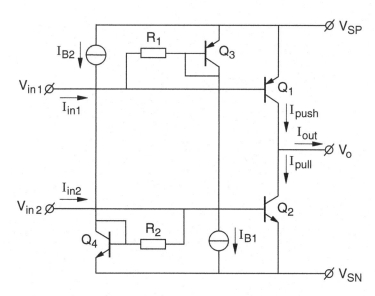

Fig. 5.23 GA feedforward biased class-AB bipolar output stage with separate resistor and diode biasing

transistor is cut off when the other is driven. This is because the base currents of the output transistors with different current gains β_P and β_N disturb the constant voltage V_B across R_1. This results in a nonlinear dynamic behavior. The resistive coupling for bipolar transistors is not strong enough so that R_1 can be driven from one side only. The minimum supply voltage is that of two diode voltages $2V_{BE} \approx 1.4 \ V$.

A different way to bias a GA output stage is to individually relate the driving voltages to the rail voltages. This is shown in a very elementary way with resistive coupling of bipolar transistors in Fig. 5.23 [15].

The idea is that the sum of the base-emitter voltages $V_{SUM} = V_{BE1} + V_{BE2}$ remains constant and equal to the sum of the voltages across the diode-connected transistors Q_3 and Q_4, $V_{SUM} = V_{BE3} + V_{BE4}$, because the driving currents I_{in1} and I_{in2} generate equal but opposite voltages across R_1 and R_2. The result is that the product of the push and pull currents remains constant, as expressed with the bipolar VF circuit of Fig. 5.7a and by Eq. 5.4. However, the base currents in the output transistors disturb this relation because they also generate currents in the biasing resistors R_1 and R_2. To reduce this disturbance the resistors must be small. On the other hand, these resistors may not be taken too small or their conductances would take away too much of the input currents, and the current gain of the output stage becomes too small. An advantage of the circuit of Fig. 5.23 is that the total supply voltage may be as low as one diode plus one saturation voltage, or about 0.9 V.

We certainly would not try to use CMOS in this circuit. Their high input impedance would be completely destroyed by the driving current of the low conductance of the resistors. But what we can do to prevent the loss of driving

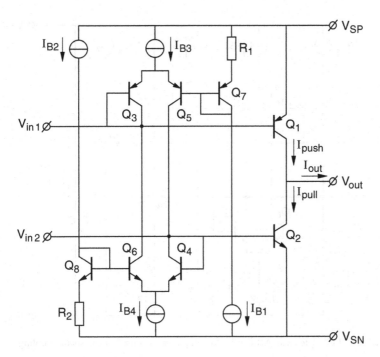

Fig. 5.24 GA feedforward biased class-AB bipolar output stage with separate diode string biasing and positive loop gain coupling

current in the separate coupling elements R_1 and R_2, is to collect the lost currents and add them again in a positive coupling loop. This principle is shown in Fig. 5.24 [10].

The transistor strings Q_1, Q_3, Q_5, Q_7 and Q_2, Q_4, Q_6, Q_8 are separate translinear loops that fix the bias current of the output transistors. To allow for a 100 mV voltage drop across the bias current sources I_{B3} and I_{B4}, a similar voltage-level shift is created across R_1 and R_2. To compensate for this level shift in the translinear loops, the output transistor Q_1 and Q_2 should be small. The currents which are lost in the diode connected biasing transistors Q_3 and Q_4 are collected by the emitters of Q_5 and Q_6 respectively, and are returned as a driving current to the complementary side. The transistors Q_3, Q_5, Q_4, Q_6 form a positive coupling loop with a current gain of slightly lower than one, which keeps this loop stable. All input current which is not used to drive one output transistor is used to drive the other. So the gain decrease due to the class-AB biasing circuit is eliminated. In fact, the positive loop-gain coupling is a way to implement a class-AB bias circuit with high input impedance. The circuit needs a supply voltage of that of one diode and two saturation voltages, which is about 1 V.

One way to eliminate the problem with the voltage drop needed for the bias current sources is to use Darlington output transistors Q_{11} and Q_{12} in combination with extra diodes in series with Q_7 and Q_8. The result is shown in Fig. 5.25 [10].

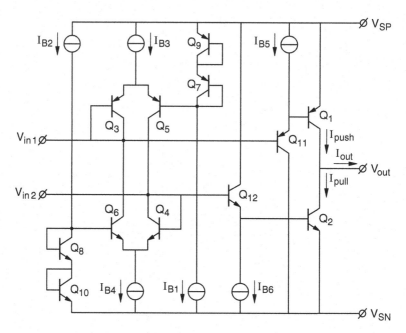

Fig. 5.25 GA FFB R-R class-AB Darlington output stage with CM transistor coupling

The result is a GA feedforward biased class-AB rail-to-rail Darlington output stage. The stage may be driven by a single input terminal. The GA Darlington output stage can be made very current efficient, because the currents in the class AB loops may be chosen β_p^2 lower than the maximum output currents; which may be a factor 1000. However, the circuit needs a minimum supply voltage of that of three diodes and a saturation voltage, which is about 2.4 V.

A very robust feedforward biasing in class-AB arises if we simplify the translinear loops such that the diode coupled transistors Q_3 and Q_4 and the Darlington transistors Q_{11} and Q_{12} are taken away, as shown in Fig. 5.26a [10].

The translinear loops Q_1, Q_3, Q_5, Q_7 and Q_2, Q_4, Q_6, Q_8 strongly fix the biasing in class-AB. The behavior of the push and pull currents I_{push} and I_{pull} in relation to the output current I_{out} is described as:

$$\left(I_{push} - \tfrac{1}{2}I_{quies}\right)\left(I_{pull} - \tfrac{1}{2}I_{quies}\right) = \left(\tfrac{1}{2}I_{quies}\right)^2$$
$$I_{out} = I_{push} - I_{pull} \tag{5.9}$$
$$I_{MIN} = \tfrac{1}{2}I_{quies} = \tfrac{1}{2}I_{B1} = \tfrac{1}{2}I_{B2} = \tfrac{1}{2}I_{B3} = \tfrac{1}{2}I_{B4}$$

with equal emitter areas for Q_1, Q_3, Q_5, Q_7 and Q_2, Q_4, Q_6, Q_8.

In the quiescent state, half the current of the current sources I_{B1} and I_{B2} flow through transistor Q_3 and the other half through Q_4. But when one of the output transistors, say Q_1, draws a large current, all of the current of the current sources I_{B1} and I_{B2} flow through Q_4. This means that the current through Q_4 is doubled.

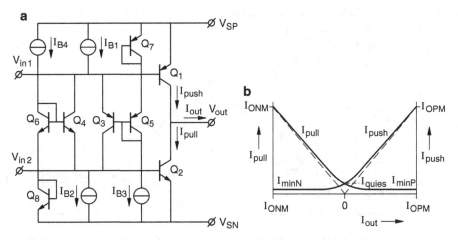

Fig. 5.26 (**a**) GA FFB R-R class-AB bipolar output stage with CM transistor coupling. (**b**) Push and pull currents I_{push} and I_{pull} as a function of the output current I_{out}

And, consequently, the current through the output transistor Q_2 is halved, and is not reduced further. This situation is in favor of the simpler VF circuit of Fig. 5.7, whose behavior was described by Eq. 5.4. There the push or pull current did approach zero at full output current.

Since the output transistors always stay into their normal working conditions, a low distortion can be achieved by this circuit.

Another advantage of this circuit is the simplicity of the coupling circuit between the upper and lower output transistors. The coupling circuit consists of a mesh of two head-to-tail connected transistors Q_3 and Q_4. This mesh shapes a positive feedback loop with a gain just below one. Therefore, the mesh has a high input resistance for CM input currents. Except for the base currents, no driving current is lost from this mash. All input current, which is not used to drive one output transistor, is automatically rerouted to the other output transistor. Firstly, this means that the class-AB coupling is so strong that the output stage can easily be driven from a single upper or lower input terminal, without losing control of the biasing. Secondly, the straight coupling through Q_3 or Q_4 from one side to the other means that the driving currents may far exceed the bias currents I_{B1} and I_{B2}. This implies a large ratio between the maximum output currents and the bias currents of the circuit.

A final advantage is the good high-frequency behavior, firstly, because the output transistors are directly accessible by their inputs, and secondly, when the input current is rerouted to the complementary output transistor, the signal has only to pass one transistor Q_3 or Q_4 in a common-base connection.

The only disadvantage of this GA output stage is that it needs at least a supply voltage of two diodes plus a saturation voltage, which amounts together to about 1.6 V.

The last described circuit also has excellent properties in CMOS technology. The GA feedforward biased CMOS output circuit with a simple common-mode transistor coupling is depicted in Fig. 5.27a [11, 12].

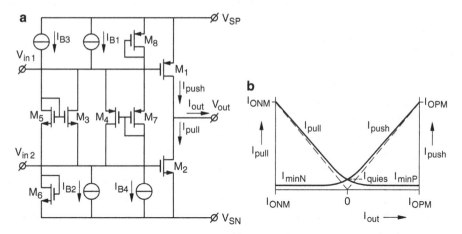

Fig. 5.27 (a) GA FFB R-R class-AB CMOS output stage with CM transistor coupling. (b) Push and pull currents I_{push} and I_{pull} as a function of the output current I_{out}

We take all $(W/L)_P$ of the P-channel transistors three times larger than the $(W/L)_N$ of the N-channel transistors to compensate for the one-third ratio of the mobilities μ_P and μ_N in order to keep the transconductances of the N-channel and P-channel transistors equal at equal currents. For simplicity, we take all $(W/L)_N$ equal for the N-channel transistors and all $(W/L)_P$ equal for the P-channel transistors, except for the output transistors which are scaled a factor α larger. If we choose the quiescent currents through the translinear loop transistors equal, we need the following relation between the bias currents: $\frac{1}{2}I_{B1} = \frac{1}{2}I_{B2} = I_{B3} = I_{B4} = I_B$. When we describe the CMOS transistors with $V_{GS} = V_{th} + \sqrt{2I_D/K}$, and $\beta = \mu C_{ox}W/L$, the following relation between the push and pull currents results:

$$\left(\sqrt{I_{push}} - 2\sqrt{I_{quies}}\right)^2 + \left(\sqrt{I_{pull}} - 2\sqrt{I_{quies}}\right)^2 = 2I_{quies}$$
$$I_{quies} = \alpha I_B, \text{ and } I_{MIN} = \left(2 - \sqrt{2}\right)^2 I_q = 0.34 I_q, \text{ at } I_{MAX} = 4 I_{quies} \qquad (5.10)$$

When one of the push or pull currents becomes four times as large as the quiescent current, the other one becomes $(2 - \sqrt{2})^2 = 0.34$ times the quiescent current I_{quies}. At this value the full bias current of $I_{B1} = I_{B2}$ flows through one of the transistors M_3 or M_4, while the other is cut off.

The smallest one of the push or pull currents will not become any smaller and stays at $0.34\ I_{quies}$, while the largest one is allowed to increase far above $4\ I_{quies}$. Again, this class-AB regulation is very robust because none of the output transistors is ever cut off, but stays largely within the normal bias conditions.

It is interesting to note that the mesh with M_3 and M_4 on one hand fixes the differential-mode voltage movement at the gates of the output transistors in robust class-AB biasing conditions, but, on the other hand, does not present any resistance to the common-mode voltage movement at the gates. Though the source impedance of M_3 and M_4 present a normal value, the head-to-tail connection of these transistors form a positive feedback loop for currents with a gain of nearly perfect unity, which cancels the influence of the source resistance on the common-mode voltage movement. Or otherwise described, no common-mode driving current can flow out of the mesh M_3, M_4. All input current is either used to drive one or the other output transistor.

The same advantages of the bipolar stage are present with the CMOS GA feedforward biased class-AB output stage. The circuit may easily be driven from one input terminal. All current which is not needed to drive one output transistor is automatically rerouted to drive the other output transistor. The high-frequency behavior is excellent, because the output transistors are directly accessible from their input, and when one input drives the other output transistor, only one transistor in a common-gate connection is in series with the signal. The only disadvantage is that the output stage needs two diode voltages and one saturation voltage as a minimum supply voltage, which is about 1.8 V.

5.3.4 Conclusion

We have discussed feedforward biased class-AB output stages. The simplest ones were the voltage follower (VF) configurations. These are the classic solution for class-AB biasing. Several circuits with Darlington output transistors have been shown. The compound (VF/GA) derivations are of particular interest because of the use of solely NPN transistors as output transistors. This makes these circuits suitable for large currents or high frequency responses. The inverting-amplifier configurations (GA) are of importance at low supply voltages, because the output can utilize nearly the full supply voltage. Several robust bipolar and CMOS GA circuits have been shown.

5.4 Feedback Class-AB Biasing (FBB)

In the preceding section, a systematic classification has been given of output stages which are feedforward-biased. We have investigated voltage-follower (VF), compound voltage-follower/inverting amplifier (VF/GA), and general-amplifier (GA) feedforward configurations. In all these stages the biasing and driving functions are processed by the same components. This often leads to compromises. In this section, output stages are discussed in which the biasing function is separated from the driving in what we will call feedback biasing (FBB). The push and pull output currents are measured and compared with a bias reference. If the biasing is not correct in a class-AB relation, the output transistors receive a correction signal by a feedback signal. In this way, we obtain more freedom to design the class-AB output stage in the GA, VF or VF/GA mode according to the derived specifications in a certain IC process.

Basically, there are two possible places to measure the push and pull currents. Firstly, a voltage measurement transistor can be connected with its base-emitter or gate-source in parallel to the base-emitter or gate source of the output transistor. Secondly, a current-measurement diode may be inserted in the collector or drain of the output transistor with a measurement transistor with its base-emitter or gate-source connected across the diode.

In any case, the currents of the measurement transistors have to be related to each other in a translinear loop which controls the class-AB biasing of the output transistors.

5.4.1 FBB Voltage-Follower Output Stages

The simplest feedback-biased VF circuits are drawn in Fig. 5.28a, b. The transistors Q_3 and Q_4 measure the sum of the base voltages of Q_1 and Q_2 and regulate this sum at a constant value so that the product of the push and pull currents remains constant, according to Eq. 5.4 with bipolar transistors.

With CMOS the sum of the square roots of the push and pull drain currents is controlled at a constant value according to Eq. 5.5. In fact, there is no difference between the feedback-biased circuits of Fig. 5.28a, b and the feedforward-biased circuits of Figs. 5.7a and 5.8a, except that the input connections are drawn immediately at the bases or gates of the output transistors instead of in between the diode-connected transistors Q_3 and Q_4 or M_3 and M_4. Because of this similarity we will not discuss VF feedback-biased stages further.

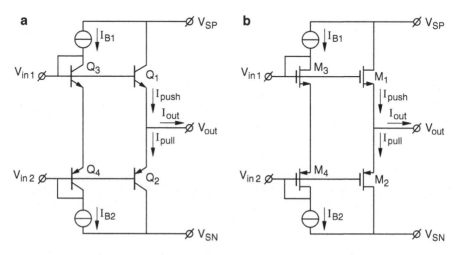

Fig. 5.28 (a) VF FBB class-AB bipolar output stage with parallel transistor measuring. (b) VF FBB class-AB CMOS output stage with parallel transistor measuring

5.4.2   FBB Compound Output Stages

A compound output stage with equal N-type push and pull output transistors is particularly important for bipolar output stages because of the much better properties of NPN transistors than PNP ones. For that reason, the examples of this paragraph are given with bipolar transistors, but the ideas extend to other transistor types as well. A compound output stage can be obtained by combining a feedforward-biased VF upper half with a feedback-biased GA complementary lower half, as is shown in the bipolar circuit of Fig. 5.29a. The diode D_2 measures the collector current of Q_2.

The transistor Q_3 and series diode D_1 measure the sum of the voltages across Q_1 and D_2 and regulate the input signals directly on the base of Q_1, and indirectly through the mirror D_3Q_4 on the base of Q_2.

The class-AB biasing is so strong that the circuit can easily be asymmetrically driven at the inverting input, as is shown in Fig. 5.29b.

The VF/GA circuit may alternatively be driven as if it were a feedforward biased stage by connecting the input symmetrically between the emitter of Q_3 and diode D_1.

If the transistor Q_3 is replaced by a diode D_3 and the feedback is removed, a strongly simplified feedforward-biased compound output stage arises, which no longer has a linear current transfer. This circuit and its current transfer characteristic are shown in Fig. 5.30a, b. A practical realization of this circuit is found in the power OpAmp μA791 of Fairchild [13] with a maximum output current of 1.2 A. The most important properties of this compound stage are that it uses push and pull output transistors of only one kind (NPN), and that it is simple. The transimpedance

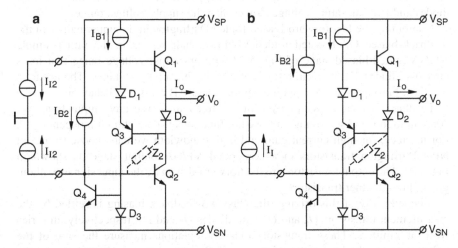

Fig. 5.29 (a) VF/GA compound mixed FFB and FBB class-AB output stage. (b) VF/GA compound mixed FF and FB biased class-AB output stage with asymmetrical driving at the inverting input

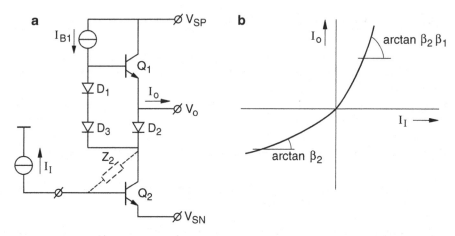

Fig. 5.30 (**a**) VF/GA compound simplified mixed FFB and FBB class-AB output stage driven at the inverting input (μA791). (**b**) Nonlinear current amplification of the circuit of Fig. 5.30a

of the circuit can be linearized by inserting a dominating Miller feedback imped-ance Z_2 across the natural collector–base impedance of Q_2.

More gain can be obtained by the application of Darlington transistor combinations.

The mixed-biased combination of a feedforward-biased Darlington VF upper half with a feedback-biased GA-boosted VF lower half shown in Fig. 5.31 is of interest. The stage has a smoother crossover behavior, a better bias stability and a higher gain at low output currents than those of the feedforward-biased equivalent of Fig. 5.14. This Darlington stage has a high ratio between maximum output-current and input-bias currents I_{B1} through I_{B4}. The output-voltage range is four diodes and two saturation voltages lower than the supply voltage range.

Combining the feedforward-biased folded Darlington in the upper half with the feedback-biased GA-boosted folded-VF lower half results in the rather simple VF/GA mixed-biased stage of Fig. 5.32 with an output-voltage range that loses only two diodes and two saturation voltages of the supply voltages. The accurate feedback biasing gives the circuit a smooth class-AB crossover behavior.

Feedback biasing provides us with the freedom to use any output transistor configuration because driving and biasing functions can be chosen independently. For instance, for high current gain and high bandwidth we can choose to use all NPN Darlington transistors in a compound VF/GA output stage as shown in Fig. 5.33. The output transistors are feedforward-driven by the input signals without going through other transistors.

Independently of this driving, the class-AB feedback biasing is settled by the measurement transistors Q_6 and Q_7 with diodes D_2 and D_3, respectively, in series with the emitters. These transistor-diode combinations measure the sum of the voltages across the base-emitter junction of Q_1 carrying the push current, and the diode D_1 carrying the pull current. The biasing current for the lower half is fed via a

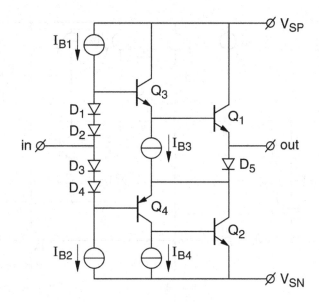

Fig. 5.31 VF/GA compound mixed-biased Darlington class-AB output stage

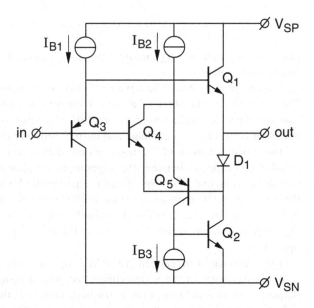

Fig. 5.32 VF/GA compound mixed-biased folded-Darlington class-AB output stage

folded CF-connected PNP transistor Q_5. The all-pass network R_f, C_f passes the low-frequency currents through Q_5, while frequencies above $\omega_f = 1/R_f C_f$ are directly fed to the lower half. The good HF properties of the feedback bias path is essential for the HF stability of the feedback loop.

The circuit may be driven asymmetrically at the upper input terminals as a voltage follower, or at the lower input terminal as an invertor. In both of these

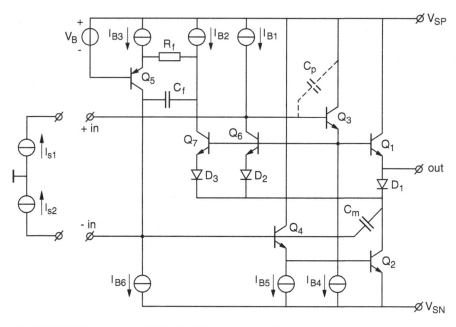

Fig. 5.33 VF/GA compound FBB All-NPN class-AB Darlington output stage

cases, the internal biasing circuitry provides a correct class-AB behavior of the
whole output stage.

The circuit may ideally be driven with two input currents having opposite signs.
These currents can be obtained at the two output terminals of a longtailed transistor
pair having a tail current source, or at the collector and emitter terminals of a
transistor connected as a .voltage and current follower (VCF)

The transimpedance of the upper output half is roughly equal to the parasitic
parallel impedance Z_p between the upper input terminal and the supply or substrate
terminals, while that of the lower half is equal to the Miller impedance Z_m between
the collector and base nodes of the lower Darlington transistor. The high-frequency
behavior and slew rate of both halves can be balanced by adding a Miller capacitor
C_m to the lower half which is as large as the natural parasitic capacitor C_p in the
upper half.

The circuit has been realized in a 30 MHz operational amplifier [14], see Sect. 7.5.

A disadvantage of the series connection of a measuring diode and the lower
output transistor is the loss of one extra diode voltage in the output voltage swing. In
order to eliminate this loss, measurement transistors Q_3 and Q_4 are placed with their
base-emitter connections in parallel to those of the output transistors Q_1 and Q_2
respectively, as in Fig. 5.34.

The currents of the measurement transistors are brought in a class-AB relation
by the translinear loop [15] with Q_5 through Q_{11} and controlled by the output
currents of Q_7, Q_8 through the mirrors Q_{12}, Q_{14} and Q_{13}, Q_{15}. Alternatively, the
controlling could be done by the output currents of Q_9 and Q_{10} through folded

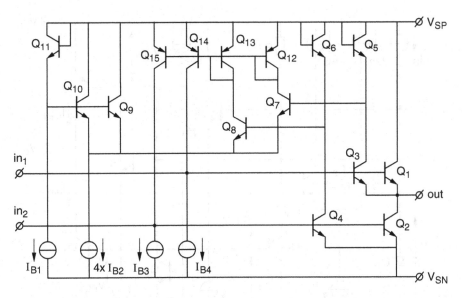

Fig. 5.34 VF/GA compound FBB All-NPN output stage with minimum selector

cascode with an all pass RC network as shown in Fig. 5.33, if the HF stability of the feedback loop with the PNP controlling transistors Q_{14} and Q_{15} appears to be insufficient. We suppose that all emitter areas are equal for the NPN transistors and equal for the PNP transistors, except that the output transistors are n times scaled up. We also suppose that all transistors Q_5–Q_{11} have an equal current in quiescent state.

The translinear loop Q_5 through Q_{11} has several segments. The diode connected transistors Q_5 and Q_6 model the base emitter voltages of the output transistors Q_1 and Q_2, respectively. The transistors Q_7 and Q_8 function like an and-gate, passing on the lowest of the two diode (Q_5, Q_6) voltages to their common emitter connection. Thus, the lowest of the two output currents is regulated. If the lowest current is half the quiescent current, Q_7 or Q_8 already take over the full current of both transistors in the quiescent-state. So the lowest of the push or pull currents never comes below half the quiescent current according to:

$$\left(I_{push} - \tfrac{1}{2}I_{quies}\right)\left(I_{pull} - \tfrac{1}{2}I_{quies}\right) = \left(\tfrac{1}{2}I_{quies}\right)^2 \tag{5.11}$$

This is equal to the expression (Eq. 5.9) of the GA feedforward biased output stage of Fig. 5.26a. This robust biasing scheme is realized in a 100 MHz precision operational amplifier [ref. 16 of Chap. 7], a 1 GHz operational amplifier, and in a 100 mA voltage and current efficient operational amplifier [23].

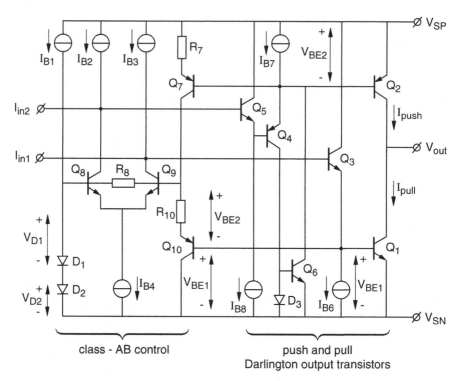

Fig. 5.35 GA FBB R-R class-AB Darlington output stage with control amplifier (NE5230)

5.4.3 FBB Rail-to-Rail General Amplifier Output Stages

Finally, we will see how the flexibility of the feedback class-AB biasing will turn the push–pull inverting-amplifier output stage into a robust rail-to-rail GA output stage which can function well at a low supply voltage.

Firstly, we will show how feedback biasing can mimic the classic class-AB control of the VF output stage with two diodes. Figure 5.35 shows how this is done [16].

The circuit has GA Darlington output transistors which are directly accessible from the input. The upper folded Darlington structure has an extra current boost Q_4, D_3, Q_6 to compensate the low current gain of the lateral upper output transistor Q_2. The base-emitter voltage V_{BE2} of the upper output transistor Q_2 is converted into a current by the base-emitter diode of Q_7 in series with R_7 and regenerated into a voltage across the base-emitter diode of Q_{10} in series with R_{10}. The base of Q_{10} is connected to the base of the lower output transistor Q_1. In this way the sum voltage $V_{BEsum} = V_{BE1} + V_{BE2}$ of the two output transistors is modeled. This sum is compared to the sum of two diode voltages $V_{Dsum} = V_{D1} + V_{D2}$ by a differential amplifier Q_8, Q_9. When the sum of the base-emitter voltages V_{BEsum} is lower than the sum of the diode voltages V_{Dsum}, the transistor Q_8 draws more current and Q_9 less current than half the tail current I_{B4}. This differentially engages both output transistors, by which the sum of the base-emitter voltages grows until $V_{BEsum} = V_{Dsum}$. With equal quiescent currents and emitter areas for all NPN transistors and diodes, as well as for all PNPs, and $R_{10} = R_7$ the push and pull currents are controlled as to obey:

$$I_{push} \cdot I_{pull} = I^2_{quies} = I^2_{B1} \qquad (5.12)$$

This behavior is equal to the VF circuit of Fig. 5.7a, as described by (Eq. 5.4).

The minimum supply voltage is two diodes and a saturation voltage, which is about 1.8 V. The circuit is utilized in the Signetics NE5230 low-voltage operational amplifier and described in Ref. [16].

A weak spot of this circuit is the high impedance at the emitter of Q_{10} when the upper output transistor is nearly cut off and the current through Q_{10} is low. This deteriorates the HF behavior of the feedback loop. To prevent this, a shunt resistor R_8 has been placed across the bases of the differential control amplifier to lower the impedance on the emitter of Q_{10}.

The class-AB control can be improved by using a minimum selector [15] Q_{10}, Q_{11} as shown in Fig. 5.36. The minimum selector is placed on top of the base-emitter voltages V_{BE1} of Q_1 and V_{BE2} of Q_2. The last voltage is modeled by the diode-connected Q_4. With equal quiescent currents and emitter areas for the NPN and PNP transistors respectively, the relation between the push and pull currents is:

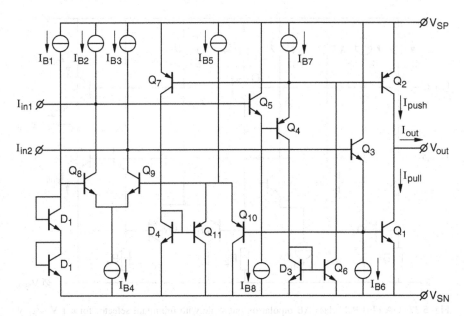

Fig. 5.36 GA FBB R-R class-AB output stage with Darlington output transistors and minimum selector (NE 5234)

$$\left(I_{push} - \tfrac{1}{2}I_{quies}\right)\left(I_{pull} - \tfrac{1}{2}I_{quies}\right) = \left(\tfrac{1}{2}I_{quies}\right)^2 = \left(\tfrac{1}{2}I_{B1}\right)^2 \qquad (5.13)$$

The low impedances anywhere at the emitters of the control circuit ensure a good HF behavior of the class-AB loop. The differential amplifier Q_8, Q_9 controls the class-AB behavior in a robust way.

The push and pull output currents do not go below $\tfrac{1}{2}I_{quies}$. The amplifier has a similar relation between the push and pull currents as the circuit of Fig. 5.26a with Eq. 5.9. The circuit of Fig. 5.36 can be found in the quad OpAmp NE5234 and is described in Ref. [17].

It is interesting to note that if, in these circuits, only one input terminal is used to drive the circuit, the regulator amplifier automatically drives the other input terminal correctly. Half of the input current is used to drive one output transistor, whereas through the collector-emitter loop of the differential amplifier the other output transistor receives the other half of the input current. Further, if one output transistor is regulated at a constant value of $\tfrac{1}{2}I_{quies}$, the other output transistor gets all the input driving current of both inputs. This guarantees a very linear signal transfer.

A low voltage version of the previous circuit is shown in Fig. 5.37. The base-emitter voltages of the output transistors are modeled in a reduced form across R_2 and R_3. The minimum selector Q_{10}, Q_{11} takes the lowest of these voltages and the differential control amplifier Q_8, Q_9 regulates this minimum equal to a reference voltage across a reference network. The circuit has the capability to work down to

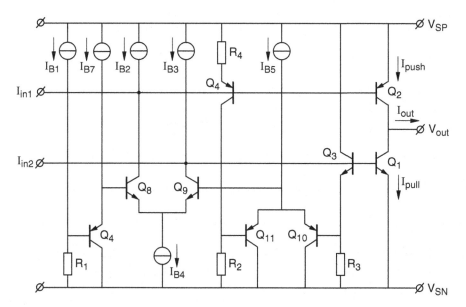

Fig. 5.37 GA FFB R-R class-AB bipolar output stage with minimum selector for a 1 V supply voltage

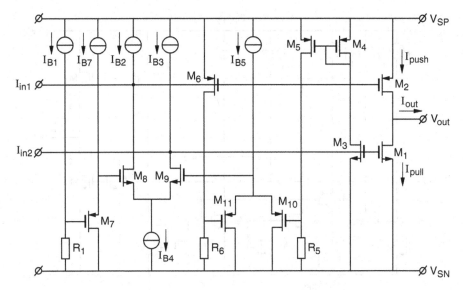

Fig. 5.38 GA FBB R-R class-AB CMOS output stage with minimum selector for 1.2 V supply voltage

one diode voltage and two saturation voltages, which amounts to about 1.0 V. The circuit is described in Ref. [17].

A similar circuit can be obtained in CMOS. A CMOS GA feedback-biased class-AB output stage is shown in Fig. 5.38.

The measuring of the lower output transistor M_1 cannot be done in the same way as in Fig. 5.37, because the threshold voltage of CMOS transistors cannot be scaled by the W/L ratio's. Only the current can be scaled down. This is done with the aid of two current mirrors M_1, M_3 and M_4, M_5.

The voltage across R_5 carries a model of the current in the lower output transistor M_1, while the voltage across R_6 carries a model of the current of the upper output transistor M_2. The minimum selector M_{10}, M_{11} takes the lowest of these voltages. The differential amplifier M_8, M_9 compares the lowest voltage with the voltage across a reference circuit M_7, R_1, which is biased by I_{B1}. For equal transconductance of the P-channel and N-channel output transistors, the W/L ratio of all P-channel transistors is taken three times that of the N-channel ones, to compensate for a lower charge mobility μ_p. The resulting behavior is roughly similar to that of the circuit of Fig. 5.27a and Eq. 5.10.

The circuit is described in Ref. [18]. It works down to a supply voltage of one diode and two saturation voltages, which amounts to about 1.2 V.

A GA feedback-biased class-AB output stage which avoids resistors in a CMOS process is shown in Fig. 5.39 [19]. It can easily be scaled with the W/L ratios of the CMOS transistors. The minimum selector is composed of two inverters M_{11}, M_{21}, M_{23}, M_{15} and M_{12}, M_{14}, followed by a maximum selector M_{18} and M_{17}, M_{16} which functions as a differential control amplifier as well. The reference is set by M_{19}.

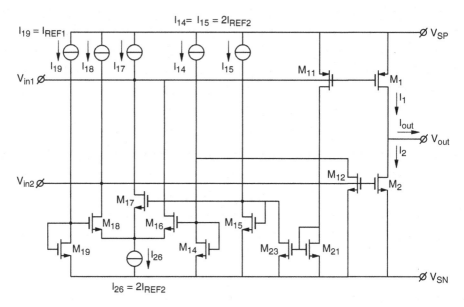

Fig. 5.39 GA FBB R-R class-AB CMOS output stage with a maximum selector for 1.2 V supply voltage

The inverters invert the minimum output currents of the output transistors to a maximum. The differential amplifier is controlled at the right-hand side by the highest of the two voltages at the gates of M_{17} or M_{16} respectively.

The quiescent current is:

$$I_q = 2 \frac{W_1 W_2}{L_1 L_2} \frac{L_{11} L_{12}}{W_{11} W_{12}} \left(2 I_{ref2} - I_{ref1} \right) \qquad (5.14)$$

The minimum current stays at about two-thirds of the quiescent current. The application of feedback-biasing allows a minimum supply voltage of that of one diode and two saturation voltages, which amounts to about 1.2 V.

A very simple minimum selector without resistors in CMOS technology is shown in Fig. 5.40 [20].

The minimum selector is composed of two measuring transistors M_{11} and M_{12}, and a mirror M_{22}, M_{24}. If M_1 carries the lowest output *current*, M_{24} is biased in triode mode and M_{11} roughly measures the minimum current of M_1. If M_2 carries the lowest current, M_{11} acts as a cascode which passes on the current of the measuring transistor M_{12} which is fed through the mirror M_{22}, M_{24}. The quiescent current and minimum currents are:

$$I_q = 2 \frac{W_1 W_2}{L_1 L_2} \frac{L_{11} L_{12}}{W_{11} W_{12}} I_{ref}$$

$$I_{min} = \frac{1}{2} I_q \qquad (5.15)$$

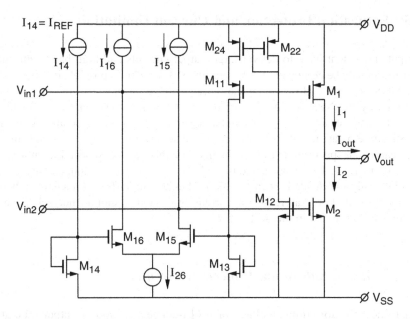

Fig. 5.40 GA FBB R-R class-AB CMOS output stage with simple minimum selector for 1.2 V supply voltage

The circuit excels by its simplicity. The minimum voltage is equal to that of one diode and two saturation voltages, which amounts to about 1.2 V.

5.4.4   Conclusion

We have discussed feedback-biased class-AB output stages in a voltage follower (VF), compound (VF/GA), and inverting amplifier (GA) configurations. The VF circuits are similar to those with feedforward-biasing. Though they are the simplest, their output voltage range loses at least two diode voltages and two saturation voltages in regard to the supply voltage range.

The VF/GA stages are of particular interest in "all-NPN" output stages, where large output currents must be drawn, or where the highest frequency response of the NPN transistors must be utilized. These circuits lose minimally one diode voltage and two saturation voltages in their output swing. The GA output stages are of particular importance at low supply voltages, as their output voltage can swing nearly from rail-to-rail, except for a saturation voltage at each rail. The feedback-biasing robustly controls these stages in class-AB and leaves much flexibility to the designer. The minimum supply voltage may be of the order of one diode voltage and two saturation voltages.

5.5 Saturation Protection and Current Limitation

Output stages that have to deliver large output currents and high bandwidth may advantageously be equipped with bipolar transistors. This is possible in bipolar and BiCMOS processes. However, if the output voltage becomes so low or high that respectively the NPN or PNP output transistor saturates, the bandwidth of these transistors may easily become so low that the amplifier starts to oscillate. Moreover, the substrate parasitic transistors become activated so that a large substrate current may cause unexpected latch-up problems. For this reason we need a saturation-protection circuit. This is presented in the Sect. 5.5.1. Bipolar output transistors can also be easily overloaded when the output has to supply heavy loads or is being short circuited. For that reason a current limitation circuit is needed. Several limiters are presented in the Sect. 5.5.2.

5.5.1 Output Saturation Protection Circuits

When the collector-emitter voltage of a GA-connected bipolar output transistor becomes lower than about 200 mV, the collector–base junction becomes forward biased. This causes several undesired effects. The situation is depicted in Fig. 5.41 for a push–pull GA/GA configuration. But the effect is also present in the lower side of a VF/GA connected all-NPN output stage.

Firstly, if one of the transistors Q_1 or Q_2 becomes saturated, the collector–base junction is activated and a large reverse current may flow back from the collector into the emitter. This must be compensated by a further increase of the forward

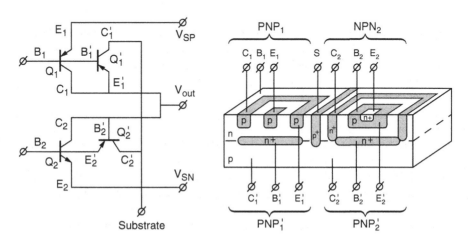

Fig. 5.41 A push–pull GA/GA output stage with a PNP_1 and NPN_2 transistor and their PNP'_1, and PNP'_2 parasitic transistors

current. The result is a decrease of β and an increase of the diffusion capacitors. The bandwidth of the output stage may decrease so much that the frequency compensation may easily become unstable. Ringing is the result.

Secondly, the parasitic substrate PNP transistor Q'_1, or Q'_2 of the concerned PNP transistor Q_1 or NPN transistor Q_2 becomes active. This may cause a large current to flow into the substrate. Particularly, in BiCMOS processes with a lightly doped substrate, a large voltage drop in the substrate may be the result so that even reverse biased diodes may turn on. This may cause unexpected latch-up problems in the circuit.

The saturation voltage may be modeled by using the Ebers-Moll model [21]. When we suppose that while saturated the ratio between the collector and base currents is in the order of the square root of the forward current gain β_F, the saturation voltage can be simply expressed as:

$$V_{CEsat} \approx I_c r_c + \frac{kT}{q} \ln\left(\frac{1}{\beta_F} \frac{I_c}{I_B}\right) \tag{5.16}$$

The first term is caused by the internal resistive collector resistance r_c. This term can be of the order of several 100 mV. The second term maybe of the order of 20–50 mV.

For detection of saturation, we could use the actual collector voltage if we are sure that the resistive collector voltage drop is sufficiently under control. This gives rise to the saturation protection circuit of the Darlington output transistor of Fig. 5.42 for example.

A disadvantage of the clamp transistor Q_3 is that the emitter can easily get too high a reverse voltage, so that it could become zenering and be destroyed. Therefore, it is better to interchange its emitter and collector connections and use it reversed. If we further connect the base of Q_3 to the base of Q_1, the collector–base junction of Q_3 perfectly matches the collector–base junction of Q_1 in voltage, dopingprofile, and process variations. The only detection error that remains is the

Fig. 5.42 Saturation protection with clamp transistor Q_3

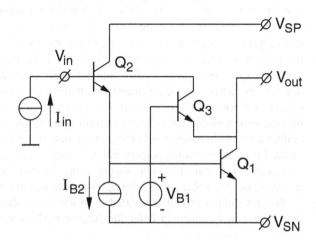

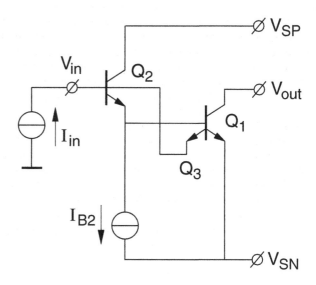

Fig. 5.43 Robust
saturation protection with a
second emitter Q_3 of the
output transistor Q_1

resistive collector voltage drop. To largely avoid this, we may use an output
transistor with a second emitter. So we obtain the robust saturation protection
circuit of Fig. 5.43 [22].

Another important feature of the protection circuit of Fig. 5.42 is that the control
loop through Q_2 and Q_1 is very short and naturally HF stable. A complementary
circuit can be used with the complementary PNP output transistor if applicable.

5.5.2   Output Current Limitation Circuits

The high bandwidth f_T and current gain β_F of the NPN transistor makes it perfectly
suitable for a high quality output stage in a bipolar or BiCMOS process. But these
very attributes also make the NPN output transistor vulnerable for overloading at
heavy load currents or at short circuit conditions. Therefore, a current limitation
circuit has to be applied. Many existing limiter circuits have control paths that are
too long, so that these circuits tend to oscillate when they have to limit the output
current. In the next three circuits, this has been avoided. There are two ways to
detect over-currents in output transistors that do not make the circuit too complicated or do not cost too much voltage loss in the output range. The first detection
method is to measure the collector current of a transistor connected with its base-
emitter contacts in parallel with those of the output transistor. The second detection
method is to insert a small resistor into the emitter lead of the output transistor and
to measure the voltage drop over that resistor. The latter is more accurate of course,
as we measure the real current through the output transistor.

The first limiter circuit is divided from the successful saturation limiter circuit
with a reversed connected detection transistor Q_4 in combination with a model

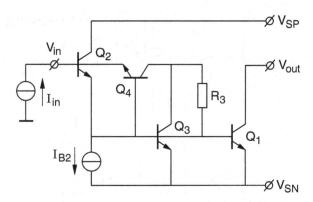

Fig. 5.44 Output current limitation circuit with a model transistor Q_3 in parallel with the output transistor Q_1 and with its collector current through a reversed connected detection transistor Q_4 and a threshold current provided by R_3

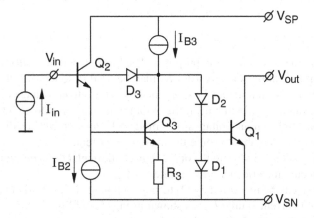

Fig. 5.45 Output current limitation circuit with a model transistor Q_3 in parallel with the output transistor Q_1 and with its collector connected through a diode D_3 to the input and a threshold bias current source I_{B3}

transistor Q_3 in parallel with the output transistor Q_1. The circuit is given by Fig. 5.44 [22].

The resistor R_3 together with the base-emitter voltage of Q_3 set the threshold current. A disadvantage of the resistor connection to the base of Q_1 is its damping action on the current gain of Q_1.

But this can be overcome by some extra elements. Another drawback of the circuit is that the reverse connected transistor Q_4 has a low reverse current gain in modern processes. This slightly weakens the limitation action.

A limiter circuit that alleviates the above mentioned disadvantages is shown in Fig. 5.45. It uses a model transistor Q_3 in parallel with the output transistor and a separate threshold bias current I_{B3} in its collector circuit. The circuit further

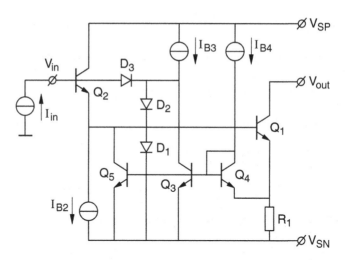

Fig. 5.46 Output current limitation circuit with an emitter-current measuring resistor R_1, a level-shift transistor Q_4, and a regulator transistor Q_3 biased with a threshold current I_{B3}

incorporates diodes for clamping purposes. The limiting action is strong enough so that a current limitation series resistor R_3 may be used in series with the emitter of Q_3. The limitation function may be sharpened by using a fixed voltage instead of R_3.

One disadvantage still remains: if the output transistor tends to go into avallange breakdown at high output voltages, the model Q_3 does not detect this. Only the current source I_{B2} can absorb some the avallange current.

If we really want to measure the output current through Q_1, an emitter resistor R_1 can be connected in series with the emitter of Q_1.

The current can be measured by Q_3 through a voltage level shifter Q_4 and biased by a threshold current source I_{B3}, as shown in Fig. 5.46 [22].

The diodes D_1 through D_3 perform clamping functions. The maximum voltage across the measuring resistor R_1 need only to be of the order of 50 mV, so that not much of the output current range is being sacrificed.

The limitation functions starts at a value of:

$$I_{out\ lim} = \frac{kT}{R_1 q} \ln\left(\frac{I_{B3}}{I_{B4}} \frac{A_4}{A_3}\right) \tag{5.17}$$

where A_4/A_3 is the ratio of the emitter areas of Q_4 and Q_3. Q_5 has further been added to increase the current by which the output transistor is being cut off. This may be needed in case of avallange breakdown of Q_1.

As a conclusion the operation characteristics of the three limiter circuits as discussed are shown in Fig. 5.47.

Further information can be found in Ref. [23].

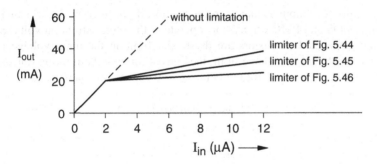

Fig. 5.47 Output current versus input current of the limiter circuits of Figs. 5.44, 5.45, and 5.46

5.6   Problems and Simulation Exercises

5.6.1   Problem 5.1

The voltage-follower output stage shown in Fig. 5.8a is biased with $I_{B1} = I_{B2} = 20\,\mu\text{A}$, with transistors sized $W/L_1 = 175$, $W/L_2 = 50$, $W/L_3 = 35$, $W/L_4 = 10$. The MOS devices are modeled by $V_{THN} = 0.5$, $V_{THP} = -0.6$, $K_N = 56\,\mu\text{A}/V^2$, $K_P = 16\,\mu\text{A}/V^2$ and the current sources are saturating at $V_{Dsat} = 0.2\text{V}$. Calculate the maximum output current so no transistor operates at zero drain current and the output voltage range which does not introduce distortion in the signal. The supply voltages are $V_{DD} = -V_{SS} = 1.5\text{ V}$.

5.6.1.1   Solution 5.1

The quiescent current of the output stage is given by the all-V_{GS} loop formed by the four MOS devices. Noting that the size of these devices obey

$$\frac{\frac{W}{L_3}}{\frac{W}{L_1}} = \frac{\frac{W}{L_4}}{\frac{W}{L_2}} \tag{5.18}$$

the solution is

$$I_{D1} = I_{D2} = 5I_{D3} = 100\mu\text{A} \tag{5.19}$$

According to Eq. 5.5, the maximum output current which still allows the both transistors to operate is

$$I_{Omax} = 4I_{quies1,2} = 400\mu\text{A} \tag{5.20}$$

The limits for output voltage can be calculated considering the fact that both pairs of NMOS and PMOS transistors operate at the same saturation voltages, but the diode connected transistors are drastically limiting the input voltage range. When approaching one voltage limit, the current through the limiting transistor is also decreasing, a convenient limit being $I_D = 0$ which corresponds to a $V_{GS} = V_{th}$

$$V_{Omax} = V_{imax} = V_{SP} - V_{Dsat} - V_{GS3}$$
$$= V_{SP} - V_{Dsat} - V_{THN} = 0.8V$$
$$V_{Omin} = V_{imin} = V_{SN} + V_{Dsat} - V_{GS4}$$
$$= V_{SN} + V_{Dsat} - V_{THP} = -0.7V$$

$$(5.21)$$

5.6.2 Problem 5.2

Figure 5.13 shows the schematic of a bipolar *IA/VF* output stage. The diodes D_1 through D_4 have an area A_{ref} and a saturation current $I_S = 10^{-16}A$. The other transistors are upsized: Q_1 and Q_2 ten times, Q_3 and Q_4 two times. The resistor R_1 and R_2 are 100 Ω while $R_3 = R_4 = 1$ kΩ. The biasing sources $I_{B1} = I_{B2} = 10\mu A$ and the transistors have large current gain and a saturation voltage $V_{sat} = 0.2$ V. Calculate the minimum supply voltage which allows the output to reach $2V_{PP}$ without distorting the signal. Consider $kT/q = 25$ mV.

5.6.2.1 Solution 5.2

In order to calculate the limits of the output voltage, the biasing currents must be known for all transistors and diodes. The biasing of Q_3 and Q_4 can be easily calculated based on the translinear loop $Q_3 - Q_4 - D_4 - D_3$

$$I_{C3} = I_{C4} = \frac{I_{SQ3}}{I_{SD3}} I_{B1} = \frac{A_{Q3}}{A_{D3}} I_{B1} = 20\mu A \qquad (5.22)$$

The biasing of Q_1 and Q_2 can be calculated by observing that in the $Q_1 - R_1 - R_3 - D_1$ loop, the ratio of sizes for both resistors and transistors is 10, so

$$I_{C1} = 10I_{D1} = 200\mu A \qquad (5.23)$$

This value satisfies the equation

$$\frac{kT}{q} \ln \frac{I_{C1}I_{SD1}}{I_{D1}I_{SQ1}} = R_3 I_{D1} - R_1 I_{C1} \qquad (5.24)$$

Alternatively a numerical solver can be used to find the solution of this equation. The maximum output signal which is not saturating any transistor is

$$V_{Omax} = V_{SP} - \max\left(V_{BE3} + V_{sat1B1}, V_{satQ3} + V_{D1} + R_3 I_3, V_{satQ1} + R_1 I_1\right) \quad (5.25)$$

Obviously, the second term in the list above has the largest value, as it adds a V_{BE} to a saturation voltage and the voltage on a resistor. This makes the value of the maximum output voltage to be

$$V_{Omax} = V_{SP} - V_{satQ3} - \frac{kT}{q} \ln\frac{I_{D1}}{I_{SD1}} - R_3 I_3 = V_{SP} - 650.8mV \quad (5.26)$$

As the requested range of the output signal is 2 V peak-to-peak, the supply voltage should be

$$V_{SP} - V_{SN} = 2V + 2*650.8mV = 3.3V \quad (5.27)$$

5.6.3 Problem 5.3

The class-AB output stage depicted in Fig. 5.20a operates at $V_{SP} = V_{SN} = 1.5\,\text{V}$ and the MOS devices are sized $W/L_3 = W/L_5 = 35, W/L_4 = W/L_6 = 10\,W/L_1 = 100$ and $W/L_2 = 350$. Resistor values are $R_1 = R_2 = 50\,\text{k}\Omega$. Considering the MOS parameters $V_{THN} = 0.5\,\text{V}$, $V_{THP} = -0.6\,\text{V}$, $K_P = 16\,\mu\text{A}/V^2$, and $K_N = 56\,\mu\text{A}/V^2$ calculate the maximum push and pull currents of the output such that no transistor operates at zero drain current.

5.6.3.1 Solution 5.3

The biasing of the output stage is set by the current leg $M_5 - R_2 - M_6$

$$V_{SP} - V_{SN} = V_{GS5} + R_2 I_{B2} + V_{GS6} \quad (5.28)$$

Replacing the V_{GS} expression as functions of drain currents, the following second order equation results

$$-V_{THP} + \sqrt{\frac{2I_{B2}}{K_P\frac{W}{L_5}}} + R_2 I_{B2} + V_{THN^+} \sqrt{\frac{2I_{B2}}{K_N\frac{W}{L_6}}} - V_{SP} + V_{SN} = 0 \quad (5.29)$$

which has the only positive solution

$$\sqrt{I_{B2}} = \frac{-b + \sqrt{b^2 - 4R_2(-V_{THP} + V_{THN} - V_{SP} + V_{SN})}}{2R_2} \tag{5.30}$$

with

$$b = \frac{1}{\sqrt{\frac{K_p}{2} \frac{W}{L_5}}} + \frac{1}{\sqrt{\frac{K_p}{2} \frac{W}{L_6}}} \tag{5.31}$$

Numerically, these equations produce

$$I_{B2} = 26\mu A \tag{5.32}$$

The quiescent current for output transistors M_1, M_2 is given by the equal voltage on R_1 and R_2 and the W/L ratios of M_1 and M_5 and for M_2 and M_6, respectively.

$$I_{D1} = I_{D2} = I_{D5} \frac{W}{L_1} \frac{L_5}{W} = 260\mu A \tag{5.33}$$

As explained with Fig. 5.8, the maximum push and pull output current at which the transistor driving the lower current is still on is

$$I_{pushmax} = -I_{pullmax} = 4I_{D1} = 1.04\text{mA} \tag{5.34}$$

5.6.4 Problem 5.4

Figure 5.27a shows a class-AB output stage with common-mode transistor coupling which allows operation at lower supply voltages compared to the common-mode resistor coupling version. Such an output stage operates at $V_{SP} = -V_{SN} = 1.0\,\text{V}$, the NMOS devices are sized $W/L_3 = 2.5$, $W/L_5 = 5$, $W/L_6 = 20$, and $W/L_2 = 100$ and all the PMOS counterparts are sized K_N/K_P larger. Considering the MOS parameters $V_{THN} = 0.5\,\text{V}$, $V_{THP} = -0.6\text{V}$, $K_P = 16\,\mu A/V^2$, and $K_N = 56\,\mu A/V^2$, calculate the output transistors quiescent current, the maximum push and pull currents and the minimum currents which are kept flowing through output devices. The biasing currents are $I_{B1} = I_{B2} = I_{B3} = I_{B4} = 10\,\mu A$.

5.6.4.1 Solution 5.4

The quiescent current of transistors M_1 and M_2 results from the translinear loop

$$V_{GS2} + V_{GS3} = V_{GS5} + V_{GS6} \tag{5.35}$$

All the threshold voltages being equal for all PMOS devices, the equation above can be translated to a drain current and sizes equation

$$\sqrt{\frac{I_{D2}}{W/L_2}} + \sqrt{\frac{I_{D3}}{W/L_3}} = \sqrt{\frac{I_{D5}}{W/L_5}} + \sqrt{\frac{I_{D6}}{W/L_6}} \tag{5.36}$$

Solution of the equation becomes

$$\sqrt{I_{D2}} = \sqrt{W/L_2} \left(\sqrt{\frac{I_{D5}}{W/L_5}} + \sqrt{\frac{I_{D6}}{W/L_6}} - \sqrt{\frac{I_{D3}}{W/L_3}} \right) \tag{5.37}$$

$$I_{D2} = 50\mu A$$

The maximum push and pull currents are equal because of the complete symmetry of the circuit related to NMOS and PMOS devices. The pull current reaches its largest value when all of the I_{B1} current is flowing through M_4, thus operating M_3 at $V_{GS3} = V_{THN}$. Starting again from the translinear loop Eq. 5.35, the drain currents equation can be rewritten for the maximum pull current

$$\sqrt{I_{pullmax}} = \sqrt{W/L_2} \left(\sqrt{\frac{I_{D5}}{W/L_5}} + \sqrt{\frac{I_{D6}}{W/L_6}} \right) \tag{5.38}$$

$$I_{pullmax} = 450\mu A$$

The minimum current, which is kept flowing through one of the output devices while the other is driving the maximum push–pull current, will also be identical for both PMOS and NMOS output transistors. For M_2, this current is reached when all of the I_{B1} current flows through M_3, reducing the available V_{GS2} to a minimum. The value of this current is

$$\sqrt{I_{D2min}} = \sqrt{W/L_2} \left(\sqrt{\frac{I_{D5}}{W/L_5}} + \sqrt{\frac{I_{D6}}{W/L_6}} - \sqrt{\frac{I_{B1}}{W/L_3}} \right) \tag{5.39}$$

$$I_{D2min} = 1.3\mu A$$

5.6.5 Problem 5.5

For the class-AB output stage in Fig. 5.40, calculate the quiescent and the minimum current through the output transistors M_1 and M_2, if the sizes of active devices are $W/L_{14} = W/L_{13} = 10$, $W/L_{15} = W/L_{16} = 20$, $W/L_{11} = 0.2\,W/L_1 = 70, W/L_{12} = 0.1\,W/L_2 = 10$, and $W/L_{22} = W/L_{24} = 70$. The biasing currents are $I_{B14} = I_{B15} = I_{B16} = 20\,\mu A$, $I_{B26} = 40\,\mu A$.

5.6.5.1   Solution 5.5

Because of the comparator built with devices M_{15} and M_{16} the feedback loop makes the devices M_{13} and M_{14} work at the same gate-source voltage, which in turn makes the current through M_{13} to be

$$I_{D13} = \frac{W}{L_{13}}\frac{L_{14}}{W}I_{D14} = 20\mu A \tag{5.40}$$

This current is also the drain current for M_{11}, as well as the drain current for M_{12} because the current mirror M_{22}: M_{24} has a current gain equal to unity. The quiescent current through the output transistor is then given by

$$I_{D2} = \frac{W}{L_2}\frac{L_{12}}{W}I_{D12} = 200\mu A \tag{5.41}$$

Identical drain current can be obtained for M_1 knowing the current through M_{11}, M_{24} and the fact that at quiescent biasing the two transistors M_{11}, M_{24} act like one transistor with double channel length compared to M_{11}.

$$I_{D1} = \frac{W}{L_1}\frac{2L_{11}}{W}I_{D11} \tag{5.42}$$

The minimum current in the output devices is limited to

$$I_{D2min} = \frac{1}{2}I_{D_2} = 100\mu A \tag{5.43}$$

5.6.6   Simulation Exercise 5.1

The feedforward-biased class-AB output stage shown in Fig. 5.48 uses V_{GS} voltage loops for AB-biasing of output transistors M_4 and M_8. A circuit shown in Fig. 5.49 can be used to put the signal currents at the input of such a stage while keeping the output at a constant voltage.

Simulate this circuit and plot the drain currents of M_4 and M_8 as a function of the input current. I_6 and I_7 in the simulation circuit must supply equal current at all times. Note how the results are affected by a change in the W/L ratios of M_5, M_6 and M_1, M_2 respectively, by doubling the widths of M_2 and M_6.

5.6.7   Simulation Exercise 5.2

For the class-AB feedback-biased output stage shown in Fig. 5.50 plot the drain currents for the output transistors M_1 and M_2 as a function of the input currents. Note the point where transistor M_{24} enters linear region by monitoring its source-drain voltage.

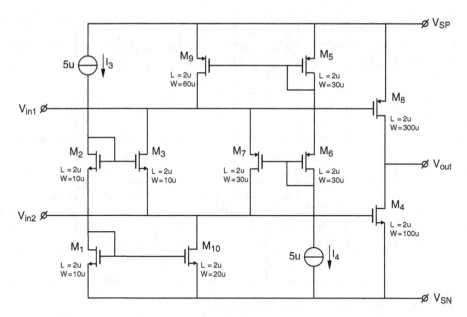

Fig. 5.48 Class-AB output stage with feedforward biasing

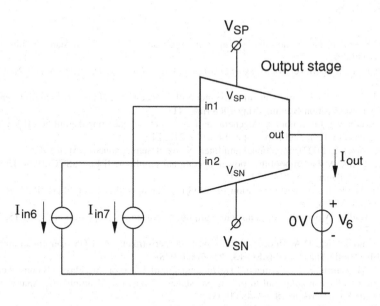

Fig. 5.49 Output stage simulation circuit for transient analysis

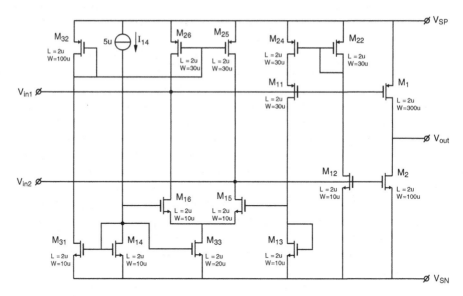

Fig. 5.50 Class-AB output stage with feedback biasing

References

1. J.E. Solomon, The monolithic op amp: a tutorial study. IEEE J. Solid-State Circuits **SC-91**, 314–332 (1974)
2. P.R. Gray, R.G. Meyer, *Analyses and Design of Analog Integrated Circuits* (Wiley, New York, 1984)
3. D. Fullager, A new high performance monolithic operational amplifier, Fairchild Semiconductor, Application Briefing, May 1968 (μA 741)
4. Y. Nishikawa, J.E. Solomon, A general-purpose wideband operational amplifier, IEEE ISSCC 73, Digest of Technical Papers, pp. 144, 145, 212, 213
5. Data sheet, LH-0021 operational amplifier, National Semiconductor, April 1972
6. F.L. Long, A dual monolithic power operational amplifier, IEEE ISSCC 1973, Digest of Technical Papers, pp. 178, 179, 221
7. O.H. Shade Jr., A new generation of MOS/Bipolar operational amplifiers. RCA Rev. **37**, 204–224 (1976)
8. R.J. Widlar, Monolithic op amp with simplified frequency compensation. IEEE **15**, 58–63 (1967)
9. D. Senderowicz, D.A. Hodges, P.R. Gray, High performance N-MOS operational amplifier. IEEE J. Solid-State Circuits **SC-13**, 760–766 (1978)
10. W.C.M. Renirie, K.J. de Langen, J.H. Huijsing, Parallel feedforward class-AB control circuits for low-voltage bipolar rail-to-rail output stages of operational amplifiers. Analog Integr. Circuits Signal Process. **8**, 37–48 (1995)
11. D.M. Montecelli, A quad CMOS single-supply op amp with rail-to-rail output swing. IEEE J. Solid-State Circuits **SC-21**, 1026–1034 (1986)
12. R. Hogervorst, J.P. Tero, R.G.H. Eschauzier, J.H. Huijsing, A compact power-efficient 3V CMOS rail-to-rail input/output operational amplifier for VLSI cell libraries. IEEE J. Solid-State Circuits **29**(12), 1505–1513 (1994)

13. P.R. Gray, A 15-W monolithic power operational amplifier. IEEE J. Solid-State Circuits **SC-7**, 478–480 (1972)
14. J.H. Huijsing, F. Tol, Monolithic operational amplifier design with improved HF behaviour. IEEE J. Solid-State Circuits **SC-11**, 323–328 (1976)
15. E. Seevinck et al., A low-distortion output stage with improved stability for monolithic power amplifiers. IEEE J. Solid-State Circuits **23**(3), 794–801 (1988)
16. J.H. Huijsing, D. Linebarger, Low-voltage operational amplifier with rail-to-rail input and output ranges. IEEE J. Solid-State Circuits **SC-20**, 1144–1150 (1985)
17. J. Fonderie, J.H. Huijsing, Operational amplifier with I-V rail-to-rail multipath-driven output stage. IEEE J. Solid-State Circuits **26**(12), 1817–1824 (1991)
18. R. Hogervorst et al., CMOS low-voltage operational amplifiers with constant-g_m rail-to-rail input stage. Analog Integr. Circuits Signal Process. **5**, 135–146 (1994)
19. R.G.H. Eschauzier, R. Hogervorst, J.H. Huijsing, A programmable 1.5 V CMOS class-AB operational amplifier with hybrid-nested Miller compensation for 120 dB gain and 6 MHz UGF. IEEE J. Solid-State Circuits **29**(12), 1497–1504 (1994)
20. K.J. de Langen, J.H. Huijsing, Compact low-voltage power-efficient operational amplifier cells for VLSI. IEEE J. Solid-State Circuits **33**(10), 1482–1496 (1998)
21. I. Getreu, *Modelling the Bipolar Transistor* (Tektronix, inc., Beaverton, 1976)
22. K.J. de Langen, J. Fonderie, J.H. Huijsing, Limiting circuits for rail-to-rail output stages of lowvoltage bipolar operational amplifiers, in *ISCAS 95*, Seattle, vol. 3, pp. 1728–1731
23. K.J. de Langen, J.K. Huijsing, *High-frequency and Low-voltage Bipolar, BiCMOS, and CMOS Operational Amplifier Techniques* (Kluwer Academic Publishers, Boston, 1999)

Chapter 6
Overall Design

Abstract The previous chapters deal with two important stages of OpAmps. With the design of the input stage the aspects of bias, offset, drift, noise, common-mode rejection, and rail-to-rail input range were covered. With the design of the output stage power efficiency, classification of the fully VF, compound VF/GA, and rail-to-rail fully GA output stages with feed forward and feedback class-AB biasing are presented. The remaining attributes of gain, high-frequency response, slew rate, and linearity have to be performed by the whole of the input, intermediate, and output stages. That is the subject of this chapter.

This chapter discusses the overall design of operational amplifiers. Firstly, we will investigate how a large gain can be achieved. Therefore, an inventory will be made of nine main overall configurations in Sect. 6.1. Secondly, a systematic overview of HF compensation techniques will be presented in Sect. 6.2. Finally, aspects of slew rate and linear distortion are surveyed in Sects. 6.3 and 6.4 respectively.

6.1 Classification of Overall Topologies

One of the most important requirements of an operational amplifier is a large voltage gain A_V and current gain A_i (defined as $A_V = -Y_t/Y_o$ and $A_i = Y_t/Y_i$ in Chap. 2). The larger the voltage and current gain, the lower the errors which are made in OpAmp applications by a non-zero input voltage and current, as calculated in Chap. 3. For obtaining a large amount of voltage and current gain, several amplifier stages can be connected in cascade. The choices of these stages and the ways they are connected determines the main topology of the operational amplifier. Therefore, we will first make an inventory of possible configurations. After that some gain-boosting and compensation techniques will be reviewed.

6.1.1 Nine Overall Topologies

A general purpose OpAmp must have a common-mode (CM) input and output voltage range which extends from nearly the negative supply-rail voltage V_{SN} to nearly the positive supply-rail voltage V_{SP}.

© Springer International Publishing Switzerland 2017   157
J. Huijsing, *Operational Amplifiers*, DOI 10.1007/978-3-319-28127-8_6

We have seen that good input stages must be connected in the GA mode, and push–pull output stages can be chosen from the fully VF, compound VF/GA or fully GA modes. These basic connections for input and output stages are shown in Figs. 6.1 and 6.2.

In the input stage the NPN bipolar and P-channel CMOS versions are shown, because these ones are generally the best of their kind. The NPN transistor has the highest bandwidth and current gain. The P-channel transistor has a floating back-gate, which can be bootstrapped by the source, so that the CMRR is high.

Moreover, the P-channel transistor generally has a lower 1/f noise than the N-channel one. But, of course, complementary versions in both technologies are also possible.

In the output stage the full VF and GA versions are already complementary in nature. The output voltage range of the full VF stage cannot reach the rail within one diode voltage and one saturation voltage of the driver transistor, which all

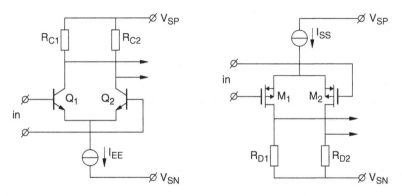

Fig. 6.1 Basic GA connections of input stages in (**a**) bipolar and (**b**) CMOS technology

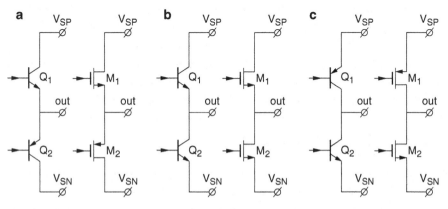

Fig. 6.2 Basic connections of the push–pull output stages in bipolar and CMOS technology: (**a**) Fully VF, (**b**) VF/GA and (**c**) GA connections

together amounts to about 1 V at each side. The full GA stage is much better in this respect. It only fails to reach the rails within one saturation voltage, which amounts to some hundreds of mV at each side. We call it a rail-to-rail output stage. The weakest spot in the full GA connection is the lateral PNP transistor Q_1 if we cannot dispose of an IC process with vertical PNPs. This transistor has a low current gain at higher current densities, so it has to be chosen very large in regard to the NPN transistor Q_2 to preserve some gain. Also its bandwidth is very low. For these reasons this transistor Q_1 may be replaced by a NPN transistor in the compound VF/GA version. We then have to accept an asymmetrical output voltage range. If the output swing must be rail-to-rail, Q_1 of the GA stage is sometimes replaced by a P-channel CMOS transistor when a BiCMOS process is at hand.

The question is now how to connect the input and output stages. A GA input transistor can only be directly connected with a GA output transistor if the latter is complementary to the former. No direct connection can be made with a VF or non-complementary GA output transistor. So we must conclude that a level-shift stage or an intermediate stage is always necessary to connect input and output stages of a general-purpose operational amplifier. Otherwise the output voltage cannot swing from close to the negative rail to close to the positive rail voltage.

An inventory of level-shift stages is given in Figs. 6.3 and 6.4.

The folded-cascode current mirror (CM) level-shift stages in Fig. 6.3a can also be regarded as folded-cascode current follower (CF) stages in the classification. They are able to subtract the differential output currents of the input stage and bring the result out at a single terminal. The bipolar version has emitter degeneration resistors to lower the current offset and noise, and thus to lower the equivalent input offset and noise voltage of the connected input stage. The bipolar folded-cascode current follower (CF) stage in Fig. 6.3b is able to connect an NPN GA input stage with any output stage at a large bandwidth. The all-pass current networks $R_F C_F$ lead the low-frequency signal components through the PNP transistors, and the high-frequency components through the NPN transistors [1]. If the turnover frequencies f_F of the $R_F C_F$ networks are higher than the cutoff frequencies f_T of the PNP transistors, there is no loss in the PNP transistors, and all of the input current is collected again at the output. This results in a flat frequency response of the current transfer from DC up to the f_T of the NPN transistors. The CMOS CF stage of Fig. 6.3c may have a large voltage gain due to the full cascoded structure.

The GA stages of Fig. 6.4 are useful as level-shift stages between a GA PNP or P-channel input stage and any output stage, and provide voltage and current gain at the same time.

It stands to reason that these stages may all be of their complementary counterparts also. Further, there is no need to mention a straight cascode CF stage after a GA stage, because such a cascode stage does not change the overall topology. Similarly, there is no need to mention a VF Darlington stage before a GA stage.

The conclusion may be drawn now, that we have only one type of input stage: the GA type; three types of output stages: the VF, VF/GA, and the GA types; and, finally, we have two types of level-shift stages: the folded-cascode current mirror (CM) or the current follower CF (both denoted as CF in Table 6.1) type, and the GA types.

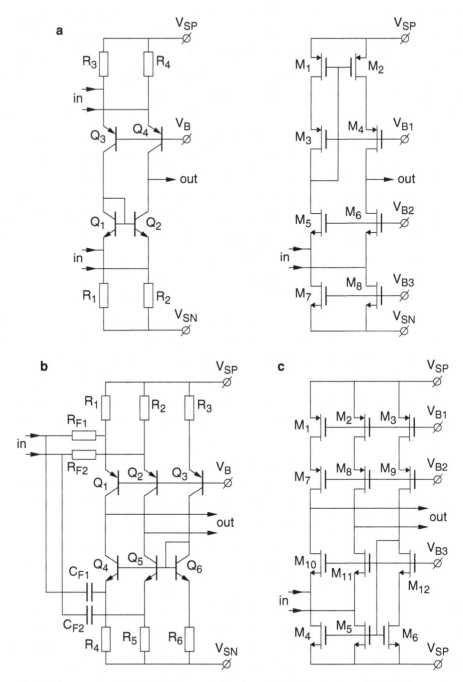

Fig. 6.3 Intermediate CF level-shift stages in bipolar and CMOS technology: (**a**) Folded-cascode current mirror (CM), and (**b/c**) folded-cascode current follower (CF)

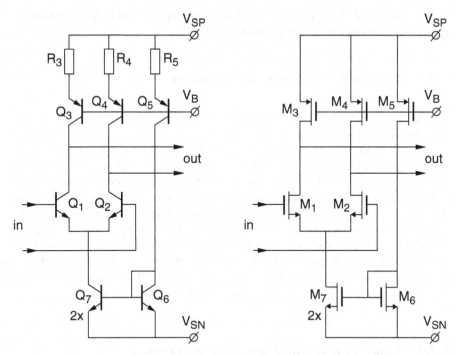

Fig. 6.4 Intermediate GA level-shift stages or intermediate amplifier stages in bipolar and CMOS technology

Table 6.1 Classification of operational amplifiers according to their main overall topology

Configuration number	Input stage	Level-shift or intermediate stage	Output stage
1	GA	CF	–
2	GA	GA	–
3	GA	CF	VF
4	GA	GA	VF
5	GA	CF	VF/GA
6	GA	GA	VF/GA
7	GA	CF	GA
8	GA	GA	GA
9	GA	GA+GA	GA

From all these possibilities we can make six three-stage combinations. We may add two possible two-stage combinations without a push–pull output stage. Also, we may add a group of multistage GA combinations with more than three GA, stages as the number of GA stages generally determines the complexity of the frequency compensation structure. Then, we have divided all operational amplifiers in nine main overall configurations. These are given in Table 6.1.

The design of these nine overall topologies together with their realization examples are further elaborated in Sects. 7.1–7.9.

6.1.2   Voltage and Current Gain Boosting

An alternative to the cascading of more GA stages is to apply voltage or current boosting to increase the gain.

Figure 6.5a shows a CF transistor M_1, used to increase the voltage gain of M_3 by the voltage-gain factor μ_1 of M_1. When we now artificially regulate the gate of M_1 with an amplifier M_2 such that the potential at the drain of M_3 remains constant, the output current is not dependent anymore on its output voltage and the output impedance at the drain of M_1 is increased [2]. The voltage gain of M_1 is boosted by the voltage-gain factor μ_2 of M_2. The total unloaded voltage gain is now $A_v = \mu_1\mu_2\mu_3$.

Figure 6.5b shows a VF transistor Q_1 used as a Darlington combination with Q_3 to increase the current gain of Q_3 by a factor β_1. We now artificially regulate the collector current of Q_1 at a constant value I_B by using Q_2 to boost the current gain of Q_1 by a factor β_2 of Q_2 [3]. The total current gain is now $A_i = \beta_1\beta_2\beta_3$.

6.1.3   Input Voltage and Current Compensation

A third method to increase the gain, besides cascading more GA stages or boosting, is compensation. Two examples are shown in Fig. 6.6a, b.

In the example of Fig. 6.6a, the relatively large source resistances $1/g_{ml,2}$ of the CMOS transistors are compensated by the equal but negative source resistances $-1/g_{m3,4}$ in a positive feedback loop [4]. The result is an overall transconductance $g_m = 1/(1/g_{ml,2} - 1/g_{m3,4})$. The transistors M3 and M4 must be taken slightly smaller

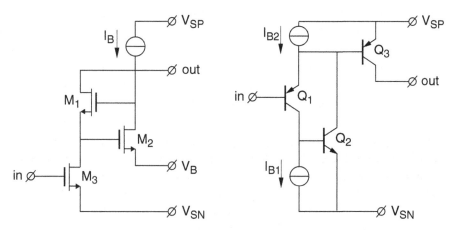

Fig. 6.5 (a) Voltage boosting of a CF transistor M_1 and (b) current boosting of a VF transistor Q_1

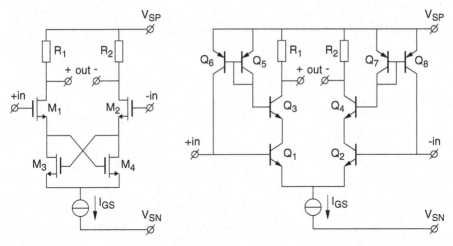

Fig. 6.6 (a) Compensation of the input voltage across of the source impedance $1/g_{m1,2}$ of $M_{1,2}$ by the negative source impedance $-1/g_{3,4}$ of $M_{3,4}$. (b) Compensation of the input base current of $Q_{1,2}$ by the negative base currents of $Q_{3,4}$

in their W/L ratio than M_1 and M_2, to make sure that the positive conductances $g_{m1,2}$ dominate the negative ones $g_{m3,4}$ so that the circuit remains stable. In the bipolar version, small degeneration resistors R_3 and R_4 have to be inserted in the emitters of Q_3 and Q_4 to keep the circuit stable. In the example of Fig. 6.6b, the base current of transistors Q_1 and Q_2 are compensated by duplicating these currents in Q_3 and Q_4 and mirroring them back into the bases of Q_1 and Q_2.

It should be clear that the above compensation methods to increase the gain by a positive feedback loop are limited by the matching accuracy of components. This is in contrast to the methods of cascading or boosting, where there is no limit in gain increase at low frequencies.

However, there are two fundamental limits to the overall gain: thermal feedback and frequency compensation. Thermal feedback on the chip going from the output stage into the input stage sets a limit to the maximum useful low-frequency voltage gain A_{V0} which can be obtained. The maximum useful value depends on the amount of dissipated power in the output stage and on the symmetry of the layout of the input stage in regard to the output stage, as was discussed with Fig. 5.1. A maximum useful value of the order of 10^5 or 10^6 can be obtained on a single chip. The other gain limitation is given by the high-frequency characteristic. This will be discussed in the next paragraph.

6.2   Frequency Compensation

The phase lag of the signal when going through the several stages of an OpAmp determines the limit to the useful gain at high frequencies. At the frequency where the phase lag exceeds 180°, the open loop gain must be dropped below unity.

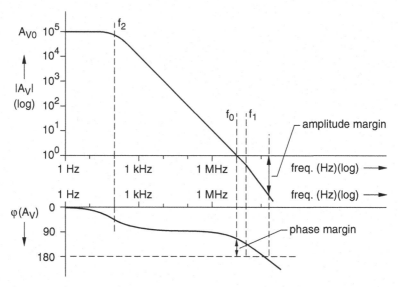

Fig. 6.7 Frequency response of an operational amplifier

Otherwise, the feedback system will become self-oscillating [5]. Moreover, a certain amplitude and phase margin must be adopted to obtain a response without peaking in the frequency domain [6] or without overshoot in the time domain [7]. In Fig. 6.7 the desired frequency characteristic of the open-loop amplifier is presented. It contains the amplitude characteristic (amplitude $|A_v|$ versus frequency) and the phase characteristic (phase $v(A_v)$ versus frequency), combined in a Bode diagram. It represents an amplifier with one dominating pole $P_2 = 2\pi f_2$ and a limiting pole $P_1 = 2\pi f_1$. Parasitic poles situated above P_1 are disregarded.

For a Butterworth pole position of the unity-gain feedback amplifier with a flat frequency response and with an overshoot in the step response of 5 %, the two pole frequencies of the open loop must be separated at a distance of two times the DC gain:

$$\frac{f_1}{f_2} \geq 2A_{v0} \tag{6.1}$$

A larger separation of the poles is also allowed. In that case the phase margin of the open loop gain is larger than 60°. But this may represent a waste of bandwidth or current.

From Eq. 6.1 it follows that the frequency $f_0 = \omega_0/2\pi$ where the amplitude characteristic crosses zero must be a factor two below the limiting pole frequency $f_1 = P_1/2\pi$, so

$$f_0 \leq \frac{1}{2} f_1 \qquad (6.2)$$

The frequency f_0 is called the 0 dB bandwidth of the open loop amplifier. The dominating-pole frequency f_2 is situated a factor equal to the DC gain A_{VO} lower than the 0 dB frequency:

$$f_2 = f_0/A_{v0} \qquad (6.3)$$

If the amplitude characteristic between f_0 and f_2 is not straight, but curved by a pole-zero doublet, the step response of the unity-gain feedback amplifier can be expected to have a slow settling overshoot or undershoot component [7]. This is undesirable in amplifiers which need a fast and accurate settling after a step signal. Hence, a general-purpose OpAmp needs a straight 6 dB per octave slope between the 0 dB frequency f_0 and the dominating frequency f_2.

An overview of how to obtain the desired frequency response will be presented now. The way this is done is called frequency compensation. The complexity of the frequency compensation is mainly determined by the number of GA stages in the loop. We start with a configuration of one GA stage and end with four GA stages.

6.2.1 One-GA-Stage Frequency Compensation

A one-GA-stage GA-CF operational amplifier is shown in Fig. 6.8b. A CF stage has been used as a cascode, to increase the voltage gain of the input GA transistor pair. The output is loaded by a capacitor C_1 and a resistor R_1, including all parasitics at this point. The tail current I_{T1} has a value of twice the bias current I_{B1}.

The DC gain is $A_{v0} = g_{m1}R_1$, with $g_{m1} = 2/(1/g_{m11} + 1/g_{m12}) \approx g_{m11}$. Hence, the transconductance of the whole stage with mirror is equal to that of one of the input transistors. If there is no resistive load besides those of the transistors, and if we suppose that the current mirror M_{23} through M_{26} is ideal, the DC gain for a bipolar circuit is maximally $A_{VOM} \approx r_{cb2}/r_{e1} \approx (r_{cb2}/r_{ce2})(r_{ce1}/r_{e1}) \approx \beta_2\mu_1$, with average values for transistors in the cascoded stages g_{m1} and g_{m2} and for a CMOS circuit $A_{VOM} = g_{m11}r_{d1}g_{m2}r_{d2} = \mu_1\mu_2$, which may be of the order of 10^5 and 10^4, respectively. In CMOS the input stage should be biased in weak inversion, as this gives the highest ratio of g_{m1}/I_{B1}.

The amplifier has a limiting frequency f_1 caused by the transit frequency f_{T2} of the cascode transistors M_{21} and M_{21}, so $f_1 = f_{T2}$. The dominating frequency f_2 is caused by the load circuit, so $f_2 = 1/(2\pi R_1 C_1)$ (Fig. 6.9).

The 0 dB bandwidth is for bipolar input transistors:

$$f_0 = \frac{g_{m1}}{2\pi C_1} = \frac{I_{B1}}{2\pi C_1 V_T} \qquad (6.4)$$

Fig. 6.8 (**a**) Block diagram
of GA-CF operational
amplifier. (**b**) Simplified
one-GA-stage GA-CF
operational amplifier

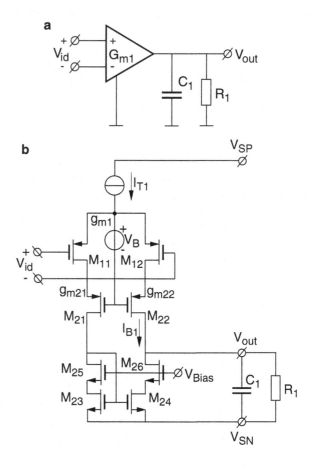

Fig. 6.9 Frequency
characteristic of a GA-CF
operational amplifier

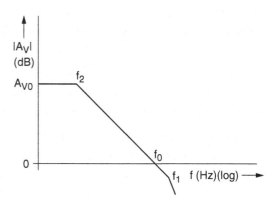

and for CMOS input transistors biased in weak inversion:

$$f_0 = \frac{g_{m1}}{2\pi C_1} \approx \frac{I_B}{2\pi C_1(V_{GS} - V_{TH})} \tag{6.5}$$

in which

$$V_T = kT/q \approx 25 \ \text{mV}$$

$$(V_{GS} - V_{TH}) \approx 60 \ \text{mV}$$

For a load capacitor of $C_1 = 10$ pF and a bias current I_{B1} of 100 µA, the bipolar circuit has a bandwidth of $f_0 = 30$ MHz, and the CMOS circuit $f_0 = 12$ MHz.

The maximum bandwidth f_0 must be kept half of that of the limiting pole frequency $f_0 = 2 f_1 = 2 f_{T2}$ by the choice of C_1. For bipolar transistors the transit frequency f_T ranges from 300 MHz up to 60 GHz, and for CMOS from 100 MHz up to 20 GHz. In fact, the load capacitance C_1 functions as well as an HF compensation capacitor.

When we divide the bandwidth by the supply power needed f_0/P_s, with $P_s = V_s I_s$, we obtain a figure of merit which is reverse to the well-known power delay product used in digital circuits. For the bipolar circuit:

$$f_0/P_s = \frac{I_{B1}}{2\pi C_1 V_T V_S 2I_{B1}} = \frac{1}{2} \frac{1}{2\pi C_1 V_T V_S} \tag{6.6}$$

and for the CMOS circuit:

$$f_0/P_s = \frac{1}{2} \frac{1}{2\pi C_1(V_{GS} - V_{TH})V_S} \tag{6.7}$$

The bandwidth over power ratio is inverse proportional to the supply voltage V_S with bipolar as well as with CMOS transistors, so the supply voltage should be as small as possible. With CMOS in weak inversion the term $(V_{GS} - V_{TH})$ may be replaced by some 20 % more than two times the thermal voltage $2V_T = 2kT/q \approx 50$ mV, which is about 60 mV. The optimum choice CMOS transistor has the smallest possible length L, and a width W so that the transistor is on the verge of weak inversion. In that case the ratio between g_m and C parasitic is the highest.

Remark: If we would eliminate the cascode transistors M_{21} and M_{22} in Fig. 6.8, there would be no limiting pole, supposing the mirror is ideal.

This would mean that we may use the circuit up to its transit frequency f_T. In that case the bandwidth over power ratio will be doubled. This is already an important conclusion. Moreover, if the OpAmp would be used in filter applications, the non-existence of internal limiting poles means that there is no extra phase shift involved. The external poles at the input and output can be made part of the intentional filter poles. This situation is highly desirable for wideband filters [8]. On the other hand, omitting the cascode transistors M_{21} and M_{22} will reduce the maximum voltage gain A_{VOM} to that of a single CMOS or bipolar transistor. Parallel compensation by an artificial negative resistance equal to the output resistance is an option [8]. But such compensation is limited by the inequality of both parasitic resistances and takes extra power again. Yet, it is worthwhile looking for system architectures that allow low-gain amplifier stages. Then such simple stages as mentioned with Figs. 3.18 and 3.19 of Chap. 3 could be used.

6.2.2 Two-GA-stage Frequency Compensation

Next we look at the two-stage GA-GA operational amplifier of Fig. 6.10. Its frequency response is given in Fig. 6.11.

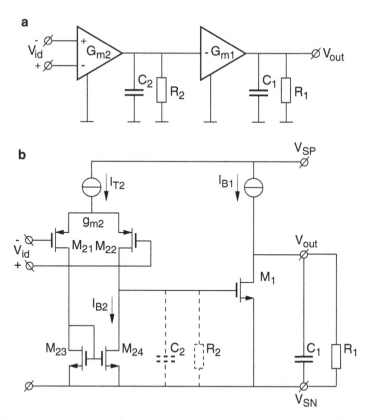

Fig. 6.10 (a) Block diagram of two-stage GA-GA operational amplifier. (b) Simplified two-stage GA-GA operational amplifier (C_2 and R_2 are usually parasitic transistor parameters)

Fig. 6.11 Uncompensated
frequency response of a
two-stage GA-GA
operational amplifier

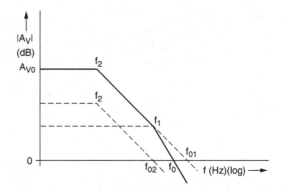

The DC voltage gain is $A_{V0} = g_{m2}R_2 g_{m1}R_1$. When there is no external load resistor the maximum gain is $A_{V0} = \mu_2\mu_1$, which value is in the order of 10^6 for the bipolar circuit and 10^4 for the CMOS version. The pole frequencies $f_1 = 1/2\pi R_1 C_1$ at the output, and $f_2 = 1/2\pi R_2 C_2$ at the output of the input stage probably do not obey the rule $f_1/f_2 \geq 2A_{v0}$. Hence, frequency compensation has to be applied.

The 0 dB bandwidth f_0 of the non-compensated amplifier is the geometric mean of the 0 dB bandwidths f_{01} and f_{02} of the two composing GA stages, which can be understood from Fig. 6.11 with its log. scales:

$$f_0 = \sqrt{f_{01}f_{02}} = \sqrt{\frac{g_{m1}g_{m2}}{2\pi C_1 2\pi C_2}} \tag{6.8}$$

Note, that the 0 dB bandwidth is independent of the resistor values. Also note, that we have drawn in Fig. 6.11 the output stage with a higher 0 dB frequency f_{01} than that of the input stage f_{02}. This is often not the case. For this moment the sequence does not change our reasoning. When it will with Miller compensation, we come back to it. Two ways of frequency compensation can be used for a two stage GA-GA amplifier: parallel compensation and Miller compensation.

6.2.2.1 Two-GA-stage Parallel Compensation (PC)

The parallel compensated amplifier is shown in Fig. 6.12, and its frequency characteristic in Fig. 6.13. The natural load elements of the input stage C_2 and R_2 are usually parasitic elements. The parallel compensation network is composed of the series connection of C_P and R_P.

The parallel compensation network $C_P R_P$ can best be placed in between the two stages. Going from lower to higher frequencies, the capacitor C_P in "parallel" with the parasitic capacitor C_2, firstly, decreases the dominating-pole frequency f_2 to f'_2. Secondly, the reduction of gain is gradually terminated by R_P above f_1. R_P must be chosen equal to $R_P = 1/2\pi f_1 C_P$.

When we choose the components such that:

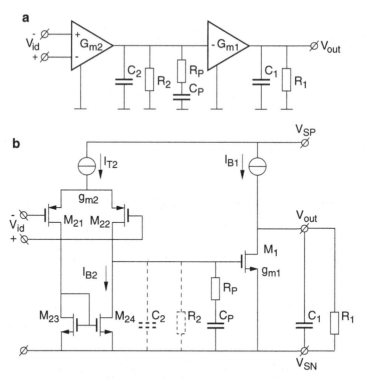

Fig. 6.12 (a) Block diagram of parallel-compensated two-stage GA-GA amplifier (C_2 and R_2 are usually parasitic elements). (b) Simplified parallel-compensated two-stage GA-GA amplifier (C_2 and R_2 are usually parasitic transistor parameters)

Fig. 6.13 Frequency characteristic of a parallel compensated two-stage GA-GA amplifier and its parallel attenuation

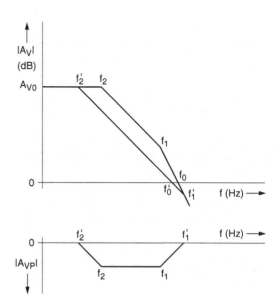

$$C_P = R_1\sqrt{2C_1C_2g_{m1}g_{ml}}$$
$$R_P = \sqrt{C_1/(2C_2g_{m1}g_{m2})}$$

(6.9)

we find the desired new straight 6 dB frequency roll-off from f'_2 up to f'_1, with a new 0 dB bandwidth f'_0 positioned a factor 2 below f'_2:

$$f'_0 = \sqrt{\frac{1}{2}\frac{g_{ml}}{2\pi C_1}\frac{g_{m2}}{2\pi C_2}}$$

(6.10)

The parallel compensation shows a remarkably good result. The new 0 dB frequency f'_0 lies only a factor $\sqrt{2}$ lower than the theoretical maximum of f_0. Particularly at heavy capacitive loads, when the bandwidth f_{01} of the output stage is much lower than the f_T of the transistors, the bandwidth f_{02} of the input stage helps to broaden the bandwidth.

When we calculate the bandwidth over power ratio for a bipolar amplifier and CMOS amplifier in weak inversion with $(V_{GS} - V_{TH}) = 60$ mV, we find respectively:

$$\frac{f'_0}{P_S} = \frac{1}{2\pi V_T}\sqrt{\frac{1}{2}\frac{I_{B1}}{C_1}\frac{I_{B2}}{C_2}}/V_S(I_{B1} + 2I_{B2})$$

(6.11)

$$\frac{f'_0}{P_S} \approx \frac{1}{2\pi(V_{GS} - V_{TH})}\sqrt{\frac{1}{2}\frac{I_{B1}}{C_1}\frac{I_{B2}}{C_2}}/V_S(I_{B1} + 2I_{B2})$$

(6.12)

We find a broad optimum for equal currents in the output and input stage around $I_{B1} = 2I_{B2}$.

When we substitute $I_{B1} = 2I_{B2}$ at the optimum, we obtain for bipolar and CMOS transistors, respectively:

$$\frac{f'_0}{P_S} = \frac{1}{4}\frac{1}{2\pi\sqrt{C_1C_2}V_TV_S}$$

(6.13)

$$\frac{f'_0}{P_S} = \frac{1}{4}\frac{1}{2\pi\sqrt{C_1C_2}(V_{GS} - V_{TH})V_S}$$

(6.14)

Unfortunately, there are a number of serious disadvantages associated with parallel compensation. The main problem is that the compensation cannot be made anywhere close to the desired value because the transconductance g_{ml} changes signal-dependently. When the output current changes from the quiescent value up to 100 times larger, the g_m varies with a factor 100. A second problem is that the compensation depends on process parameters which are different from those to be compensated. This means that the choices given by Eq. 6.9 are not accurately met and that a pole-zero doublet will occur resulting in a slow settling component.

For class-A operation and for applications where a large slow settling signal component is not a problem, parallel compensation can be considered. However, the relative large parallel compensation capacitor C_p that is needed may take an excessive large chip area. These are the reasons that other means of compensation have to be explored.

6.2.2.2 Two-GA-stage Miller Compensation (MC)

Miller compensation is the other possibility. Figure 6.14 shows a two-stage GA-GA amplifier with Miller compensation [9]. And Fig. 6.15 shows the frequency response of it.

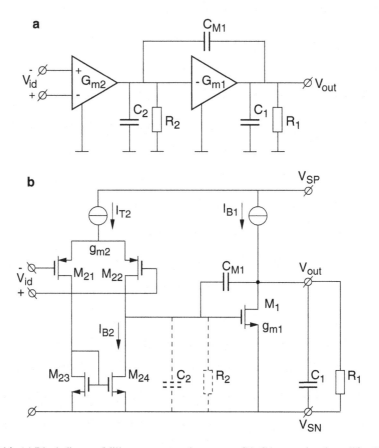

Fig. 6.14 (a) Block diagram Miller-compensated two-stage GA-GA operational amplifier (C_2 and R_2 are usually parasitic elements). (b) Simplified Miller-compensated two-stage GA-GA operational amplifier (C_2 and R_2 are usually parasitic transistor parameters)

Fig. 6.15 Pole-splitting by Miller compensation of a two-stage GA-GA amplifier

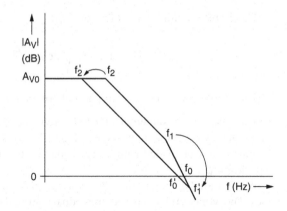

The output stage with a Miller capacitor C_{M1} around it behaves as an integrator with a transimpedance $1/2\pi fC_{M1}$. It integrates the output current I_{02} of the input stage and presents it as the output voltage V_{out}.

The transfer of the Miller-compensated output stage is:

$$\frac{V_{out}}{I_{02}} = \frac{1}{2\pi fC_{M1}} \qquad (6.15)$$

The voltage gain of the whole amplifier, including the input stage with a transconductance g_{m2}, becomes:

$$\frac{V_{out}}{V_{in}} = \frac{g_{m2}}{2\pi fC_{M1}} \qquad (6.16)$$

The new 0 dB bandwidth f'_0 of the amplifier is determined for $V_{out}/V_{in} = 1$, as:

$$f'_0 = \frac{g_{m2}}{2\pi C_{M1}} \qquad (6.17)$$

The Miller capacitor, on the one hand, reduces the dominant pole frequency f_2 to f'_2 by its integrator function, but, on the other hand, provides feedback by which the bandwidth increases from f_1 to f'_1. It looks as if the pole frequencies f_1 and f_2 are split apart, as is clearly shown in Fig. 6.15. The maximum obtainable bandwidth f'_0 for a 60° phase margin must lie a factor 2 below the new pole frequency f'_1, or a factor $/2$ below the uncompensated 0 dB frequency f_0, just as with parallel compensation.

So the maximum 0 dB bandwidth f'_0 is:

$$f'_0 \leq \frac{1}{\sqrt{2}}f_0 = \sqrt{\frac{1}{2}f_{01}f_{02}} = \sqrt{\frac{1}{2}\frac{g_{m1}}{2\pi C_1}\frac{g_{m2}}{2\pi C_2}} \qquad (6.18)$$

This implies a minimal choice for C_{M1}:

$$C_{M1} \leq \sqrt{2C_1C_2 \frac{g_{m2}}{g_{m1}}} \tag{6.19}$$

This result is as good as with parallel compensation.

Remark: We have assumed that the 0 dB frequency f_{01} of the output stage lies above the frequency f_{02} of the input stage in Fig. 6.11. This may not always be the case, as the output may be loaded with a large capacitor C_1, and is also loaded by the Miller capacitor C_{M1}.

With parallel compensation the sequence of f_{01} and f_{02} did not matter, but now it does. The Miller effect is based on feedback around one stage [9]. Feedback is only effective if the loop gain around that stage is higher than unity. At frequencies where the loop gain is unity or lower the Miller effect is not present anymore. This is the case when we try to split the output pole frequency f_1 beyond the point f_{01} where the gain of the output stage is lower than unity. The situation is drawn in Fig. 6.16.

The splitting of the output pole frequency f_1 stops at the limiting pole frequency f_{01}.

$$f_1' \approx f_{01} = \frac{g_{m1}}{2\pi C_1} \tag{6.20}$$

Further lowering of f_2 by further increasing C_{M1} must be carried on until the 0 dB frequency is two times lower than f'_1 for a phase margin of 60°. Hence, the 0 dB frequency f'_0 must be chosen:

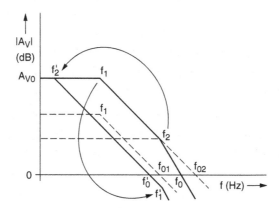

Fig. 6.16 Frequency characteristics of a Miller-compensated two-stage GA-GA amplifier with a 0 dB frequency of the output stage f_{01} lower than that of the input-stage pole frequency f_{02}

$$f'_0 = \frac{1}{2}f'_1 \approx \frac{1}{2}f_{01} \tag{6.21}$$

The lowering of f'_2 by increasing C_{M1} without increasing f'_1 beyond f_{01} is inefficient. We could as well have started with a lower bandwidth f_{02} of the first stage to save power. For the moment it can be concluded that the 0 dB bandwidth of the whole amplifier is limited by half that of the output stage $2f_{01}$. This is the reason why we are counting the stages from the output back to the input. For applications where f_{01} is lower than f_{02} due to heavy capacitive output loads we can follow the methods described with Figs. 6.30, 6.31, 6.32, 6.33, 6.34, 6.35, and 6.36. This ends the remark.

The Eqs. 6.17 and 6.21 determine the minimum choice of C_{M1}:

$$C_{M1} = \frac{g_{m2}}{2\pi f'_0} \tag{6.22}$$

The Miller compensation capacitor C_{M1} must be calculated at the lowest current through M_1 which is I_{B1} where g_{m1} has the lowest value and at the highest expected value of the load capacitance C_1.

The dominating pole frequency becomes:

$$f'_2 = \frac{f'_0}{A_{v0}} \tag{6.23}$$

with $A_{V0} = g_{m2}R_2g_{m1}R_1$.

The cutoff frequency f_T of the output transistor M_1 may pose a further restriction on the bandwidth. The Miller feedback is attenuated by the voltage divider $C_{M1}/(C_{M1}+C_2)$. By this value the limiting pole frequency f'_1 and the bandwidth f'_0 are lowered:

$$f'_0 = \frac{1}{2}f'_1 = \frac{1}{2}\frac{g_{m1}}{2\pi C_1}\frac{C_{M1}}{C_{M1}+C_2} \tag{6.24}$$

For bipolar transistors with $C_{M1}=C_2=C_{BE1}$, and for CMOS transistors with the width W set so that $C_{M1}=C_2=C_{GS1}$, and the length L at the minimum, the bandwidth is practically limited to:

$$f'_0 = \frac{1}{2}f'_1 = \frac{1}{2}\frac{g_{m1}}{2\pi C_1}\frac{C_{M1}}{C_{M1}+C_2} \approx \frac{1}{4}f_T \tag{6.25}$$

Coming back to the case of Figs. 6.11 and 6.15 with $f_{02}<f_{01}$, the bandwidth f_{02} of the input stage need not be higher than $f_{02} = \frac{1}{2}f_{01} = f'_0$. The result is a bandwidth over power ratio of the whole amplifier for bipolar and CMOS transistors of respectively:

$$\frac{f_0'}{P_S} = \frac{1}{2} \frac{1}{2\pi(C_1 + 2C_2)V_T V_S} \tag{6.26}$$

$$\frac{f_0'}{P_S} = \frac{1}{2} \frac{1}{2\pi(C_1 + 2C_2)(V_{GS} - V_{TH})V_S} \tag{6.27}$$

When we compare these results with those of parallel compensation we find a somewhat lower bandwidth over power ratio. With $C_1 = C_2$ the ratio is $3/\sqrt{2} \approx 2.1$ lower, and with $C_1 = 10C_2$ the ratio is $12/\sqrt{20} \approx 2.7$ lower than that of parallel compensation. Although, this is a disadvantage of Miller compensation, there are many advantages.

The main advantage of the Miller-compensated amplifier over the parallel-compensated version is the straight 6 dB/oct. roll-off of the frequency characteristic, without pole-zero doublets from pole-zero cancellation techniques. This makes the amplifier suitable for fast settling step responses without slow settling components.

Another important advantage of Miller compensation is that all gain which is attenuated by the compensation capacitor C_{M1} is used to lower the output impedance and the linear distortion of the output stage. With parallel compensation the attenuated gain is lost for any use!

A final advantage is that the compensation capacitor C_{M1} is much smaller compared to that which is needed with parallel compensation. That means a much smaller chip.

The obvious choice with Miller compensation for unity-gain feedback and relatively heavy capacitive load is to utilize most of the supply power in the output stage.

It should be kept in mind that if the amplifier need not be unity-gain feedback, but has for instance a closed-loop gain of 10, the supply current of the output transistor can be taken ten times lower, and so the supply power will strongly be reduced.

A disadvantage of Miller compensation is that a zero appears in the right-half complex plane at a frequency f_z':

$$f_z' = \frac{g_{m1}}{2\pi C_{M1}} \tag{6.28}$$

The additional zero frequency f_z' in the right half of the complex plane causes a reduction of the phase margin for values of f_z' above but still close to f_1'. If f_2' is below f_1', even nearly a reversal of the phase occurs. This effect can easily be understood if we consider that the Miller capacitor C_{M1} functions as a feed forward path across the output transistor. At frequencies of f_z' lower than the limiting pole frequency $f_1' = g_{m1}/2\pi C_1$ the output transistor is no longer effective and the transistor does not inverse anymore.

Many techniques have been used to overcome this effect. The best ways are shown in Figs. 6.17, 6.18, and 6.19.

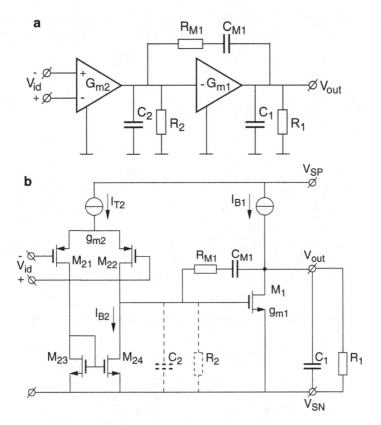

Fig. 6.17 (a) Block diagram of resistive Miller-zero cancellation with a resistor $R_{M1} = 1/g_{m1}$ in series with C_{M1} (C_2 and R_2 are usually parasitic elements). (b) Simplified circuit with resistive Miller-zero cancellation with a resistor $R_M = 1/g_{m1}$ in series with C_{M1} (C_2 and R_2 are usually parasitic transistor parameters)

Resistive Miller Zero Cancellation with a resistor $R_{M1} = 1/g_m$ in series with C_{M1} is shown in Fig. 6.17. The resistor R_{M1} must attenuate the feed forward path through C_{M1} by the same amount as by which the g_{m1} of the output transistor provides gain.

The value of R_{M1} must be matched with $1/g_{m1}$. For large variations in output current resistive Miller Zero Cancellation is poor. For a high current R_{M1} must be small to match a high g_{m1}. This means that at low output current the cancellation is not effective, just when it is most needed at a low g_{m1}. So this solution functions well when the output transistor is biased in class-A, but is less useful for class-AB biasing.

Another popular solution is Active Miller Zero Cancellation using a cascode transistor M_{32} in series with the Miller loop, as shown in Fig. 6.18 [10, 11]. The zero cancellation is most needed with CMOS, because of a relatively low g_{m1}. With CMOS a cascode is often already present in the input stage for increasing its gain. In Fig. 6.18 the mirror in the input stage is cascoded. The cascode has two advantages:

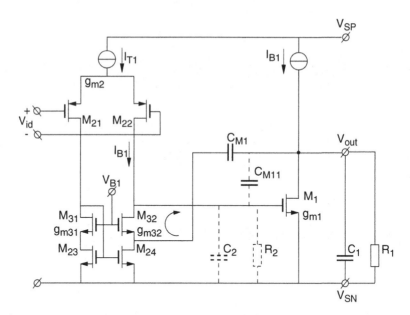

Fig. 6.18 Active Miller-zero cancellation with a cascode in series with the Miller loop

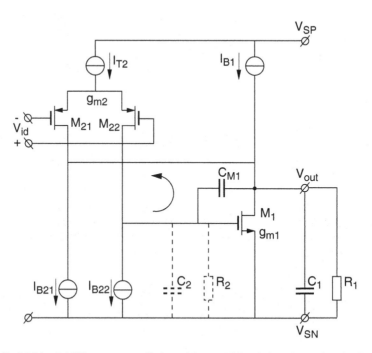

Fig. 6.19 Multipath Miller-zero cancellation with an additional feed forward path of opposite polarity

Firstly the Miller feedback circuit has more loop gain, viz. C_{M1}/C_2, instead of that of the conventional configuration which amounts $C_{M1}/(C_{M1}+C_2)$.

This means that the limiting pole frequency f'_1 can be increased by roughly a factor 2 from $\frac{1}{2}f_{o1}$ to 1 f_{o1}, with $C_{M1}=C_2$, or even more [10, 11]. Hence the bandwidth of the whole amplifier can be a factor 2 larger at the same supply current or even more.

Secondly, the Miller capacitor feeds back to a point at the emitter or source of M_{32} with a much smaller voltage movement than at the base or gate of the output transistor M_1. This improves the linearity of the OpAmp in the frequency range where C_{M1} is active.

However, at high output currents the loop gain can become so high, because of a high g_{m1}, that the phase margin within the Miller loop, having a second pole at the input of M_{32} with a value of $f_3 = g_{m32}/C_{M1}$, becomes too low. This is particularly the case with CMOS transistors. In that case, C_2 can be intentionally enlarged. Alternatively, we can choose a nested combination of a part of C_{M1} through the cascode transistor M_{32} and another part C_{M11} directly to the gate of M_1 [11]. A ratio of 2 between C_{M1} and C_{M11} is usually taken. Note that the cascode stage is capacitive balanced by $C_{P1} = C_{M1}$ so that there is no pole-zero doublet in the cascode stage at high frequencies.

Finally, the Miller zero can be cancelled by adding a feed forward path across C_{M1}, as shown in Fig. 6.19. This precisely cancels the current at the right-hand side of C_{M1} which causes the zero of the feed forward path through C_M. The equal but opposite currents are precisely available at the outputs of M_{21} and M_{22} of the differential pair. This technique is called Multipath Miller Zero Cancellation [12, 13].

A disadvantage of this technique is that this feed forward path is not always available. Nor can its voltage always swing from rail-to-rail.

6.2.3   Three-GA-Stage Frequency Compensation

Figure 6.20 shows the basic circuit of a three-GA-stage operational amplifier.

The way of frequency compensation of a three-GA-stage operational amplifier is to: firstly, compensate the two-stage output and intermediate stages as was done in previous sections by parallel or Miller compensation; and secondly, compensate the whole amplifier again with parallel or Miller compensation as if the intermediate and output stages were one new output stage.

Though parallel compensation may lead to a slightly higher bandwidth over power ratio, its pole-zero cancellation is very dependent on parameter spreading of the IC process and current variations (Fig. 6.21).

Moreover, parallel compensation leads to large compensation capacitors. For these reasons we will only choose the overall Miller compensated structure of an already Miller compensated intermediate and output stage. This is called nested Miller compensation [3, 14] (Fig. 6.20).

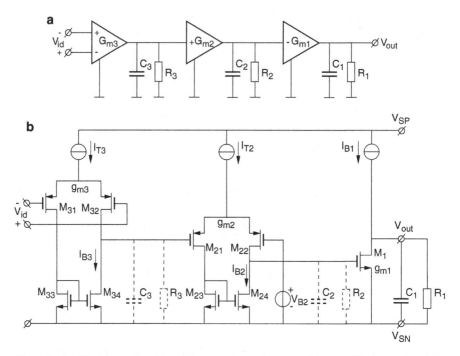

a

b

Fig. 6.20 (a) Block diagram of three-GA-stage operational amplifier. (b) Simplified three-GA-stage operational amplifier

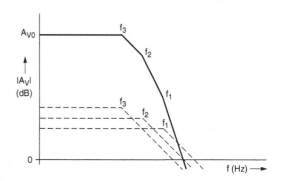

Fig. 6.21 Frequency characteristic of the three-GA-stage amplifier of Fig. 6.20

6.2.3.1 Three-GA-Stage Nested Miller Compensation (NMC)

A nested Miller compensated amplifier is drawn in its basic structure in Fig. 6.22 [3, 14].

We firstly split apart f_1 and f_2 by a regular Miller compensation capacitor C_{M1} to new positions f'_1 and f'_2. Next, we split again f'_2 and f_3 apart by a nested Miller compensation capacitor C_{M2} to final positions f''_2 and f'_3. The second splitting by C_{M2} has been made possible by the choice of a non-inverting intermediate stage g_{m2}.

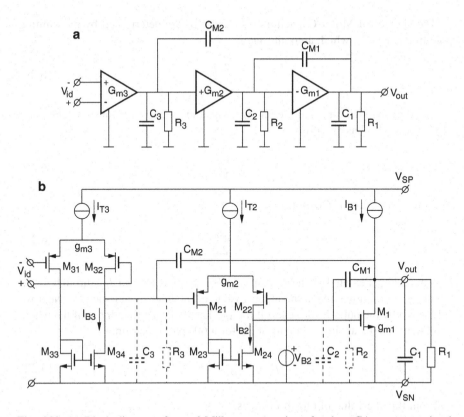

Fig. 6.22 (**a**) Block diagram of nested Miller compensation of a three-GA-stage operational amplifier. (**b**) Simplified nested Miller compensation of a three-GA-stage operational amplifier

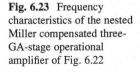
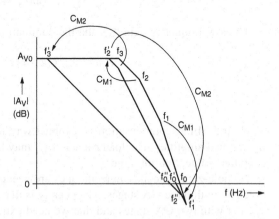

Fig. 6.23 Frequency characteristics of the nested Miller compensated three-GA-stage operational amplifier of Fig. 6.22

The combination of a non-inverting intermediate stage g_{m2} and inverting output stage g_{m1} results in an inverting two-stage amplifier combination, which can be compensated by a nested Miller capacitor C_{M2} again. The frequency characteristic is shown in Fig. 6.23.

The choices of Miller Capacitors C_{M1} and C_{M2} are determined by the limiting pole frequency f'_1, which amounts again:

$$f'_1 = \frac{g_{m1}}{2\pi C_1} \tag{6.29}$$

The 0 dB bandwidth f'_0 of the intermediate and output stage must be taken half of f'_1 for a 60° phase margin:

$$f'_0 = \frac{1}{2}f'_1 \tag{6.30}$$

From this the value of C_{M1} follows as:

$$C_{M1} = \frac{g_{m2}}{2\pi f'_0} \tag{6.31}$$

C_{M1} must be calculated at the lowest value of the bias current I_{B1} through M_1 at which g_{m1} is minimal and at the highest value of the load capacitance C_1. Next we split again by the outer Miller capacitor C_{M2}. The 0 dB bandwidth f''_0 of the whole amplifier must be again taken half of f'_0 for a 60° phase margin:

$$f''_0 = \frac{1}{2}f'_0 = \frac{1}{4}f'_1 \tag{6.32}$$

From this the value of C_{M2} follows as:

$$C_{M2} = \frac{g_{m3}}{2\pi f''_0} \tag{6.33}$$

The low-frequency gain A_{V0} and the dominating pole frequency f'_3 are:

$$A_{v0} = g_{m3}R_3 g_{m2}R_2 g_{m1}R_1 \tag{6.34}$$

$$f'_3 = f''_0/A_{v0} \tag{6.35}$$

Nested Miller compensation is a robust way to compensate a three-GA-stage operational amplifier. The load capacitor C_1 may be chosen smaller than the upper supposed value, but not larger.

No pole-zero cancellation techniques are needed. An abundance of gain can be obtained with three GA stages. The penalty is that the bandwidth is half that of the version with two GA stages and that we need extra current for the input stage.

An optimal choice for the currents of the three stages is roughly determined by:

$$f_{03} = \frac{1}{2}f_{02} = \frac{1}{4}f_{01}, \quad \text{or} \tag{6.36}$$

$$\frac{g_{m3}}{2\pi C_3} = \frac{1}{2}\frac{g_{m2}}{2\pi C_2} = \frac{1}{4}\frac{g_{m1}}{2\pi C_1} \tag{6.37}$$

So the contribution to the total supply current of the current of the input stage is relatively low. And the bandwidth over power ratio is about two times lower than that of the two-GA-stage amplifier.

6.2.3.2 Three-GA-Stage Multipath Nested Miller Compensation (MNMC)

A way to regain the factor 2 again is to use the multipath nested Miller compensated (MNMC) circuit of Fig. 6.24 [14, 15].

The circuit has a second input stage with a transconductance of g_{m32}, additional to the first input stage with g_{m31}. The second input stage forms an independent parallel path across the first input stage and intermediate stage. The output of the second input stage is connected to the input of the output stage. In fact, we have an

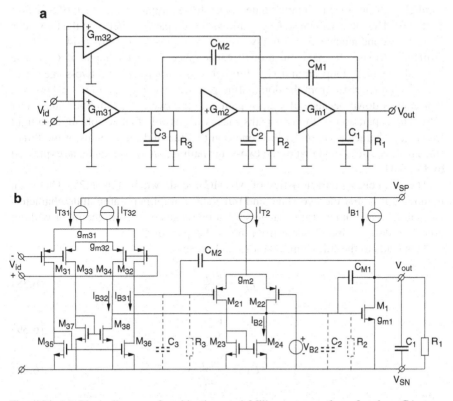

Fig. 6.24 (a) Block diagram of multipath nested Miller compensation of a three-GA-stage operational amplifier. (b) Simplified multipath nested Miller compensation of a three-GA-stage operational amplifier

Fig. 6.25 Frequency
characteristics of a
multipath nested Miller
compensated (MNMC)
three-GA-stage operational
amplifier

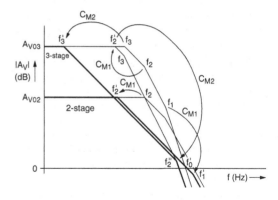

additional two-stage amplifier in parallel to a three-stage amplifier. At low frequencies the three-stage amplifier dominates by its large low-frequency gain. But at high frequencies the two-stage amplifier dominates. In the middle range both paths have equal transfer functions, without adding their transfers. Adding would have resulted in a pole-zero doublet. Adding would have happened if we had a simple capacitive parallel path across the intermediate stage, for example as in the NE5534 (see Sect. 7.6). The intermediate stage g_{m2} allows for a separation between the two paths if its transconductance is relatively weak, as expressed by Eq. 6.44, or if we artificial increase the parasitic ground capacitance C_3 at the input of G_{m2}. The internal feedback loop around G_{m2} through C_{M2} suppresses the two-stage's frequency characteristic from the dominating pole frequency f'_3 up to f''_2, where that loop breaks down because of lack of gain, and the two-stage amplifier takes over [16]. The separation of the two paths makes it possible that we obtain a straight frequency characteristic under the condition that the mid-range gains of the three-stage path g_{m31}/C_{M2} and that of the two-stage path g_{m32}/C_{M1} are equal, as expressed by Eq. 6.41.

The frequency characteristics of MNMC are drawn in Fig. 6.25. The main requirement is that the two-stage and three-stage amplifier's amplitude characteristics touch each other at mid range so that one takes over from the other without pole-zero doublet, independent from the load capacitor C_1.

This leads to the following relations and choices:

$$f'_1 = \frac{g_{m1}}{2\pi C_1} \tag{6.38}$$

$$f'_0 = \frac{1}{2} f'_1 \tag{6.39}$$

$$C_{M1} = \frac{g_{m32}}{2\pi f'_0} \tag{6.40}$$

$$f_0'' = f_0' = \frac{1}{2}f_1' \qquad (6.41)$$

$$C_{M2} = \frac{g_{m31}}{2\pi f_0''} \qquad (6.42)$$

$$f_3' = f_0''/A_{v0} \qquad (6.43)$$

$$g_{m2} < \frac{1}{3}g_{m32} \qquad (6.44)$$

The limiting pole frequency f_1' is set again by the output stage according to Eq. 6.38. The 0 dB bandwidth f_0' is chosen half of f_1'. The additional input stage with g_{m32} and f_0' is compensated as if it were the first stage of a two-stage amplifier combination with C_{M1} according to Eq. 6.40.

The original input stage with g_{m31} is now compensated with C_{M2} according to Eq. 6.42, at the same 0 dB frequency f_0' as g_{m32} with C_{M1} has been done. This implies that the compensated characteristics of the two and three-stage amplifiers touch each other for a wide frequency range. At the high frequency end the two-stage amplifier extends up to the limiting pole frequency f_1'. At low frequencies the three-stage amplifier goes down to $f_3' = f_0'/A_{v0}$. The overall characteristic is straight from f_1' down to f_3' with a 6 dB/oct roll-off. There is no summing of the two characteristics in the middle range because of the independent nature of the two characteristics separated by a weak g_{m2}. The two-stage amplifier must have a 0 dB frequency $f_0' \approx g_{m32}/2\pi\, C_{M1}$ equal to that of the three-stage amplifier $f_0' \approx g_{m31}/2\pi C_{M2}$.

The matching of $g_{m31}/C_{M2} = g_{m32}/C_{M1}$ determines the pole-zero cancellation in the overall characteristic. Accurate matching can be realized because C_{M2} and C_{M1} can be made by the same type of integrated MOS capacitors, and g_{m31} and g_{m32} are the result of two equal stages, biased with equal current sources.

The accurate controllable matching of the two characteristics in IC technology is a major advantage of the multipath nested Miller compensation technique over other pole-zero cancellation schemes.

The value of g_{m2} of the intermediate stage is still undetermined. We want the frequency characteristic to be determined by the second input stage g_{m32} at the 0 dB frequency f_0' and not by the intermediate stage g_{m2}. For this reason we take $g_{m2} < g_{m32}$, as prescribed by Eq. 6.44. Later, at Fig. 6.30 we will see that a quenching capacitor C_3 at the input of g_{m2} to ground can be used to solve this issue.

The result is a three-stage OpAmp with a large amount of gain and nearly the same bandwidth over power ratio as the two-stage Miller compensated operational amplifier. The extra costs are the addition of the second input stage and its addition to the supply power.

6.2.4   Four-GA-Stage Frequency Compensation

When we need more gain we can add a fourth stage or even more stages. The question is how to compensate these stages reliably. Besides parallel compensation, which becomes very impractical with four or more stages, the nested Miller compensation method can be extended. However, without multipath, we loose a factor 2 in bandwidth each time we nest. With many multipaths the circuit becomes complex. For instance, with a four-stage amplifier, already three input stages are needed in front of the circuit of Fig. 6.24. To simplify the nesting, and not lose a factor 2 each nest, we can use hybrid nested Miller compensation.

6.2.4.1   Four-GA-Stage Hybrid Nested Miller Compensation (HNMC)

The circuit of a hybrid nested Miller compensated (HNMC) four-GA-GA-GA-GA stage operational amplifier is shown in Fig. 6.26 [13, 17]. The circuit is maximally simple for a four-stage amplifier because no differential stages are needed for the second and third stage.

The driving of the output stage can nearly be done from rail-to-rail. No cascodes are needed because there is an abundancy of gain. This means that this HNMC is suitable for the lowest possible supply voltage with one V_{GS} plus one V_{SAT} [17].

Three invertors are connected in cascade: The first and second stages, counted from the output, are compensated by the Miller capacitor C_{M1}. The new pole positions f_1 and f_2 are split to f'_1 and f'_2.

The third and fourth stages with f_3 and l_4 are split by the Miller capacitor C_{M3} into the new pole positions f'_3 and f'_4. The frequency characteristics are drawn in Fig. 6.27. The two two-stage Miller-compensated amplifiers $g_{m1,2}$ and $g_{m3,4}$ are cascaded into one amplifier with two dominating poles f'_2 and f'_4. The phase of the three cascaded inverting amplifiers is just right so that the poles f'_2 and f'_4 can be split by an overall Miller capacitor C_{M2} into the final pole positions f''_2 and f''_4. The following choices have been made:

The limiting-pole frequency f'_1 is:

$$f'_1 = \frac{g_{m1}}{2\pi C_1} \tag{6.45}$$

The 0 dB frequency f'_0 of the driver and output stage multiplied by the loop gain C_{M2}/C_{M3} of the intermediate inverting stage M_3 must be half f'_1, so:

$$f'_0 \frac{C_{M2}}{C_{M3}} = \frac{1}{2} f'_1 \tag{6.46}$$

This leads to the choice of C_{M1}:

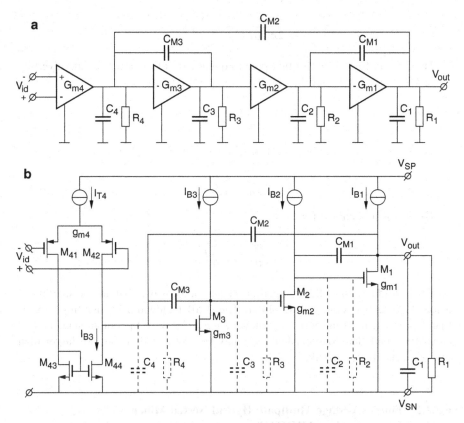

Fig. 6.26 (**a**) Block diagram of hybrid nested Miller compensated four-GA-stage operational amplifier. (**b**) Simplified hybrid nested Miller compensated four-GA-stage operational amplifier

Fig. 6.27 Frequency characteristics of hybrid nested Miller compensated four-GA-stage operational amplifier

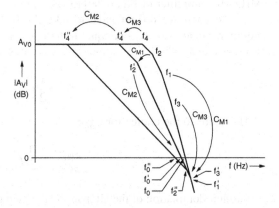

$$\frac{g_{m2}}{2\pi C_{M1}} \frac{C_{M2}}{C_{M3}} = \frac{1}{2} f'_1 \qquad (6.47)$$

This is only true if the limiting pole frequency f'_3 of the intermediate inverting stage M_3 is equal or larger than f'_1, which means a choice for g_{m3}:

$$\frac{g_{m3}}{2\pi C_{M2}} \geq f'_1 \qquad (6.48)$$

The overall bandwidth f''_0 must be taken a quarter of f_1, hence:

$$f''_0 = \frac{1}{4} f'_1 \qquad (6.49)$$

This leads to a choice for C_{M2}:

$$\frac{g_{m4}}{2\pi C_{M2}} = \frac{1}{4} f'_1 \qquad (6.50)$$

The hybrid nested Miller compensation is a robust way to handle the stability of a four-GA-stage operational amplifier. The 0 dB bandwidth is four times lower than that of the limiting-pole frequency $f'_1 = g_{m1}/2\pi C_1$, of the output stage and its capacitive load. The load capacitor C_1 may be taken smaller, but not larger than given by the above formulas.

6.2.4.2   Four-GA-Stage Multipath Hybrid Nested Miller Compensation (MHNMC)

We can improve the bandwidth of the HNMC amplifier with a factor 2 if we add a multipath input stage according to the multipath hybrid nested Miller compensated (MHNMC) amplifier of Fig. 6.28 [13, 17].

In that case, we do not have to take the 0 dB bandwidth f''_0 of the whole amplifier a quarter of f'_1 but equal to half f'_1. So the overall 0 dB bandwidth f''_0 becomes equal to half f'_1, without losing a factor 2:

$$f''_0 = \frac{1}{2} f'_1 \qquad (6.51)$$

This leads to the choice for C_{M2}:

$$\frac{g_{m41}}{2\pi C_{M2}} = \frac{1}{2} f'_1 \qquad (6.52)$$

Further, domination of the HF path by the gain path should be prevented at high frequencies. This means that the gain through the driver stage must be lower than that of the direct path. This leads to the choice:

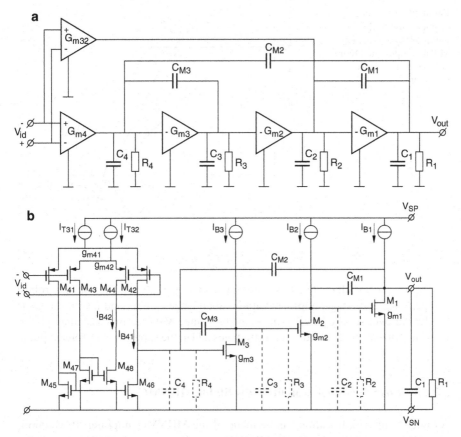

Fig. 6.28 (a) Block diagram of multipath hybrid nested Miller compensated four-GA-stage operational amplifier. (b) Simplified multipath hybrid nested Miller compensated four-GA-stage operational amplifier

$$\frac{C_{M2}}{C_{M3}} \frac{g_{m2}}{2\pi C_{M1}} \leq \frac{1}{3} \frac{g_{m42}}{2\pi C_{M1}} \tag{6.53}$$

This also means that the limiting-pole frequency f'_3 of the intermediate stage may be lower:

$$\frac{g_{m3}}{2\pi C_{M2}} \leq \frac{1}{3} f'_1 \tag{6.54}$$

The gain reduction of the driver and intermediate stage can be prevented by using a quenching capacitor to ground, as we will see with Fig. 6.31.

Finally, we must match the 6 dB roll-off of the gain path with that of the HF path, in order to avoid a pole-zero doublet. This leads to the important choice of:

Fig. 6.29 Frequency
characteristic of the
multipath hybrid nested
Miller compensated four-
GA-stage operational
amplifier

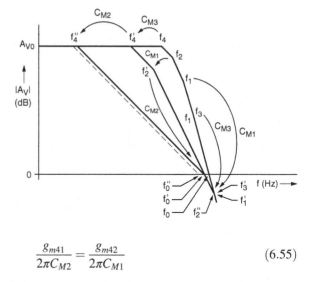

$$\frac{g_{m41}}{2\pi C_{M2}} = \frac{g_{m42}}{2\pi C_{M1}} \qquad (6.55)$$

The frequency characteristics are shown in Fig. 6.29. The result is a robust compensated four-stage amplifier with an abundance of gain and nearly the same bandwidth over supply power ratio as the two-stage Miller compensated amplifier. Only the currents of the third and fourth stages add to the supply power consumption.

6.2.4.3 Four-GA-Stage Conditionally Stable MHNMC

An interesting simplification can be made of the MHNMC amplifier. If we eliminate C_{M2}, a conditionally stable MHNMC amplifier arises [13]. This can be of advantage in audio amplifiers to increase the loop gain of the feedback amplifier up to 20 kHz for a lower distortion, while a slow settling component can be tolerated. The gain is allowed to roll off above the 20 kHz at a higher rate than 6 dB/oct. In this case it is 12 dB/oct. However, at the 0 dB frequency this must be slowed down again to 6 dB/oct. The independent nature of the multipath makes that this can be robustly realized, only depending on the multipath parameters. The frequency characteristic of the conditionally stable MHNMC amplifier without C_{M2} can be derived from Fig. 6.29.

Starting with the DC gain flat from the right until f'_4, then with 6 dB/oct down until f'_2, following with 12 dB/oct down until it hits the curve of the compensated two-stage amplifier $g_{m42}/2\pi C_{M1}$, which equals the overall characteristic of the MHNMC amplifier. The overall characteristic is determined in this range by the second input stage g_{m42} and the output stage which take the frequency response down by 6 dB/oct through the 0 dB line.

6.2.5 Multi-GA-stage Compensations

Many variations and combinations can be made in the HF compensation of operational amplifiers. With MNMC and MHNMC more than four stages can reliably be compensated without too much loss in bandwidth over power ratio [13].

6.2.6 Compensation for Low Power and High Capacitive Load

Low power consumption in combination with high capacitive load becomes more important in many cases than a straight 6 dB per octave open-loop frequency response with fast settling close-loop behavior. Particularly in low-frequency applications, low-dropout power regulators and headphone amplifiers with unpredictable high capacitive load a low quiescent current consumption is very important.

In the previous Miller compensated amplifiers a phase margin of 60° was required for fast settling. This implied an output stage with enough current to allow a capacitive load C_l at the limiting pole frequency f'_l, according to:

$$f'_l = \frac{g_{ml}}{2\pi C_l} \tag{6.56}$$

A first relieve is that we may lower the limiting pole frequency of the output transistor and thus its current consumption if there is a high closed-loop gain A_{v0}. If we know that the closed-loop gain is not unity but higher, the bandwidth is lower and we can lower the limiting-pole frequency of the output stage in a two-stage Miller compensated amplifier.

6.2.6.1 Active Miller Compensation

But what happens with the stability when we do not obey the rules at all? Let we again look at the two-stage active Miller compensated amplifier of Fig. 6.30a.

Suppose we give the output stage of the OpAmp enough current for a phase margin S of 60° at a capacitive load C_l of 30 pF and a limiting pole frequency f'_l of 2 MHz, so that the bandwidth f'_0 for unity gain stability is 1 MHz according to Eq. 6.18. If we now increase the load capacitor C_l the phase margin will decrease. But if we increase the load capacitor further, the phase margin will increase again. The situation is sketched in Fig. 6.30b.

According to [18] the minimum phase margin S_{Min} obeys the following rule:

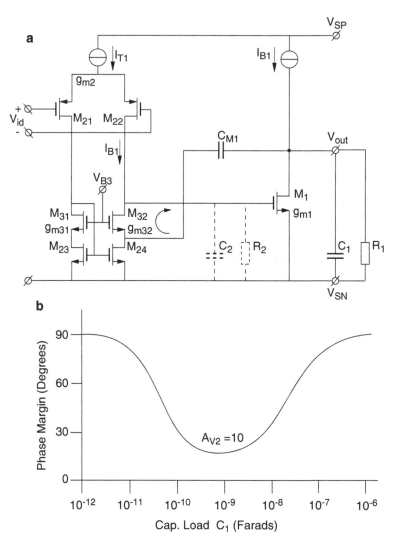

Fig. 6.30 (a) Two-stage active Miller compensated OpAmp. (b) Phase Margin versus load capacitance for an active Miller-compensated OpAmp with a voltage gain A_{v2} of 10 of the input stage

$$S_{Min} = \frac{4}{g_{m2}R_2} \frac{C_{M1}}{C_2} \qquad (6.57)$$

If the voltage gain $A_{v2} = g_{m2}R_2$ of the input stage is 10 and if the ratio C_{M1}/C_2 is 50, the phase margin will not get below about 20°. This result is even independent of g_{m1} and hence independent of the current consumption of the output transistor M_1. The bandwidth will reduce, though.

6.2.6.2 RC or Distributed RC Compensation Network

The reduction of the voltage gain of the input stage is a disadvantage, as this reduces the loop gain and the accuracy. To allow more gain in the input stage a resistor R_{M1} can be inserted in series with C_{M1}, as shown in the Low-Drop-Out amplifier [19] of Fig. 6.31a. The resulting phase margin with a resistor R_{M1} in series with C_{M1} is shown in Fig. 6.31b.

The best result is obtained when R_{M1} is chosen $0.5xR_2$. Still there are two points where the phase margin is low, about 20°. The phase margin can be further improved by using a distributed $R_{M1}C_{M1}$ network. This result is also shown in

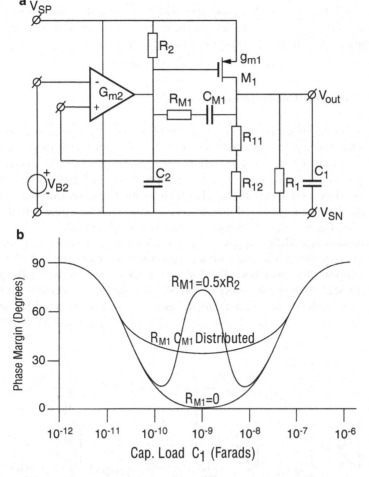

Fig. 6.31 (a) Low-Drop-Out regulator with a resistor R_{M1} in series with C_{M1} or with a purposefully distributed $R_{M1}C_{M1}$ network. (b) Phase margin as a function of the load capacitance of a Low-Drop-Out regulator of (a) with a resistor $R_{M1} = 0.5xR_2$ in series with C_{M1} or with a distributed $R_{M1}C_{M1}$ network

Fig. 6.31b. Theoretically, the phase margin does not sink below 45°. The main idea is to obtain a frequency characteristic with 3 dB per octave roll-off over a wide frequency range.

In our search for a combination of high gain and stability at high capacitive load or to lower current consumption of the output stage we now look at the three-stage OpAmp. By itself the nested Miller compensated OpAmp can not be made stable for a wide range of load capacitances at a low bias current of the output transistors. The reason is that, if the limiting pole frequency $f'_1 = g_{m1}/2\pi C_1$ falls within the frequency band where the closed-loop gain of the amplifier is higher than one, the nested Miller compensation around the output stage breaks down.

However, if we are able to reduce the three-stage amplifier into a two-stage amplifier for high frequencies then the compensation can fall back on the two-stage stability presented above. Two solutions will be presented in the following: one with pole-zero damping of the intermediate stage and one with high-frequency quenching of the intermediate stage in a multipath topology.

6.2.6.3 Damping Compensation Network

Several solutions with damping of the intermediate stage by pole-zero cancellation schemes have been proposed to stabilize three-stage amplifiers with high capacitive load at low quiescent currents. Important ones are described in [20, 21]. But they are designed for a certain range of the load capacitances and may become instable when loaded by a small capacitance. In [22] a solution is presented where the gain of the intermediate stage is reduced and made broadband by a pole-zero damping network R_D, C_D, and g_{mD}. That solution is depicted in Fig. 6.32.

The inner-nested Miller capacitor C_{M1} is made inactive by the low impedance of the damping network. The outer Miller capacitor C_{M2} takes the place of a single Miller compensation capacitor. Only for small capacitive loads, where the gain of the output stage is high, the inner-nested Miller capacitor C_{M1} comes into functioning again. Hence the circuit largely mimics a two-stage amplifier at high frequencies that can be made stable for all capacitive loads either by limiting the

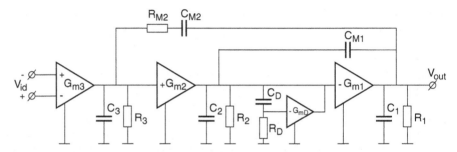

Fig. 6.32 Three-stage nested Miller compensated OpAmp with damping network

gain of g_{m3}, or a resistor R_{M2} in series with C_{M2}, or by the use of a distributed $R_{M2}C_{M2}$ network, as described before. The following choices have been made:

$$f_0'' = \frac{g_{m3}}{2\pi C_{M2}} = f_0' = \frac{g_{m2}}{2\pi C_{M1}} = \frac{1}{2}f_1' = \frac{1}{2}\frac{g_{m1}}{2\pi C_1} \quad (6.58)$$

$$\frac{1}{2}\frac{1}{2\pi R_D C_2} = \frac{g_{m2}}{2\pi C_{M1}} = \frac{1}{2}f_1' \quad (6.59)$$

6.2.6.4 Quenching Capacitor Network

The next solution is the multipath nested Miller compensated three-stage amplifier that can be made stable by it self. The multipath two-stage path makes the three-stage path stable when we quench the three-stage path by a quenching capacitor CQ to ground in parallel with C_3 [23], so that the two-stage path is dominating at high frequencies, see Fig. 6.33a, b.

For the two-stage path the same measures have to be taken as explained above. These either pose a limitation on the voltage gain of G_{m32}, or let us insert a resistor R_{M1} in series with C_{M1}, or use a distributed $R_{M1}C_{M1}$ network.

If the transconductance g_{m31} is large the quenching capacitor C_Q may also become large for high capacitive loads. This disadvantage is not present in the four-stage hybrid nested Miller compensated amplifier of Fig. 6.34.

The most important result of the quenched multipath three-stage solution is that we do not anymore have to lower the transconductance of the intermediate stage g_{m2} as required earlier in formula 6.44. We just use a small quench capacitor C_Q in any multipath three-stage amplifier to allow a large g_{m2} of the intermediate stage and make it more stable at high load capacitances.

A circuit for a multipath nested three-stage OpAmp that has a minimum phase margin of 20° for all capacitive loads with a relative small voltage gain of g_{m32} is shown in Fig. 6.33b.

In search for more gain the Multipath Hybrid Nested Miller compensation of a four-stage OpAmp is a good candidate for stability at all capacitive loads if we use a quenching capacitor C_Q [23] across C_{M3}, as shown in Fig. 6.34.

The circuit has a minimum phase margin of 20° for all capacitive loads if the voltage gain of g_{m42} is restricted. The phase margin can be improved by using a resistor R_{M1} in series with C_{M1} or a distributed $R_{M1}C_{M1}$ network. The quenching capacitor is now part of the Hybrid Nested Miller Compensation scheme across g_{m3} which amplifies the influence of the quenching capacitor. Therefore it can be taken smaller for the same task than the quenching capacitor to ground in the three-stage amplifier. Stability requirement for normal capacitive loads was already indicated by Eq. 6.53. But now C_{M3} is further increased by C_Q to obtain stability for all capacitive loads and for a larger g_{m2}.

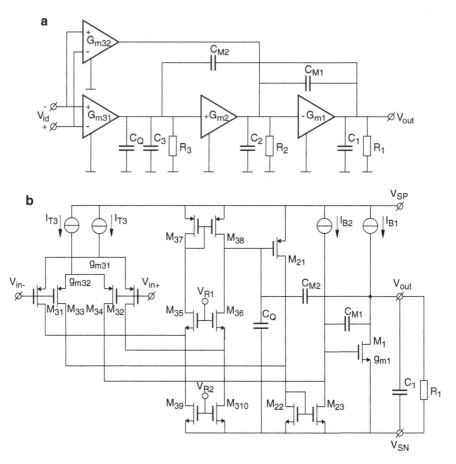

Fig. 6.33 (a) Block diagram of a multipath nested Miller compensation with quench capacitor C_Q that can be made stable for all capacitive loads. (b) Circuit for a multipath nested three-stage OpAmp with a quenching capacitor C_Q to ground at the input of the intermediate stage M_{21} of the three-stage gain path

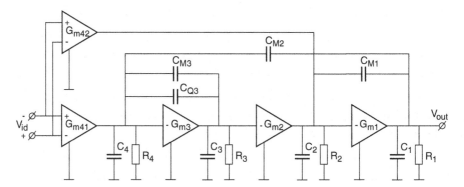

Fig. 6.34 Block diagram of a multipath hybrid nested four-stage OpAmp with a quenching capacitor C_Q across C_{M3} that is stable for all capacitive loads

6.2.6.5 Reversed Nested Miller Compensation (RNMC) for Low Power and High Capacitive Load

A most interesting way of compensation is the forward nested Miller compensation (RNMC) in which high load capacitances can be facilitated at low quiescent current and low power [13] under certain conditions. In the original patent of 1985 [24] this method was called Forward Nested Miller Compensation (FNMC). With reverse or forward nesting a stage close to the input stage is used as the center of nesting, while the nesting is extended in the direction to the output. An example is RNMC around the inverting intermediate stage of a three-GA-stage amplifier. The situation is sketched in Fig. 6.35a.

The strength of this architecture for heavy capacitive loads is based on two attitudes: Firstly, to the capacitive load it looks like a two-stage amplifier with one Miller capacitor C_{M1}. As we have seen earlier this one can be made suitable for a large capacitive load in regard to the quiescent current of the output stage. Secondly, the intermediate stage makes extra loop gain around the output stage to boost its limiting pole frequency if the ratio C_{M1}/C_{M2} larger than 1. It can be easily seen that the intermediate stage is connected in a feedback loop around the output stage as an inverting amplifier with input impedance C_{M1} and feedback impedance C_{M2}.

The limiting pole frequency $f'_2 = g_{m2}/2\pi C_2$ is now situated at the output of the intermediate stage. This pole is not loaded by the output capacitance. Hence, the limiting pole frequency can be high at a low current consumption of the intermediate stage. The limiting pole frequency is:

$$f'_2 = g_{m2}/2\pi C_2 \tag{6.60}$$

For 60° phase margin the outer Miller capacitor C_{M1} must be taken as given in Eq. 6.61:

$$f'_1 = \frac{1}{2}C_{m1}g_{m1}/C_{m2}2\pi C_1 \tag{6.61}$$

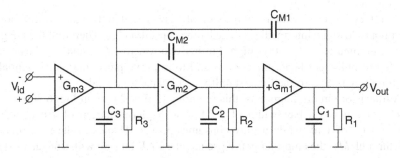

Fig. 6.35 Block diagram of a reverse or forward nested Miller compensated three-stage OpAmp

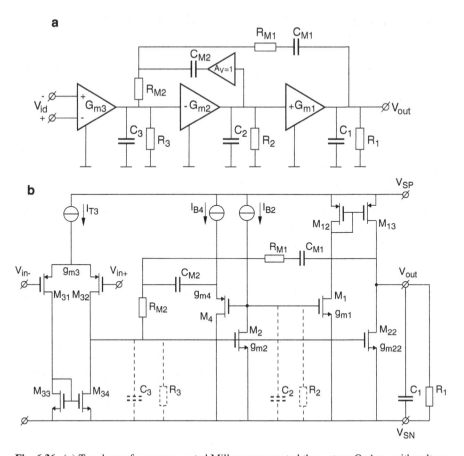

Fig. 6.36 (a) Topology of a reverse nested Miller compensated three-stage OpAmp with voltage buffer and nulling resistors R_{M1} and R_{M2}. (b) Circuit diagram of a reverse nested Miller compensated three-stage OpAmp with voltage buffer and nulling resistors R_{M1} and R_{M2}

In [25] the topology of a three-stage RNMC is presented in which refinements are introduced to facilitate high capacitive load at low current consumption. The topology is shown in Fig. 6.36a, b.

A voltage buffer A_v with a voltage gain of 1 is inserted in the inner Miller loop in order not to load the intermediate stage by the capacitor C_{M2}. Therefore the HF gain of the intermediate stage stays high. Further two resistors R_{M1} and R_{M2} have been inserted to null right-hand-plane zeros and to improve phase margin. Potentially, this topology is able to facilitate stability at all capacitive load.

A transistor circuit diagram of this topology [26] is shown in Fig. 6.36b.

The circuit has a bandwidth of 3 MHz with phase margin of 70° at a load capacitor C_1 of 1 nF while it only consumes 33 μA. Thanks to the feed forward path through M_{22} the amplifier gets a pseudo class-AB biasing with a negative slew-rate as large as the positive slew-rate of nearly 2 V/μs.

6.2.7 Conclusion

We have seen that frequency compensation is needed with amplifiers having more than one stage. Parallel compensation leads to the highest bandwidth over supply-power ratio. However, strong pole-zero doublets can be expected, because of unmatched IC process parameters. This makes parallel compensation unfit for amplifiers used to obtain a fast step response without slow settling components. Miller compensation is very robust and has by nature no pole-zero doublets. However, with normal Miller compensation the bandwidth is limited by that of the output stage and its capacitive load. Derivations of normal Miller compensation (MC), such as NMC, MNMC, HNMC, and MHNMC for amplifiers with more than two-GA-stages can successfully be compensated without loosing too much of the bandwidth over power ratio. For a combination of high capacitive load and low power consumption several techniques can be used, such as a distributed RC network in combination MC of a two-stage amplifier, or damping control of the intermediate stage of a three-stage MC amplifier, or quenching of the gain path of a HNMC four-stage amplifier, or finally RNMC which relieves the amplifier of its limiting pole frequency at the output.

6.3 Slew Rate

The currents and voltages in an operational amplifier are limited. Each stage has its own limitation. The two-stage amplifier of Fig. 6.37 has two current limitations: one at the output of the output stage I_{B1}, and one at the output of the input stage I_{B2}. These currents limit the speed at which the output voltage V_{out} can change. This speed is called slew rate $S_r = dV_{out}/dt$.

When the limitation at the output dominates, the output voltage V_{out} cannot slew faster than:

$$S_{rl} = \left(\frac{dV_{out}}{dt}\right)_{max} = I_{B1}/C_1 = V_{2max}g_{m1}/C_1 = V_{2max}2\pi f_{01} \qquad (6.62)$$

for linear or sinusoidal waveforms, with V_{2max} as the maximum linear approximated voltage swing at the input of M_1, which amounts to $V_{2max} = V_T = kT/q$ for bipolar, or $V_{2max} = (V_{GS} - V_{TH})$ for CMOS, and $f_{01} = g_{m1}/2\pi C_1$ is the 0 dB frequency of the output stage.

However, the above is seldom the case, because when the output is biased in class-AB, it has basically no current limitation.

Moreover, when feedback is applied the voltage gain is strongly reduced. The limitation will be almost certain in the input stage. Its output is attenuated by the parallel or Miller compensation network, which is dotted in Fig. 6.37.

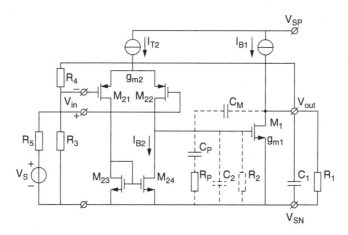

Fig. 6.37 Two-stage operational amplifier with parallel or Miller compensation and feedback resistor

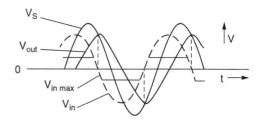

Fig. 6.38 Time responses of various signals in the circuit of Fig. 6.37 showing slewing with $\beta = R_3/(R_3 + R_4) = 1$, at unity gain

If the frequency of the input signal is raised above the dominating pole frequency f_d of the compensation network, the signal is attenuated proportionally to the frequency, while the output voltage remains constant because of the feedback.

The increase of the output current of the input stage goes on until the maximum current of $I_{B2} = I_{M2}$ is reached. At higher frequencies, the returning part $\beta = R_3/(R_3 + R_4)$ of the output voltage V_{out} cannot follow the source voltage V_S any more. Figure 6.38 shows the various response voltages in Fig. 6.37 with $\beta = 1$, at unity gain feedback. The input stage is blocking if the input voltage $V_{in} = V_S - \beta V_{out}$ is larger than $V_{in\ max}$, which is $V_{in\ max} \approx 2\,V_T = 2\ kT/q$ for a bipolar differential input pair and $V_{in\ max} \approx 2(V_{GS} - V_{TH})$ for a CMOS differential input pair.

So normally the slew rate S_r is dominated by the input stage, and can then be calculated for parallel and Miller compensation as:

$$S_r = \left(\frac{dV_{out}}{dt}\right)_{max} = \frac{I_{T2}}{C_p}\,g_{m1}R_1 = \frac{I_{T2}}{C_M} = V_{in\ max}\,2\pi\,f_0' \qquad (6.63)$$

for parallel compensation with $f_0' = g_{m2}g_{m1}R_1/2\pi C_p$, or for Miller compensation with $f_0' = g_{m2}/2\pi C_M$, respectively.

We can now compare the slewing of the output stage with that of the input stage. Normally we choose f'_0 of the whole amplifier half of f_{01} of the output stage so that the slewing of the input stage dominates. This is also valid for a well-designed operational amplifier with more than two stages.

The relation between slew rate and bandwidth makes it possible to calculate one from the other. For a bipolar input pair $V_{in\ max} = 2\ kT/q = 50$ mV. For CMOS, $V_{in\ max} = 2(V_{GS} - V_{TH})$ may be from 100 mV up to 1 V.

$$(S_r)_{bipolar} \approx 0.3\,f'_0 \qquad\qquad (6.64)$$

$$(S_r)_{CMOS} \approx (0.6\ \text{to}\ 6)f'_0 \qquad\qquad (6.65)$$

Apparently, CMOS transistors have a higher slew rate than bipolar transistors, and more so when they are in stronger inversion. However, this is only seemingly true. The same results as those with CMOS are obtained with bipolar transistors if we use degeneration resistors in series with the emitters to lower the g_{m2}/I_{T2} ratio. A compromise must be chosen between slew rate and offset and noise.

A general approach to increase the slew rate is to make the bias currents signal dependent with a class-AB input stage. This will be discussed in an example in Sect. 7.1.

6.4 Nonlinear Distortion

We encounter two types of nonlinear distortion when using operational amplifiers: firstly, distortion caused by the input stage close to slewing, and secondly, distortion caused by the output stage.

To investigate the distortion caused by the input stage close to slewing, we have to have a closer look at the transconductance g_{m2} for large signals of the input stage of Fig. 6.37. This is depicted in Fig. 6.39.

When we looked at slewing, we treated the input stage as if it had a straight transconductance g_{m2} over the maximum useful input range from $-V_{in\ max}$ up to $+V_{in\ max}$ at which values the output current I_2 changed from $-I_{T2}$ up to $+I_{T2}$. In reality we find a lying "S" shaped function. For a bipolar and a CMOS pair we have, respectively:

$$I_2 = I_{T2}tangh(V_{in}q/2kT) \qquad\qquad (6.66)$$

$$I_2 \approx I_{T2}tangh(V_{in}/(V_{GS} - V_{TH})) \qquad\qquad (6.67)$$

With $V_{GS} - V_{TH} \approx 60$ mV for CMOS input transistors in weak inversion.

It is interesting to note that by the balanced nature of the input pair all even order distortion components have disappeared. Only odd harmonic distortion remains. At

Fig. 6.39 Transconductance g_{m2} of the input stage for large signals

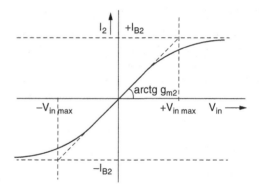

plus and minus half $V_{in\ max}$, which corresponds with half the slew rate, the third-order distortion voltage component $(V_{in\ 3rd})_{sr/2}$ referred to the input voltage is about 10 % of V_{in} or 2.5 mV for bipolar, and 5 % of V_{in} or 5–50 mV for CMOS transistors. This distortion component can be seen as an additional voltage source $V_{in\ D}$ in series with the differential input voltage V_{in}.

Though the input distortion voltage is larger with CMOS than with bipolar transistors, the relative frequency at which half the slew rate occurs is much larger with CMOS than with bipolar transistors. This over-compensates the larger distortion with CMOS, as we will see in the following calculation.

The signal-to-distortion ratio is:

$$\frac{S}{D} = \frac{V_S}{V_{inD}} = \frac{V_{out}}{AV_{inD}} \tag{6.68}$$

For an output voltage of $V_0 \approx 1$ V, and a closed loop gain $A = 1$, the S/D ratio is in the order of 400 for bipolar transistors and between 200 and 20 for CMOS at the frequency where the amplifier is excursed at half the slew rate.

The bandwidth $f_{sr/2}$ at half the slew rate is:

$$f_{sr/2} = \frac{1}{2}S_r/2\pi V_{out} = \frac{1}{2}f_0'V_{in\ max}/V_{out} \tag{6.69}$$

For an output voltage of 1 V, the bandwidth $f_{sr/2}$ is $f_0'/40$ for bipolar transistors, and $f_0'/20$ to $f_0'/2$ for CMOS. Above this frequency the distortion rises sharply, but below $f_{sr/2}$ the third-order distortion $V_{in\ 3rd}$ reduces with about the square of the ratio $(fV_{out}/2f_0'V_{in\ max})$ [13]:

$$V_{\in 3rd} \approx (V_{in3rd})_{sr/2}\left(fV_0/\frac{1}{2}f_0'V_{in\max}\right)^2 \tag{6.70}$$

We can now take this in consideration for the signal-to-distortion ratio S/D, and obtain:

$$\frac{S}{D} = \frac{V_S}{V_{inD}} = \frac{V_{out}}{AV_{inD}} \approx \frac{1}{A} \frac{V_{in\,max}^2}{V_{out}(V_{in3rd})_{sr/2}} \left(\frac{\frac{1}{2}f_0'}{f}\right)^2 \tag{6.71}$$

For bipolar and CMOS transistors this can be roughly estimated when slewing below half the slew rate:

$$\left(\frac{S}{D}\right)_{bipolar} \approx \frac{1}{A} \frac{1}{V_{out}} \left(\frac{\frac{1}{2}f_0'}{f}\right)^2 \tag{6.72}$$

$$\left(\frac{S}{D}\right)_{CMOS} \approx \frac{1}{A} \frac{2\ to\ 20}{V_{out}} \left(\frac{\frac{1}{2}f_0'}{f}\right)^2 \tag{6.73}$$

The result is that for a signal-to-distortion ratio of better than 0.1 %, at $V_{out} = 1$ V, and $A = 1$, we have a bandwidth of about $f'_0/60$ for bipolar transistors, and between $f'_0/40$ and $f'_0/13$ for CMOS transistors.

The slack characteristic of CMOS transistors in strong inversion is in favor of the steep one of bipolar transistors for distortion. The more the CMOS transistors are in strong inversion, the better. But again, with emitter degeneration of bipolar transistors we can obtain the same results as with CMOS (Fig. 6.40).

Unequal slew rate for up and down CM input voltage movements is caused by the parasitic capacitance at the tail of the input stage. This can generate excessive distortion in unity-gain feedback amplifiers.

The distortion caused by the output stage can be modeled by the transconductance of the output transistor g_{m1}. The output current I_0 is distorted as a function of the drive voltage V_2 at the output of the second stage.

The distortion to signal ratio D/S cannot generally be expressed by a function of the ratio of the output current to the maximum output current I_{B1}:

$$\frac{D}{S} = f(I_0/I_{B1}) \tag{6.74}$$

because this function may have several types of nonlinearity.

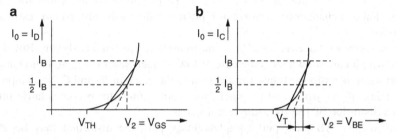

Fig. 6.40 Transconductance of (**a** *to the left*) CMOS transistors and (**b** *to the right*) bipolar transistors

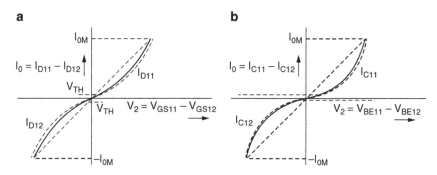

Fig. 6.41 Transconductance of (**a** *to the left*) CMOS class-AB push–pull complementary pair, and (**b** *to the right*) of a bipolar pair

Fig. 6.42 The current transfer characteristic of (**a** *to the left*) a single bipolar output transistor, and (**b** *to the right*) a class-AB push–pull complementary bipolar pair

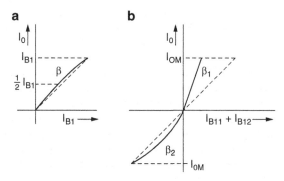

For CMOS with $I_D = \frac{1}{2}\beta(V_{GS} - V_{TH})^2$ the distortion at a maximum sinusoidal output current can be estimated at 10 %, and is of the second order. For bipolar transistors with $I_c = I_s exp(V_{BE}/V_T)$ the distortion can be estimated at 20 % totally, including even and odd order components. For a complementary transistor pair in a push–pull class-AB configuration with a high ratio between the maximum output current I_{OM} and the quiescent output current I_{Oquis}, the second order harmonic distortion disappears but the odd harmonics increase. This is depicted in Fig. 6.41.

The distortion with CMOS transistors is estimated at 10 %, when we have adapted the W/L ratios of both transistors so that their Ks are equal for the P channel and N channel transistor. For bipolar transistors the distortion is estimated at 30 %.

Until now we have assumed that the transistors are voltage driven. But if we drive with a current, which is the case at low frequencies, or if there is no parallel compensation, a bipolar transistor determines the value of R_2 and C_2 at its input. If we replace R_2 for $\beta_1 r_{e1}$, and $I_{c1} = \beta_1 I_2$, the bipolar transistor becomes fairly linear with a distortion of 1 % for a single transistor.

For a class-AB push–pull complementary pair, the distortion may be 20 % because of a large difference between the current gain β_1 and β_2 of the PNP and NPN transistors. The situation is drawn in Fig. 6.42.

The main advantage of feedback with an operational amplifier is that the distortion of the output stage is strongly reduced by the return difference or loop gain $A_V\beta_V$, in which A_V is the open loop gain of the amplifier and β_V the attenuation of the feedback network in the loop.

The closed-loop gain A_V is determined by the feedback network

$$A_V = \frac{A_{VC}}{1 + A_{VC}\beta_V} = \frac{1}{\beta_V}\frac{A_{VC}\beta_V}{1 + A_{VC}\beta_V} \approx \frac{1}{\beta_V} \tag{6.75}$$

As per inst $\beta_V = R_3/(R_3 + R_4)$

The sensitivity $S_{A_{VC}}^{A_V}$ of the closed-loop gain A_V to changes in the open loop gain A_{VC} is reduced by the loop gain $A_{VC}\beta_V$, as is calculated in Eq. 6.76.

$$S_{A_{VC}}^{A_V} = \frac{\delta A_V}{\delta A_{VC}}\frac{A_{VC}}{A_V} = \frac{1}{1 + A_{VC}\beta_V} \approx \frac{1}{A_{VC}\beta_V} \tag{6.76}$$

With the same value as that of the loop gain the distortion is reduced. Figure 6.43 shows the loop gain $A_{VC}\beta_V$ of the compensated and A_{VNC} of the non-compensated amplifier.

At low frequencies the loop-gain is $A_{VO}\beta_V = g_{m2}R_2g_{m1}R_1\beta_V$. This is normally a large value and the distortion is mostly reduced by it. At low frequencies bipolar transistors can be regarded as current driven along with their specific distortion. The reduction of the distortion at high frequencies depends on the way the amplifier is compensated.

With parallel compensation using C_P and R_P the return difference is unity $A_{VC}\beta_V = 1$ at the bandwidth of the feedback amplifier $f_B = f'_0\beta_V$.

Hence, at this frequency the full distortion of the output transistor appears at the output. The shaded area of the loop gain in Fig. 6.43 is lost and cannot be used for reducing the distortion. At a frequency ten times lower than l_B the distortion is only reduced with a factor 10.

Fig. 6.43 Frequency characteristics of the closed-loop gain (A_{VC}) and open-loop gain (A_{VO}) two-stage amplifier of Fig. 6.37

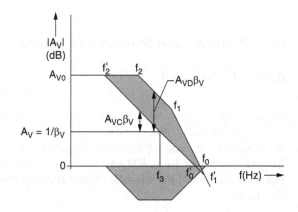

With Miller compensation using C_M the linearity is much better than with parallel compensation. The reason is that the Miller capacitor internally provides feedback across the output transistor reducing its distortion. It can be regarded as if the gain of the non-compensated amplifier is still available to reduce the distortion. At the bandwidth $f_B = f'_0 \beta_V$ of the feedback amplifier, the return difference is still $A_{VNC} \beta_V = f_2/f'_2$, which might easily be of the order of 100. At a frequency ten times lower than f_B the distortion is reduced by a factor 1000. With a closed-loop bandwidth near f'_0 at unity gain, the advantage of the Miller compensation over parallel compensation is lower.

With nested Miller compensation around the output stage of a three-stage amplifier, the advantage is even more pronounced as the uncompensated gain is larger than that of a two-stage amplifier.

There is one distortion component which is added with Miller compensation. The Miller capacitor is connected to the diode voltage at the input of the output transistor as a virtual ground. The nonlinear voltage characteristic of this diode appears in series with the output. The nonlinearity of this diode voltage is only reduced by the return difference $A_{VC} \beta_V$ of the compensated amplifier. To eliminate this effect, the Miller capacitor can be connected to the input of a cascode transistor inserted between the input and output stage, as shown in Fig. 6.18.

6.4.1 Conclusion

We have seen that the distortion of the operational amplifier originates from the input stage and the output stage. The distortion of the input stage is lower than 1 % in regard to a signal voltage of 1 V. It is reduced by the ratio of the signal frequency f and the frequency f_{sr} at which the amplifier slews. The distortion of the output stage is in the order of 10 %. It is reduced by the compensated loop gain with parallel compensation and with the uncompensated open loop with Miller compensation. Miller compensation and particularly nested Miller compensation strongly reduces the distortion of the output stage.

6.5 Problems and Simulation Exercises

6.5.1 Problem 6.1

Using NMOS devices for the one-stage amplifier in Fig. 6.8, sized $W/L_{11,12} = 200\,\mu/2\,\mu$, $W/L_2 = 50\,\mu/1\,\mu$, draw the Bode amplitude characteristic if the devices have $V_{THN} = 0.5$ V, $K_N = 56\,\mu A/V_2$, $\lambda_N = 0.1$ V$_{-1}$, and $f_{T2} = 30$ MHz. Biasing currents are $I_{M1} = 100\,\mu A$, $I_{B1} = 50\,\mu A$ and the load capacitor is $C_1 = 10$ pF. How does the Bode characteristic change if a load resistor $R_1 = 500\,k\Omega$ is added in the circuit?

6.5.1.1 Solution 6.1

The DC gain of the amplifier can be calculated for no external resistive load as the product of input stage g_m and the output impedance of the cascoded transistor M_{11}

$$A_{DC} = \frac{1}{2} g_{m11} r_{ds11} g_{m2} r_{ds2} = \frac{1}{2} \mu_{11} \mu_2 \tag{6.77}$$

For saturated MOS transistors, the voltage gain μ can be calculated

$$\mu_{11} = \frac{1}{\lambda_N \sqrt{2 I_{D11} K_N W / L_{11}}} = 75$$

$$\mu_2 = \frac{1}{\lambda_N \sqrt{2 I_{D2} K_N W / L_2}} = 53 \tag{6.78}$$

The DC voltage gain of a one-stage NMOS amplifier becomes

$$A_{DC} = \frac{1}{2} g_{m11} r_{ds11} g_{m2} r_{ds2} = \frac{1}{2} \mu_{11} \mu_2 = 2000 \tag{6.79}$$

The second nondominant pole of the amplifier is given by the transit frequency of M_2

$$f_1 = f_{T2} = 30\,\text{MHz} \tag{6.80}$$

The unity gain bandwidth is set by the ratio of the input stage g_m and the load capacitor C_1

$$f_0 = \frac{1}{2} \frac{g_{m11}}{2\pi C_1} = 11.9\,\text{MHz} \tag{6.81}$$

at $g_{m11} = 750\,\mu\text{S}$. Note that the second pole frequency is more than two times larger compared to the unity gain bandwidth, so the phase margin will be higher than $60°$. The dominant pole frequency is obtained as the division of unity gain bandwidth to DC voltage gain value

$$f_2 = \frac{f_0}{A_{DC}} = 5.6\,\text{kHz} \tag{6.82}$$

The changes produced by the presence of a load resistor are related to a reduced output impedance, which in turn reduces to voltage gain and the unity gain bandwidth. If the load resistance is much smaller than the amplifier output impedance

$$R_1 = 500\,\text{k}\Omega \gg r_{ds11} g_{m2} r_{ds2} = 10.25\,\text{M}\Omega \tag{6.83}$$

then the two equations above, which are dependent on DC voltage gain, are affected
by the load resistance alone

$$A_{DC} = \frac{1}{2} g_{m11} R_1 = 187$$

$$f_2 = \frac{f_0}{A_{DC}} = 63.6 \,\text{kHz}$$

(6.84)

The decrease in voltage gain is dramatic because a single stage MOS amplifier
obtains the biggest part of its voltage gain by using a high output impedance

6.5.2 Problem 6.2

A two-stage amplifier using parallel compensation is depicted in Fig. 6.12. Consid-
ering an all-NMOS circuit, with transistors sized $W/L_{21,22} = 100 \,\mu/2 \,\mu$ and $W/L_1 = 200 \,\mu/2 \,\mu$, and loaded with $C_1 = 20$ pF, $R_1 = 50$ kΩ, compensate the circuit by
calculating the values of C_P, R_P. Also calculate the unity gain bandwidth of the
compensated amplifier and the bandwidth over dissipated power ratio if the circuit is
biased at $V_{SP} - V_{SN} = 2$ V. The transistors have $V_{THN} = 0.5$ V and $K_N = 56 \,\mu\text{A/V}^2$ and
are biased with $I_{M2} = 50 \,\mu\text{A}$, $I_{B1} = I_{B2} = 25 \,\mu\text{A}$. The internal components are $C_2 = 1$ pF
and $R_2 = 1$ MΩ.

6.5.2.1 Solution 6.2

According to Eq. 6.9, the parallel compensation network should be designed with

$$C_P = R_1 \sqrt{2 C_1 C_2 g_{m1} g_{m2}}$$

$$= R_1 \sqrt{2 C_1 C_2 \sqrt{2 K_N \frac{W}{L_1} I_{B1}} \frac{1}{2} \sqrt{2 K_N \frac{W}{L_2} 2 I_{B2}}} = 118 \,\text{pF}$$

(6.85)

$$R_P = \sqrt{\frac{C_1}{C_2 g_{m1} g_{m22}}} = 8.4 \,\text{k}\Omega$$

With this compensation the unity gain bandwidth of the amplifier is, according to
Eq. 6.10,

$$f_0' = \sqrt{\frac{g_{m1}}{2\pi C_1} \frac{g_{m22}}{2\pi C_2}} = 18.8 \,\text{MHz}$$

(6.86)

Noting that $I_{B1} = I_{B2}$ and all transistors operate at the same $V_{GS} - V_{THN}$, Eq. 6.14
can be used to find the bandwidth over power ratio

$$\frac{f_0'}{P_S} = \frac{1}{8\pi\sqrt{2C_1C_2}(V_{GS} - V_{THN})(V_{SP} - V_{SN})} = 4.7 \times 10^{10} \qquad (6.87)$$

6.5.3   Problem 6.3

Using the same device parameters and resistor and capacitor values as in Problem 6.2, calculate the Miller capacitor C_{M1} and the Miller zero cancellation resistor R_{M1} for maximum unity gain bandwidth of the amplifier shown in Fig. 6.17. What is the frequency of the dominant pole?

6.5.3.1   Solution 6.3

Since the relationship between the uncompensated amplifier's poles is important, with the Miller cancellation techniques the values of these poles should be calculated first

$$f_{01} = \frac{g_{m1}}{2\pi C_1} = 6\,\text{MHz}$$
$$f_{02} = \frac{g_{m2}}{2\pi C_2} = 30\,\text{MHz} \qquad (6.88)$$

Noting that $f_{01} < f_{02}$, the attainable unity gain bandwidth will be

$$f_0' = \frac{1}{2}f_{01} = 3\,\text{MHz} \qquad (6.89)$$

and the corresponding Miller capacitor value becomes

$$C_{M1} = \frac{g_{m2}}{2\pi f_0'} = \frac{g_{m22}}{4\pi f_0'} = 10\,\text{pF} \qquad (6.90)$$

according to Eq. 6.22. The Miller zero cancellation resistor results from Eq. 6.24 which states the value of the Miller zero frequency

$$R_{M1} = \frac{1}{g_{m1}} = 1.3\,\text{k}\Omega \qquad (6.91)$$

The dominating pole frequency produced by Miller compensation becomes

$$f_2' = \frac{f_0'}{A_{V0}} = \frac{f_0'}{g_{m2}R_2 g_{m1}R_1} = 426\,\text{Hz} \qquad (6.92)$$

The bandwidth over power ratio for this amplifier is

$$\frac{f_0'}{P_S} = \frac{1}{8\pi(C_1 + 2C_2)(V_{gs} - V_{th})(V_{SP} - V_{SN})} = 1.7 \times 10^{10} \tag{6.93}$$

6.5.4 Problem 6.4

The three-GA-stage amplifier in Fig. 6.22 is designed with NMOS transistors sized $W/L_{31,32} = 100\,\mu/2\,\mu$, $W/L_{21,22} = 10\,\mu/1\,\mu$, and $W/L_1 = 100\,\mu/1\,\mu$ biased at $I_{M3} = 2I_{B3} = 50$ μA, $I_{M2} = 2I_{B2} = 20$ μA, and $I_{B1} = 50$ μA. The internal nodes are loaded with $C_3 = 0.5$ pF, $R_3 = 300$ kΩ, $C_2 = 1$ pF, $R_2 = 200$ kΩ while the output is loaded with $C_1 = 10$ pF, $R_1 = 100$ kΩ. Calculate the nested Miller compensation network, the unity gain bandwidth frequency and the dominant pole frequency. The NMOS devices have $V_{THN} = 0.5$ V and $K_N = 56$ μA/V^2.

6.5.4.1 Solution 6.4

The 0 dB bandwidth f'_0, of the Miller compensated intermediate and output stages should be, according to Eq. 6.30,

$$f_0' = \frac{1}{2}f_1' = \frac{1}{2}\frac{g_{m1}}{2\pi C_1} = \frac{1}{2}\sqrt{\frac{2K_N\frac{W}{L_1}I_{B1}}{2\pi C_1}} = 6\,\mathrm{MHz} \tag{6.94}$$

which takes a Miller capacitor C_{M1} sized

$$C_{M1} = \frac{g_{m2}}{2\pi f_0'} = \frac{g_{m22}}{4\pi f_0'} = 1.3\,\mathrm{pF} \tag{6.95}$$

The unity gain bandwidth of the nested Miller compensated amplifier will be half f'_0.

$$f_0'' = \frac{1}{2}f_0' = 3\,\mathrm{MHz} \tag{6.96}$$

which needs a nested Miller capacitor C_{M2}

$$C_{M2} = \frac{g_{m3}}{2\pi f_0''} = \frac{g_{m32}}{4\pi f_0''} = 10\,\mathrm{pF} \tag{6.97}$$

The dominant pole frequency is obtained by combining Eqs. 6.34 and 6.35

$$f_3' = \frac{f_0'}{A_{v0}} = \frac{4f_0''}{g_{m32}R_3 g_{m22}R_2 g_{m1}R_1} = 70\,\mathrm{Hz} \tag{6.98}$$

6.5.5 Problem 6.5

A multipath nested Miller compensated amplifier designed with the schematic shown in Fig. 6.24 and using NMOS transistors sized $W/L_{31,32} = W/L_{33,34} = 100\,\mu/2\,\mu$, $W/L_{21,22} = 10\,\mu/1\,\mu$ and $W/L_1 = 100\,\mu/1\,\mu$ is driving a load capacitor $C_1 = 10$ pF with an output impedance of $R_1 = 100$ kΩ. The biasing currents are $I_{M32} = I_{M31} = 2I_{B32} = 2I_{B31} = 50$ μA, $I_{M2} = 2I_{B2} = 20$ μA and $I_{B1} = 50$ μA. The internal nodes are loaded with $C_3 = 0.5$ pF, $R_3 = 300$ kΩ, $C_2 = 1$ pF, $R_2 = 200$ kΩ. Calculate the nested Miller compensation network C_{M1} and C_{M2}. The NMOS devices have $V_{THN} = 0.5$ V and $K_N = 56$ μA/V$_2$.

6.5.5.1 Solution 6.5

$$f_0' = \frac{1}{2}f_1' = \frac{1}{2}\frac{g_{m1}}{2\pi C_1}\frac{1}{2}\sqrt{\frac{2K_N \frac{W}{L_1}I_{B1}}{2\pi C_1}} = 6\,\text{MHz} \tag{6.99}$$

The 0 dB bandwidth f_0' after first compensation with C_{M1}
  which in turn makes the Miller capacitor C_{M1} to be

$$C_{M1} = \frac{g_{m32}}{2\pi f_0'} = \frac{g_{mT34}}{4\pi f_0'} = 1.3\,\text{pF} \tag{6.100}$$

Because of the existing multipath, the overall unity gain bandwidth

$$f_0'' = f_0' = 6\,\text{MHz} \tag{6.101}$$

$$C_{M2} = \frac{g_{m31}}{2\pi f_0''} = \frac{g_{mT32}}{4\pi f_0''} = 10\,\text{pF} \tag{6.102}$$

will be equal to f_0' after the second compensation by C_{M2} which according to Eq. 6.42 needs a Miller capacitor C_{M2}
  The dominant pole frequency is given by Eq. 6.43
  In order to obtain a phase margin larger than 55°, the condition in Eq. 6.44 should be checked

$$f_3' = \frac{f_0''}{A_{v0}} = \frac{4f_0''}{g_{mT34}R_3 g_{m22}R_2 g_{m1}R_1} = 140\,\text{Hz} \tag{6.103}$$

$$g_{m2} = \frac{1}{2}g_{m22} = 51\,\mu\text{A/V} < \frac{1}{3}g_{m32} = \frac{1}{6}g_{mT34} = 62.5\mu\text{A/V} \tag{6.104}$$

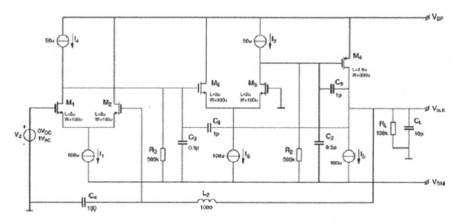

Fig. 6.44 Three-stage Miller compensated amplifier

6.5.6 Simulation Exercise 6.1

The three-GA-stage operational amplifier in Fig. 6.44 is shown in a setup for AC differential gain simulation. The amplifier is Miller compensated by C_8 and C_9. Remove the Miller capacitors and simulate the circuit to calculate the phase margin, unity gain bandwidth and dominant pole frequency, then repeat the previous step with the Miller capacitors in the circuit. Considering the fact that some transistor models do not model accurately the moderate inversion region of a MOS device, resize the transistors to place them in strong inversion by checking the $V_{GS} - V_{TH}$ to be at least 0.2 V. Considering a threshold voltage variation of 40 mV (the actual V_{TH} variation can be calculated dividing this value by transistor width) for all devices, compensate the amplifier for 60° phase margin in the worst case combination of the mismatch mentioned above.

6.5.7 Simulation Exercise 6.2

The operational amplifier shown in Fig. 6.45 can provide a large DC gain due to the dual input stage. Compare the gain and phase characteristics of this amplifier with the ones from the previous exercise, after removing the Miller capacitors in both circuits and after making the output stage transistors M_4 of equal size $W/L_4 = 300/1.6$. The three-stage gain characteristic is now shown by the results from the previous circuit and can be compared with the results for the multipath version. Add the Miller capacitors and compare the two results again. Note the bandwidth increase for the multipath circuit.

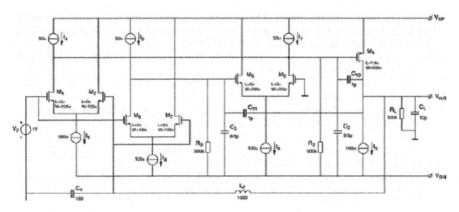

Fig. 6.45 Multipath nested Miller compensated amplifier

References

1. J.H. Huijsing, F. Tol, Monolithic operational amplifier design with improved HF behavior. IEEE J. Solid-St. Circ. **SC-11**, 323–328 (1976)
2. K. Bult, G.J.G.M. Gielen, A fast settling CMOS op amp for SC-circuits with 90 dB gain. IEEE J. Solid-St. Circ. **SC-25**(12), 1379–1383 (1990)
3. J.H. Huijsing, D. Linebarger, Low-voltage operational amplifier with rail-to-rail input and output ranges. IEEE J. Solid-St. Circ. **SC-20**(6), 1144–1150 (1985)
4. R. Caprio, Precision differential voltage-current converter. Electron. Lett. **9**, 147–148 (1973)
5. H. Nyquist, Regeneration theory. Bell Syst. Tech. J. **11**, 126–147 (1932)
6. H.W. Bode, *Network Analysis and Feedback Amplifier Design* (D. van Nostrand Company Inc., New York, NY, 1949)
7. B.Y. Kamath, R.G. Meyer, P.R. Gray, Relationship between frequency response and settling time of operational amplifiers. IEEE J. Solid-St. Circ. **SC-9**(6), 347–352 (1974)
8. B. Nauta, A CMOS transconductance-C filter technique for very high frequencies. IEEE J. Solid-St. Circ. **27**(2), 142–153 (1992)
9. J.M. Miller, Dependence of the input impedance of a three-electrode vacuum tube upon the load in the plate circuit. Sci. Paper Bur. Stds. **15**(351), 367–385 (1920)
10. R. Hogervorst et al., A compact power-efficient 3 V CMOS rail-to-rail input/output operational amplifier for VLSI cell libraries. IEEE J. Solid-St. Circ. **29**(12), 1505–1513 (1994)
11. R. Hogervorst, J.H. Huijsing, *Design of Low-Voltage Low-Power Operational Amplifier Cells* (Kluwer Academic Publishers, Boston, MA, 1996)
12. R.G.H. Eschauzier, J.H. Huijsing, An operational amplifier with multipath Miller zero cancellation for RHP zero removal, in Proceedings ESSCIRC 1993, Editions Frontières, Gif-sur-Yvettes, France
13. R.G.H. Eschauzier, J.H. Huijsing, *Frequency Compensation Techniques for Low-Power Operational Amplifiers* (Kluwer Academic Publishers, Boston, MA, 1995)
14. M.J. Fonderie, J.H. Huijsing, *Design of Low-Voltage Bipolar Operational Amplifiers* (Kluwer Academic Publishers, Boston, MA, 1993)
15. M.J. Fonderie, J.H. Huijsing, Operational amplifier with 1-V rail-to-rail multipath-driven output stage. IEEE J. Solid-St. Circ. **26**(12), 1817–1824 (1991)
16. K.J. de Langen, J.H. Huijsing, *Compact Low-Voltage and High-Speed CMOS, BiCMOS and Bipolar Operational Amplifiers* (Kluwer Academic Publishers, Boston, MA, 1999), p. 200

17. R.G.H. Eschauzier, R. Hogervorst, J.H. Huijsing, A programmable 1.5 V CMOS class-AB operational amplifier with Hybrid Nested Miller compensation for 120 dB gain and 6 MHz UGF. IEEE J. Solid-St. Circ. **29**(12), 1497–1504 (1994)
18. R.J. Reay, G.T.A. Kovacs, An unconditionally stable two-stage CMOS amplifier. IEEE J. Solid-St. Circ. **30**(5), 591–594 (1995)
19. A. Bakker, Low-dropout voltage regulator with improved stability for all capacitive loads, U.S. Patent 6,373,233 B2, 16 April 2002
20. K.N. Leung, P.K.T. Mok, W.H. Ki, J.K.O. Sin, Three-stage large capacitive load amplifier with damping-factor-control frequency compensation. IEEE J. Solid-St. Circ. **35**(2), 221–230 (2000)
21. X. Peng, W. Sansen, Transconductance with capacitances feedback compensation for multistage amplifiers. IEEE J. Solid-St. Circ. **40**(7), 1514–1520 (2005)
22. V. Dhanasekaran, J.S. Martinez, E. Sanchez-Sinenco, Design of three-stage class-AB 16 Ohm headphone driver capable of handling wide range of load capacitance. IEEE J. Solid-St. Circ. **44**(6), 1734–1744 (2009)
23. J. Hu, J.H. Huijsing, K.A.A. Makinwa, A three-stage amplifier with quenched multipath frequency compensation for all capacitive loads. IEEE Int Symp Circ Syst **2007**(27–30), 225–228 (2007)
24. J.H. Hiujsing, (as misspelled in US Patent 4,559,502), Multi-stage amplifier with capacitive nesting for frequency compensation, U.S. Patent 4,559,502, filed by Signetics Corporation (Sunnyvale, CA), 17 Dec 1985
25. D. Marano, G. Palumbo, S. Pennisi, Improved power-efficient RNMC technique with voltage buffer and nulling resistors for low-power high-load three-stage amplifiers. JCSC **18**(7), 1321–1331 (2009)
26. A.D. Grasso, D. Marano, G. Palumbo, S. Pennisi, Improved reversed Nested Miller frequency compensation technique with voltage buffer and resistor. IEEE Trans. Circ. Syst. **54**(5), 382–386 (2007)

Chapter 7
Design Examples

Abstract We have made a classification of operational amplifiers in Chap. 6. Nine main topologies have been listed as in a periodic system. In this chapter we will present practical design examples of the nine topologies. Hence, it is easy to recognise them, an to choose the one we need. Many famous designs are incorporated.

For a good input stage only the GA type is suited, as shown in Chap. 4. For level-shift or intermediate stages there are two possible types, the CM or CF (both denoted as CF), and the GA stage, as shown in Chap. 6. Three push–pull output stages exist, the fully VF, VF/GA, and fully GA stages, as shown in Chap. 5. The combination of these possibilities create two two-stage configurations without a push–pull output stage, six three-stage configurations, and one multistage configuration with four GA stages or more. These nine main topologies are inventoried again in Table 7.1.

We will give practical examples of each of these nine overall topologies in the following nine paragraphs and end with conclusions. We will see that the first configuration is mainly useful for capacitive loads. Configurations 3–6 are mostly useful in bipolar technology. While the configurations 7–9 with rail-to-rail output stage are particularly useful for low-voltage CMOS technology, but may also be used in some low-voltage bipolar and BiCMOS applications.

7.1 GA-CF Configuration

The most simple topology for an operational amplifier is the GA-CF configuration.

7.1.1 Operational Transconductance Amplifier

The simple class-A complete general OpAmp is called operational transconductance amplifier (OTA). The OTA has besides the GA-input stage, a current-mirror (CM) or a folded-cascode current-follower (CF) stage for providing an

© Springer International Publishing Switzerland 2017   215
J. Huijsing, *Operational Amplifiers*, DOI 10.1007/978-3-319-28127-8_7

Table 7.1 Classification of nine main overall topologies for operational amplifiers

Configuration number	Input stage	Level-shift or intermediate stage	Output stage
1	GA	CF	–
2	GA	GA	–
3	GA	CF	VF
4	GA	GA	VF
5	GA	CF	VF/GA
6	GA	GA	VF/GA
7	GA	CF	GA
8	GA	GA	GA
9	GA	GA + GA	GA

Fig. 7.1 (**a**) Class-A operational transconductance amplifier (OTA) with GA-CM configuration and (**b**) frequency characteristic of the GA-CF configuration

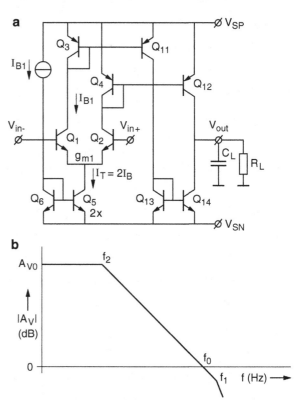

output-voltage swing independent of the CM input voltage. This type of amplifier is mainly used to drive capacitive loads C_L. The transconductance, which is provided by the GA input stage, is too low to provide a large DC gain with a resistive load R_L. A typical structure in bipolar technology is shown in Fig. 7.1a.

The OTA of Fig. 7.1a has a transconductance G_m equal to the transconductance of the input stage g_{m1} which is proportional to the bias current I_{B1}. The series connection of the two emitters of the input stage halves the transconductance of the input transistors $g_{ml,2}$. The current mirror Q_{13}, Q_{14} doubles the transconductance again, so that G_m of the whole OTA equals that of one of the input transistors $g_{ml,2}$. The emitter-area ratios have been chosen such that all transistors have a quiescent current of I_{Bl}. With $V_T = kT/q$ the total G_m becomes:

$$G_m = g_{ml} = g_{ml,2} = \frac{I_{Bl}}{V_T} \qquad (7.1)$$

Formula 7.1 shows that the transconductance G_m is constant over temperature if the bias current I_{Bl} is proportional to the absolute temperature T (*PTAT*). The limiting pole frequency f_l results from the mirrors. Their transit frequency f_l is half that of a single transistor f_T because of the double base-emitter capacitor $2c_{be}$ parallel to one diode resistor $r_e = 1/g_m$. The PNP transistors usually limit the bandwidth by their low f_{TP}. So the limiting pole frequency f_l is:

$$f_l \approx \frac{1}{2} f_{TP} \qquad (7.2)$$

The load capacitor C_L must be so large that the zero-dB frequency f_0 lies a factor 2 below f_l for a 60° phase margin:

$$f_0 = \frac{G_m}{2\pi C_L} \leq \frac{1}{2} f_1 \approx \frac{1}{4} f_{TP} \qquad (7.3)$$

The dominating pole frequency f_2 is:

$$f_2 = \frac{1}{2\pi R_L C_L} \qquad (7.4)$$

The DC voltage gain is:

$$A_{V0} = G_m R_L \qquad (7.5)$$

The GA-CM configuration is often used without load resistor R_L to drive a capacitive load.

The maximum obtainable gain without R_L is limited by the parallel ($\|$) output resistance r_0 of the PNP and NPN current mirrors ½ r_{cep} and ½ r_{cen}:

$$A_{V0M} = G_m R_o = G_m \left(\frac{1}{2} r_{cep} \middle\| \frac{1}{2} r_{cen} \right) \approx G_m \frac{1}{4} r_{ce} \approx \frac{1}{4} \mu \qquad (7.6)$$

The value of $\mu = g_m r_{ce} = r_{ce}/r_E$ may be of the order of 10^3 or 60 dB. When we would have cascoded the output stage with bipolar transistors, the output impedance r_o will be limited by the collector–base resistance $r_{cb} \approx \beta r_{ce}$, and the maximum obtainable gain would become:

$$A_{V0M} = G_m \left(r_{cbp} \| r_{cbn} \right) \approx G_m \frac{1}{2} r_{cb} \approx \frac{1}{2} \beta \mu \qquad (7.7)$$

7.1.2   Folded-Cascode Operational Amplifier

To obtain a high output impedance in CMOS, a folded cascode can be placed in cascade, as shown in the circuit of Figs. 7.2a, b.

The CMOS folded-cascode amplifier of Fig. 7.2a has a transconductance $G_m = g_{m1,2}$ equal to that of one input transistor $g_{m1,2}$. The series connection of the two input transistors halves the transconductance. But the mirror-connected transistors M_{11}, M_{12} sum the differential currents of the input transistors, so that the transconductance is doubled again. We have chosen relative wide input transistors, so that these become biased in weak inversion at I_{B1}. That provides the highest G_m at the given tail current of $2I_{B1}$. This means that all offset and noise sources of the input stage and the following stage are minimally reflected in the input offset voltage V_{inoff}. In weak inversion the CMOS transistors function like bipolar transistors in an exponential way. But the transconductance is n times smaller. The transconductance of the whole amplifier is:

$$G_m = g_{m1,2} = \frac{I_{B1}}{nV_T} \tag{7.8}$$

with $V_T = kT/q$, and nV_T is chosen about 60 mV at room temperature, which is in moderate inversion close to weak inversion.

The W/L ratios of all current-source transistors M_3, M_{17}, and M_{18} are chosen such that their output currents are all equal to $2I_{B1}$. This provides a maximum symmetrical output current at a minimal supply power. The current sources need to have low noise, low offset, and a high output resistance. Therefore, relatively long CMOS transistors have been chosen for these current sources such that they become biased in strong inversion at an effective gate-source voltage $V_{GS} - V_{TH}$ of about 300 mV. The G_m is:

$$G_m = g_{m1,2} = K(V_{GS} - V_{TH}) = \sqrt{2KI_{B1}} = \frac{2I_{B1}}{(V_{GS} - V_{TH})} \tag{7.8a}$$

with: $K = \mu C_{ox} W/L$.

Some times the mirror connection is chosen at the bottom at the gates of the lower N-channel transistors M_{11} and M_{12} as in Fig. 7.2a. This gives the highest frequency response because the gate capacitors of the N-channels are smaller than that of the P-channels. The P-channel transistors are usually chosen three times larger to compensate the g_m for a three times lower mobility. With the current mirror connection at the lower side, the current sources M_{17} and M_{18} at the upper side must provide two times the bias current of one input transistor, to fully excurse the output. As a consequence, the mirror-connected transistors M_{11} and M_{12} must carry three times the bias current of one input transistor. Only then the maximum positive and negative output current, equal to two times that of one input transistor, can be provided at the output to the load.

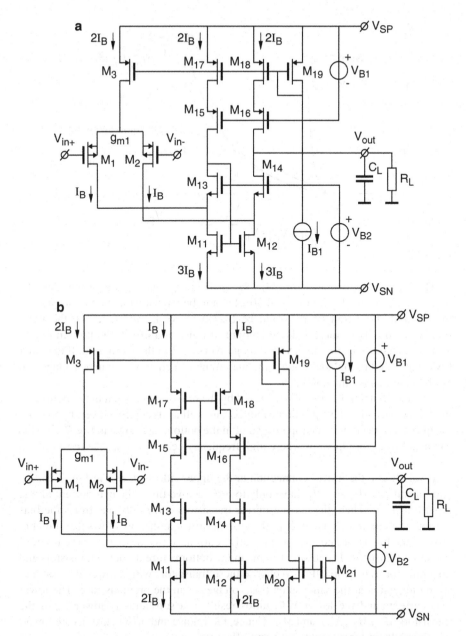

Fig. 7.2 (a) Class-A folded-cascode operational amplifier with GA-CF configuration. (b) Class-A folded-cascode operational amplifier (OpAmp) with GA-CF configuration and the current-mirror connection at the P-channel side (c) Comparison of the G_m of a CMOS transistor in weak inversion, strong inversion, and degenerated weak inversion

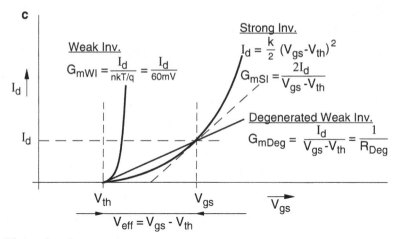

Fig. 7.2 (continued)

The bottom mirror transistors M_{11}, M_{12} carry three times more current than the input transistors do. Their noise and offset would be reflected three times stronger at the input if these transistors would be biased in weak inversion like the input transistors are. To avoid this large amount of noise and offset, the bottom mirror transistors have to be biased in very strong inversion. But there is a limit. If the input CM range need to include ground, the maximum effective gate-source voltage can not be much larger than 300 mV.

The additional noise and offset caused by the mirror connection in the bottom of the folded cascode CF stage can be reduced, as well as its supply power, by making the mirror connection at the top instead of in the bottom, as shown in Fig. 7.2b. This can be allowed for OpAmps where high frequency behavior is not the most critical issue.

If we choose the mirror connection at the upper side as drawn in Fig. 7.2b, the current sources M_{11} and M_{12} have only to provide two times the bias current of one input transistor. Then, the current mirror transistors carry only one times the bias current of an input transistor. Still, the current mirror at the top allows the output to provide the maximum positive and negative current of two times the bias current of one input transistor. The lower currents in the bottom current-source transistors and upper mirror transistors allow to choose these transistors with a lower W/L ratio in strong inversion at the same saturation voltages as in the previous case. The lower bias currents and the lower W/L ratios result in a substantially lower g_m of the transistors M_{11}, M_{12}, M_{17}, and M_{18}. Hence, their noise and offset currents are lower, which results in lower input noise and offset voltages.

An alternative to strong inversion is the use of resistive degenerated transistors in weak inversion. That might take more area in the layout. But the degeneration resistors can be accurately matched and their G_m is two times smaller than that of transistors in strong inversion at the same bias current and voltage loss. This results in a lower noise and offset. This alternative can also be applied to the lower current sources M_{11} and M_{12} and upper current mirror M_{17} and M_{18}. The alternatives are shown in Fig. 7.2c.

The limiting pole frequency f_1 in the folded cascode of Fig. 7.2a results from the folded current follower M_{14} and mirror M_{11}, M_{12}. The transit frequency of the N-channel mirror $f_1 = \frac{1}{2} f_{TN}$ roughly limits the bandwidth:

$$f_1 = \frac{1}{2} f_{TN} \tag{7.9}$$

The load capacitance C_L must be large enough so that the zero-dB frequency f_0 lies a factor 2 below f_1:

$$f_0 = \frac{g_{ml}}{2\pi C_L} \leq \frac{1}{2} f_1 = \frac{1}{4} f_{TN} \tag{7.10}$$

The dominating pole frequency f_2 is:

$$f_2 = \frac{1}{2\pi R_L C_L} \tag{7.11}$$

The maximum DC voltage gain (without load resistance R_L) is limited by the parallel ($\parallel$) output resistance R_o of the cascodes M_{14} and M_{16} and can be estimated at $R_o \approx \mu_{14} r_{ds12} \parallel \mu_{16} r_{ds18} \approx \frac{1}{2}\mu\ r_{ds}$:

$$A_{VOM} = G_m R_o \approx g_m \frac{1}{2} \mu r_{ds} \approx \frac{1}{2} \mu_P \mu_N \tag{7.12}$$

The maximum value of the voltage gain μ_P or μ_N for P-channel or N-channel CMOS transistors, respectively, may vary from 30 to 300, depending on the chosen W/L ratio and on the current level.

7.1.3 Telescopic-Cascode Operational Amplifier

In cases where the common-mode input voltage range need not be large, for instance in an inverting integrator (see Fig. 3.16), a telescopic cascode may be used instead of a folded cascode output stage. This OpAmp is shown in Fig. 7.3.

A very important feature is that the telescopic OpAmp only needs half the supply current of the folded version for the same maximum output current and G_m. Moreover, the extra noise and offset of the extra currents sources in the folded cascode stage are not present here. A disadvantage, though, is that the minimum supply voltage is one saturation voltage higher than that of the folded version. For the folded version of Fig. 7.2a the minimum supply voltage is: $V_S = V_{SP} - V_{SN} = V_{diode11} + V_{sat15} + V_{sat17}$, while the minimum supply voltage for the telescopic version is $V_S = V_{sat9} + V_{sat1,2} + V_{sat3,4} + V_{diode5}$.

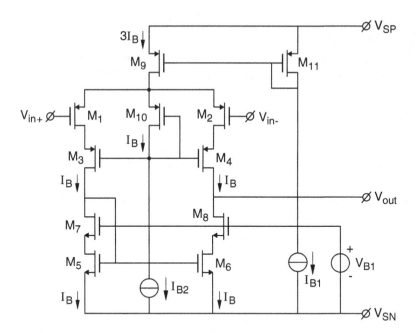

Fig. 7.3 Telescopic cascoded class-A operational amplifier with GA-CF configuration

A disadvantage of the telescopic OpAmp is that the common-mode input voltage range of the telescopic version can not include the negative rail voltage at all. Nor can the common-mode input voltage cover much of the output voltage range. The output CM voltage can not come higher than a threshold voltage minus two saturation voltages above the input CM voltage.

7.1.4 Feedforward HF Compensation

The main advantage of the preceding simple GA-CM or CF circuits is that the zero-dB bandwidth lies only a factor 4 below the f_T of the transistors. Ways to improve the bandwidth are based on bypassing the mirrors or cascodes by an $R_F C_F$ all pass network [1, 2]. This eliminates the PNPs from the HF path in the OTA in Fig. 7.4 so that the limiting pole frequency f_1 is equal to $\frac{1}{2} f_{TN}$ of an NPN mirror. Note that the HF feedforward network has been cross-coupled to match the negative transfer of the mirrors.

Moreover, the HF feedforward network provides a direct HF path from the collector of Q_1 to the output without a limiting-pole frequency.

This same HF path, but not cross-coupled, is provided in the CMOS folded cascode amplifier of Fig. 7.5. The partly elimination of the dominating pole makes that the zero-dB bandwidth f_0 may approach $\frac{1}{2} f_T$, or even f_T, by the HF-bypassing of the internal connection. Note that the elimination is only at one half and that the

Fig. 7.4 OTA with
GA-CM configuration and a
cross-coupled HF
feedforward network
$R_{F1}C_{F2}$ and $R_{F2}C_{F1}$

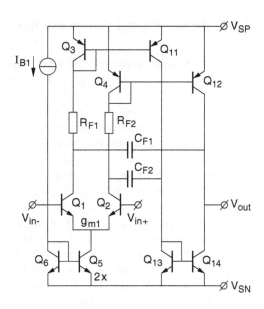

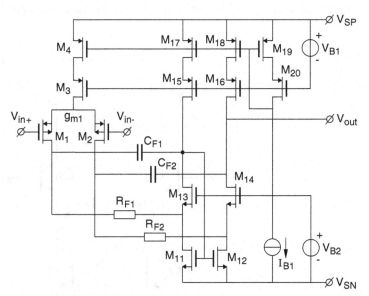

Fig. 7.5 Folded-cascode amplifier with GA-CF configuration and a straight HF feedforward
network $R_{F1}C_{F1}$ and $R_{F2}C_{F2}$

output load also influences the existence of a pole-zero doublet in the frequency
characteristic.

A disadvantage of the simple circuits of Figs. 7.1 and 7.2 is the relatively low
transconductance G_m and DC gain A_{V0}. Many approaches have been made to increase
the G_m or A_{V0}. Some of these approaches will be discussed in the following sections.

7.1.5 *Input Voltage Compensation*

The most important attribute of an input stage is its transconductance g_m. Bipolar transistors provide the largest g_m at a certain quiescent current in nature. Their $g_m = I_E/V_T$ with $V_T = kT/q$ is determined by the Boltzmann statistics for distribution of charge carriers in a semiconductor. There is no motive for charge carriers in a junction to react more sensitively to a voltage than according to the Boltzmann statistics law. Tunneling or other quantum effects may break this law. Yet, there is a demand to improve the g_m without enlarging the quiescent current. A network approach to do something is to compensate one source or emitter resistance with another of the same value but with a negative sign by positive feedback. A basic circuit is drawn for a bipolar and CMOS circuit in Figs. 7.6 and 7.7 [3].

The cross-coupled transistors Q_3 and Q_4 create a positive feedback loop. The currents of Q_1 and Q_3 are equal to I_{C1}, and those of Q_2 and Q_4 equal to I_{C2}, if we disregard the base currents. This means that the base emitter voltages of Q_1 and Q_3 are equal: $V_{BE1} = V_{BE3}$, and even so those of Q_2 and Q_4: $V_{BE2} = V_{BE4}$. The sum of the base-emitter voltages of Q_1 and Q_4 are equal to the sum of those of Q_2 and Q_3. Hence the input voltage appears unattenuated across R_3 and R_4. If R_3 and R_4 are chosen zero, the circuit may oscillate uncontrollably due to parasitic capacitances.

Fig. 7.6 Bipolar input stage with enlarged g_m by compensation

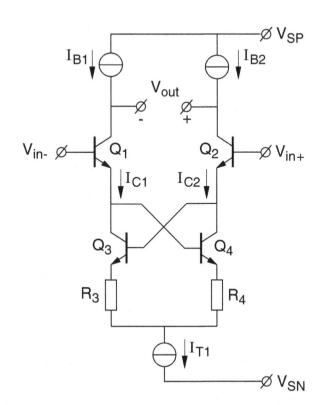

Fig. 7.7 CMOS input stage
with enlarged g_m by
compensation and
$W/L_{3,4} < W/L_{1,2}$

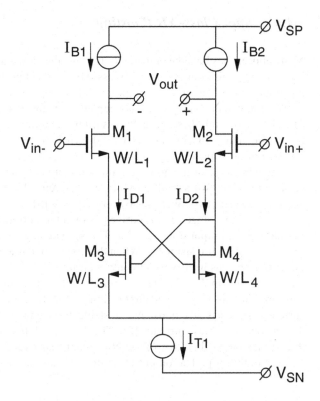

The value of $1/R_{3,4}$ may not be chosen larger than 30 times $g_{ml,2} = 1/r_{el,2}$ to insure
stability. With non-zero R_3 and R_4 the transconductance G_m is:

$$G_m = \frac{\Delta I_{out}}{V_{in}} = \frac{I_{C1} - I_{C2}}{V_{in}} = \frac{-2}{R_3 + R_4} = \frac{-1}{R_{3,4}} \quad (7.13)$$

Note that the input stage does not inverse the signal like it does in a conventional
input stage. This effect appears as a negative value of g_m. The positive feedback in
$Q_{3,4}$ does not work above the transit frequency f_T of the transistors, so the bandwidth cannot be increased by this method. But the g_m at lower frequencies can be
enlarged by a factor 30, and so can the voltage gain A_{vo}.

The CMOS circuit in weak inversion can also be kept stable by series resistors.
But by choosing the $W/L_{3,4}$ ratio of M_3, M_4 smaller than W/L_{12} of M_1, M_2 the
stability may also be assured in strong inversion

$$g_m = \frac{\Delta I_{out}}{V_{in}} = \frac{1}{1/g_{ml,2} - 1/g_{m3,4}} = g_{ml,2}\left(1 - \sqrt{\frac{W/L_{1,2}}{W/L_{3,4}}}\right)^{-1} \quad (7.14)$$

When $W/L_{3,4}$ equals $W/L_{1,2}$ the circuit is uncontrolled. We may not choose
$W/L_{3,4}$ less than 5 % smaller than $W/L_{1,2}$ to assure stability over variations in
transistor parameters. This means an increase of g_m of 20 times.

7.1.6 Input Class-AB Boosting

Many approaches have been made to improve the slew rate and settling time of the
GA-CM configuration by using a class-AB input stage. Its output current may be
much larger than the quiescent current. A first approach is shown in Fig. 7.8a [4].

The input pair M_1M_2 is cross-coupled with the complementary pair M_3 M_4.
Transistors M_5 and M_6 serve as level shifters. The translinear loop through M_1–M_6
biases the transistors in class-AB as shown in Fig. 7.8b. A slew rate has been
reported of 80 V/µs at a supply-power consumption of 1 mW.

A disadvantage of the above circuit is that it needs a supply voltage larger than
three diodes and two saturation voltages. A circuit that can function at one diode
and two saturation voltages is shown in Fig. 7.9 [5].

The first input stage $M_{11}M_{12}$ drives its own tail current through $M_{15}M_{16}M_{17}$ and
also drives the output transistor M_1. The second stage $M_{21}M_{22}$ does the same and
drives the output transistor M_2. When the first stage has a negative input voltage the
tail current boosts itself with the output current.

The current in M_{11} increases strongly, while the current in M_{12} stays nearly
constant. In this way much larger signals can be processed than the quiescent
current. The class-AB boost has a stable loop gain independent of the device
dimensions and bias currents [5]. The P-channel transistors have a W/L ratio
three times larger than that of the N-channel transistors to compensate the g_m for
a lower mobility µ. The output transistors M_1 and M_2 are scaled Bx to provide a

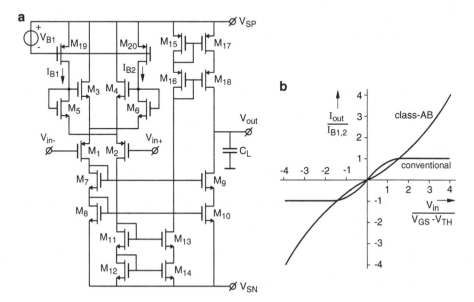

Fig. 7.8 (a) Class-AB cascode current-mirrored (CCM) amplifier and (b) transconductance of the
class-AB input stage

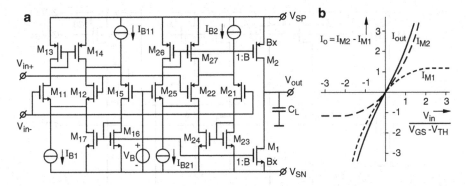

Fig. 7.9 (a) OTA with class-AB biasing and (b) transconductance of the class-AB OTA of (a)

large output current. The ratio between maximum output current I_{OM} and quiescent current I_Q is limited by the supply voltage available to M_{11} and M_{21} according to

$$\frac{V_{OM}}{V_Q} = \frac{1}{4}\left(1 + \frac{V_{SP} - V_{SN}}{V_{GS11,21} - V_{TH}}\right)^2 \tag{7.15}$$

The circuit was used to drive a large capacitive load at a reasonable slew rate and settling time.

Many other approaches to class-AB input stages have been made to combine a low quiescent power consumption with a high current drive capability, slew rate, and settling time.

7.1.7 Voltage-Gain Boosting

The open-loop low-frequency voltage gain of the GA-CF operational amplifier of Fig. 7.2a, b can be increased by voltage gain boosting [6, 7]. The idea is to boost the voltage gain of the cascode transistors M_{14} and M_{16} by measuring their source voltages and regulating them at a constant value by actively controlling their gate voltages. The circuit is drawn in Fig. 7.10.

The amplifier without gain boosting had a voltage gain according to Eq. 7.12: $A_{VOM} = \frac{1}{2}\mu_p\mu_n$. This gain is now boosted with the voltage gain of the differential amplifiers M_{21}, M_{22} and M_{31}, M_{32}, which is $\frac{1}{2}g_m r_{ds} = \frac{1}{2}\mu_P$ or $\frac{1}{2}\mu_N$ multiplied by the voltage gain of the folded cascodes M_{25}, M_{35}, which is μ_n or μ_P. The total DC gain then becomes:

$$A_{VOM} = \frac{1}{4}\mu_p^2\mu_n^2 \tag{7.16}$$

A voltage gain has been reported of 90 dB in combination with a zero-dB frequency of 116 MHz at a load capacitor of 16 pF and a power consumption of 52 mW [6].

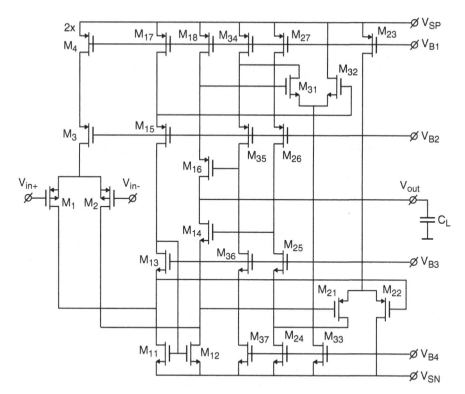

Fig. 7.10 Voltage-gain boosting in a class-A GA-CF operational amplifier

7.1.8   *Conclusion*

The GA-CF configuration has a high bandwidth f_0 with a value in the order of a factor 4 lower than the transit frequency f_T of the transistors. The limited g_m and the class-A biasing makes that no significant resistive loads can be driven at a reasonable gain. The main application of this type of operational amplifier is to drive capacitive loads, as in integrator filters, switched capacitor filters, and sample and hold circuits. A capacitive load does not endanger HF stability, but improves it.

We have seen that the GA-CF configuration can be improved in various ways. Feedforward compensation can improve the bandwidth by a factor 2. Class-AB biasing of the input stage results in a high slew rate and fast settling. Transconductance compensation with an equal but negative value results in a higher low-frequency g_m but no larger bandwidth. Voltage boosting may greatly enlarge the DC voltage gain with pure capacitive loads.

The GA-(CM or) CF configuration will further be used as a basis for the GA-CF-VF configuration in Sect. 7.3.

7.2 GA-GA Configuration

The GA-GA configuration is the first step to profit from a second transconductance gain factor after the input stage. As a result, resistive loads R_L can be driven at a reasonable voltage gain, besides capacitive loads C_L. A direct consequence of two cascaded GA stages is that we have two poles and that one pole has to be compensated by an HF compensation network, which is often a Miller capacitor.

7.2.1 Basic Bipolar R-R-Out Class-A Operational Amplifier

A basic GA-GA bipolar operational amplifier is shown in Fig. 7.11.

The frequency characteristic of the basic GA-GA bipolar operational amplifier of Fig. 7.11 is given in Fig. 7.12.

The limiting pole frequency with the highest specified load capacitor C_L and the transconductance g_{m1} at Q_{11} is roughly:

$$f'_1 = \frac{g_{m1}}{2\pi C_L} \tag{7.17}$$

The zero-dB bandwidth f'_0 must be chosen half f'_1 for a 60° phase margin by choosing the right value for C_{m1}:

$$f'_0 = \frac{g_{m2}}{2\pi C_{M1}} = \frac{1}{2} f'_1 = \frac{1}{2} \frac{g_{m1}}{2\pi C_L} \tag{7.18}$$

Fig. 7.11 Basic bipolar R-R-out class-A GA-GA configuration in bipolar technology

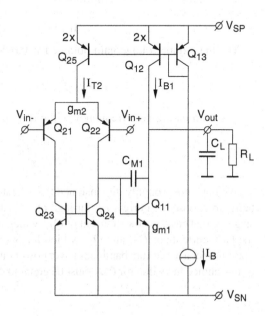

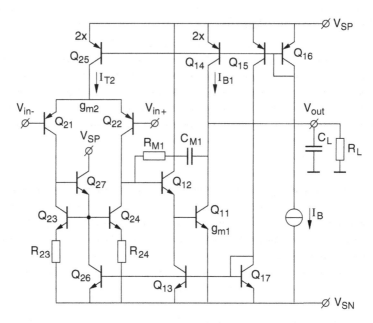

Fig. 7.12 Improved basic bipolar R-R-out class-A GA-GA configuration in bipolar

The value for C_{M1} must be chosen at a value:

$$C_{M1} = 2g_{m2}\frac{C_L}{g_{m1}} = C_L\frac{I_{T2}}{I_{B1}} \tag{7.19}$$

The voltage gain of the amplifier is:

$$A_{V0} = g_{m2}\beta_{11}R_L \tag{7.20}$$

The maximum voltage gain obtained without R_L is:

$$A_{VOM} \approx g_{m2}\beta_{11}\left(r_{ce11}\|r_{ce12}\right) \approx \frac{1}{2}\beta\mu \tag{7.21}$$

The dominating pole frequency without R_L is:

$$f_2' = \frac{f_0'}{A_{VOM}} \tag{7.22}$$

We have now roughly dimensioned the operational amplifier of Fig. 7.11. We are better to choose the current I_{B1} through Q_{11} equal to the tail current I_{T2} of the input stage. In that case, the base current of Q_{11}, which loads Q_{22}, compensates the offset of the base currents of Q_{23} and Q_{24}, which loads Q_{21}. This choice is also close to the optimum for a maximum bandwidth over power ratio as expressed in Eq. 6.27. Because of this choice the value for C_{M1} must be equal to C_L, as follows from Eq. 7.14.

7.2.2 Improved Basic Bipolar R-R-Out Class-A Operational Amplifier

Now we will try to improve the basic amplifier of Fig. 7.11 as shown in Fig. 7.13. The main improvement is to use a Darlington second stage. This enlarges the gain by roughly a factor β_{12} to:

$$A_{V0} \approx g_{m2}\beta_{12}\beta_{11}R_L \tag{7.23}$$

The extra Darlington transistor Q_{12} also requires some extra measures. An important consequence is that there is the extra pole frequency f_{12} in the Miller loop at the emitter of Q_{12}, which is loaded by the base-emitter capacitor C_{be11} of Q_{11}. Its frequency is $f_{12} = 1/(r_{bb11} + r_{e12})2\pi C_{be11}$. In 1975, James Solomon touched this problem in his famous OpAmp tutorial [8]. To place this pole frequency f_{12} at least at f_1' the current through Q_{12} must at least be $I_{B12} = I_{B1}f_1'/f_{T1} = I_{B1}C_{be2}/C_L$, and the bulk base resistance r_{bb11} of Q_{11} must be made small by a fingered structure of the emitter of Q_{11}.

To eliminate the extra phase shift by the right-half-plane zero $f_z' = g_{m1}/2\pi C_{M1}$ (Eq. 6.24), a resistor R_{M1} is placed in series with C_{M1}. Its value is chosen such that the feedforward path through C_{M1} is canceled by the g_m of Q_{11}:

$$R_{M1} = r_{e11} = 1/g_{m11} \tag{7.24}$$

To balance the asymmetrical load current i_{B12} at one side of the input stage, an equal load i_{B27} of the Wilson mirror transistor Q_{27} is used at the other side. The bias

Fig. 7.13 Frequency characteristic of the GA-GA operational amplifier of Fig. 7.11

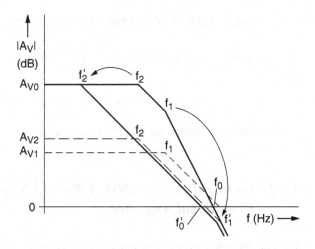

Fig. 7.14 Basic class-A
R-R-out GA-GA
configuration in CMOS

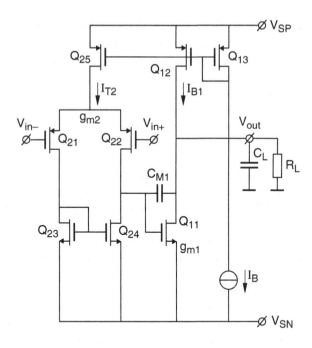

currents through Q_{12} and Q_{27} must be equal to prevent an induced input voltage offset. To further prevent the full offset and noise of Q_{23} and Q_{24} being added to the input offset and noise voltage, the emitter resistors r_{e23} and r_{e24} of these transistors have to be degenerated by R_{23} and R_{24}.

7.2.3   Basic CMOS R-R-Out Class-A Operational Amplifier

In CMOS the situation is different. The voltage gain of the basic circuit of Fig. 7.14 is:

$$A_{VOM} = g_{m2} \left(2r_{ds22} \| r_{ds24} \right) g_{ml} R_L \approx \mu_2 g_{ml} R_L \qquad (7.25)$$

The maximum voltage gain without R_L is $\mu_2 \mu_1$.

7.2.4   Improved Basic CMOS R-R-Out Class-A Operational Amplifier

The current gain can not be improved by a Darlington in CMOS, but the voltage gain can be enlarged by cascoding the output transistors. This is shown by the circuit of Fig. 7.15.

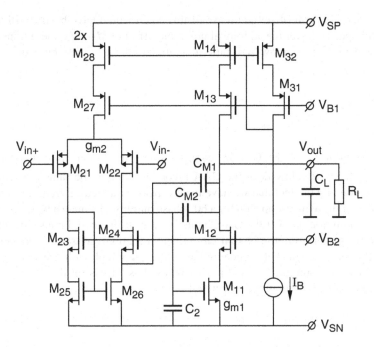

Fig. 7.15 Improved class-A R-R-out GA-GA configuration in CMOS

Inserting the cascodes M_{12} and M_{13} improves the maximum voltage gain without R_L with a factor μ_3 to:

$$A_{VOM} = \mu_2\mu_1\mu_3 \qquad (7.26)$$

With CMOS low-capacitive loads are often applied. This increases the influence of C_M and its corresponding right-half plane zero f'_z, as discussed with Fig. 6.18.

To solve this problem we split C_M into C_{M1} and C_{M2} and connect the left-hand side of C_{M1} to the source of the cascode transistor M_{24}. This cuts the feedforward path through C_{M1} so that f'_z is eliminated. However, the limited g_{m24} of M_{24} causes another zero at $f_z = g_{m24}/2\pi C_{M1}$ in the Miller loop, as at high frequencies the current through C_{M1} is hindered to flow through M_{24} and does not counteract the driving current from M_{22}. Several measures can be taken to repair this zero. A good compromise is to use the $C_{M1}/2$ smaller split-off part C_{M2} as a regular Miller capacitor.

The connection of C_{M1} to the cascode transistor M_{24} introduces a voltage gain C_{M1}/C_2 in the Miller loop, with C_2 as the parasitic capacitance at the gate of M_{11}. This loop consists of three poles, one at each output of M_{24}, M_{11}, and M_{12}. The pole at the gate-source capacitance C_2 of M_{11} dominates. If the HF stability of the loop is insufficient, C_2 can artificially be enlarged.

If we had not used the cascoded current mirror to connect the left-hand side of C_M but rather a separate folded cascode chain in parallel to the input stage as in Fig. 7.2b (which can be done to lower the minimum supply voltage from $3V_{SAT} + 1V_{TH}$ to

$2V_{SAT} + 1V_{TH}$ [9]), all offset and noise of this chain would have been added to the input voltage. To lower the additional noise and offset of M_{25}, M_{26} these transistors must have a low g_m at a low W/L ratio in relation to those of the input pair.

7.2.5 Conclusion

The bandwidth f'_0 of the GA-GA configuration with Miller compensation lies roughly a factor 2 below f'_0 of the GA-CF configuration at the same bias current. This is caused by the fact that we have to stay away by a factor 2 from the zero-dB frequency of the Miller loop gain for a 60° phase margin. The main advantage of the GA-GA configuration over the GA-CF one is a β or β^2 increased transconductance of the bipolar or Darlington bipolar circuit, and a μ improved transconductance of the CMOS circuit. This helps keeping the voltage gain high with resistive loads.

The GA-GA configuration will further be used as a basis for the description of the GA-GA-VF configuration in Sect. 7.4.

7.3 GA-CF-VF Configuration

The lack of a reasonable transconductance of the GA-CF configuration of Sect. 7.1 and the reduction of bandwidth of the GA-GA configuration of Sect. 7.2 makes us look to the combination of the GA-CF configuration with an output VF buffer for a larger current gain, while maintaining as much of the high bandwidth of the GA-CF topology as possible.

7.3.1 High-Speed Bipolar Class-AB Operational Amplifier

An example of a high-speed bipolar circuit is shown in Fig. 7.16. It has a class-AB VF output stage with folded Darlington transistors [10].

The current gain $\beta_n \beta_p$ of this VF stage is linear and fairly symmetric because each push and pull path has an NPN and a substrate PNP transistor in cascade, the total transconductance is:

$$g_{mtot} = g_{m3} \beta_n \beta_p = g_{m3} \beta^2 \qquad (7.27)$$

The limiting pole frequency f_l lies at the output, and has a value of:

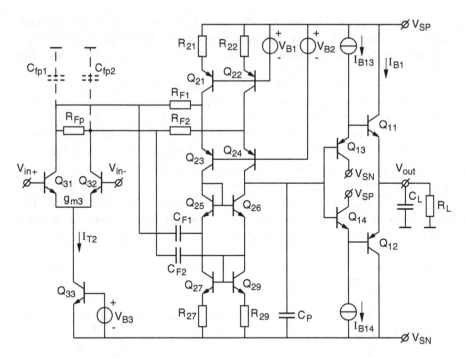

Fig. 7.16 High-speed GA-CF-VF configuration with a class-AB folded Darlington output stage

$$f_1 = \frac{g_{ml}}{2\pi C_L} = \frac{2I_{Bl}q}{2\pi C_L kT} = \frac{I_{Bl}}{\pi C_L V_T} \tag{7.28}$$

with: $g_{ml} = g_{ml1} + g_{ml2}$.

The two transconductances of the output transistors g_{ml1} and g_{ml2} are in parallel to each other. This helps to keep the quiescent current I_{Bl} a factor 2 lower for a certain bandwidth than with a single VF transistor.

Moreover, the phase shift of the VF output stage is reduced by the feedforward path of its base-emitter capacitances. This means a zero in the left-hand plane at the f_T of these transistors. We do have to keep the bulk base resistors r_{bb} low by a fingered layout of the emitter. The zero-dB bandwidth f_0' of the whole amplifier must be chosen again at half of f_1 by the choice of C_P.

$$f_0' = \frac{g_{m3}}{2\pi C_p} = \frac{1}{2} f_1 = \frac{g_{ml}}{4\pi C_L} = \frac{I_{Bl}}{2\pi C_L V_T} \tag{7.29}$$

$$C_p = C_L = \frac{2g_{m3}}{g_{ml}} \tag{7.30}$$

The quiescent currents through the Darlington transistors Q_{13} and Q_{14} should be at least $I_{B13} \geq I_{B11}C_{bel1}/C_L$, and $I_{B14} \geq I_{B12}C_{bel2}/C_L$, to avoid a second limiting pole in the Darlington output stage, with I_{B11} and I_{B12} the bias currents of the output transistors Q_{11} and Q_{12}, respectively.

A disadvantage of the class-AB VF output buffer is that its nonlinear diode voltage characteristic is outside the grip of the compensation capacitor C_P. Therefore, at high frequencies below f'_0, where the external loop gain is low, the distortion will be relatively high.

Another disadvantage of the Darlington VF output buffer is that it might peak or even oscillate at high frequencies when the amplifier is heavily capacitively loaded. Solomon calls this the output bump [8]. The situation is depicted in Fig. 7.17.

The PNP output emitter follower Q_2 shows a negative input conductance $-G_T = 2\pi f^2 C_L / f_{T2}$. While the Darlington emitter follower Q_4 shows an inductive output $L = R_{B4}/2\pi f_{T4}$. Together with the interstage capacitor C_p the circuit gives rise to a resonance LC circuit at a resonance frequency of $f_p = (1/2\pi)\sqrt{LC_p}$ with a poor damping. The effect is shown in the amplitude characteristic of Fig. 7.18. The design parameters must be chosen such that the top of the peak remains sufficiently below 0 dB, by choosing a fingered layout for the emitter of Q_4 to lower R_{B4}.

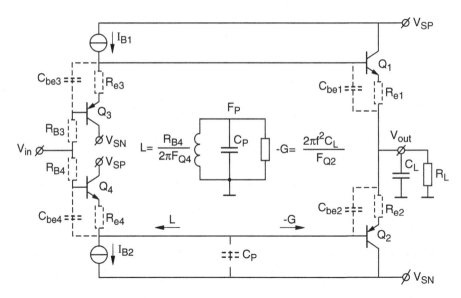

Fig. 7.17 Frequency peaking of the VF Darlington output stage

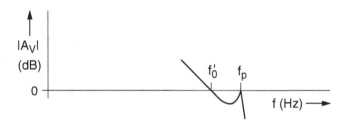

Fig. 7.18 Amplitude characteristic of the frequency peaking in a Darlington VF output buffer

The input and folded-cascode stages have been sufficiently described in Sect. 7.1. It will be noted that into the all-pass current network $R_F C_F$ behind the input stage a parallel resistor R_{fp} has been inserted. This is used to compensate at low frequencies the attenuation caused by the parasitic substrate capacitors C_{fp1} and C_{fp2} at the left-hand side of C_{F1} and C_{F2} which take away about 20 % of the HF current. The resulting pole-zero doublet has a value of 1 or 2 %. It is important that the remaining pole-zero doublet is small as this appears in the overall frequency characteristic of the whole amplifier and results in a slow settling component in the step response of the OpAmp.

The turn-over frequency of $R_F C_F$ must be made more than three times lower than the f_T of the PNP cascode. The bandwidth of the level shift stage is equal to that of the f_T of the NPN transistors, loaded with the parasitic parallel capacitors C_{fp1} and C_{fp2} of the capacitors C_{F1} and C_{F2}, respectively.

The all-pass network and its pole-zero doublet can be completely eliminated when the bipolar process allows for vertical PNPs with high f_T. This is of advantage for a short settling time. A settling time to 0.1 % of 100 µs was reported [10] with a slew rate of 375 V/µs, and a GBW of 60 MHz. CMOS versions of the GA-CF-VF configuration have been made in early years, but the relatively low g_m of the VF output buffer results in a low limiting pole frequency. Moreover, the large voltage loss across the diodes of the VF buffer makes the CMOS variant less attractive.

7.3.2   High-Slew-Rate Bipolar Class-AB Voltage-Follower Buffer

A very special simplification of the GA-CF-VF configuration is the high slew-rate voltage follower of Fig. 7.19 with an GA-VF configuration. Because the input is always connected to the output, there is no need for a level shift stage. When we replace the compensation capacitor C_P with C_C between the collector and emitter of Q_{21}, the compensation still functions, but the slew rate is not reduced by C_C because it is bootstrapped with the CM signal voltage. The slew rate is limited by the parasitic capacitance C_P.

An asymmetric slew-rate can be observed in all OpAmps connected as voltage followers by the effect of the parasitic tail capacitor C_T of the input stage. If the input voltage goes down, the capacitor decreases the tail current I_{T2} and the slew rate decreases. But if the input voltage goes up, the slew rate increases. The downwards slew rate of the voltage follower is limited to:

$$S_{rdown} = I_{B2} / (C_p + C_T) \qquad (7.31)$$

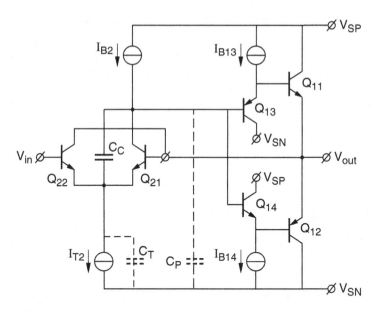

Fig. 7.19 High-slew-rate voltage follower in GA-VF configuration

7.3.3 Conclusion

The high bandwidth of the GA-CF configuration combined with a high bipolar transconductance VF output buffer raises an excellent bipolar amplifier with an GA-CF-VF configuration. The bandwidth and slew rate can be chosen high. The settling time is particularly high when the bipolar process allows for complementary vertical PNPs with high f_T, so that the all-pass network and its pole-zero doublet can be eliminated. But still a high bandwidth can be obtained. CMOS versions of the GA-CF-VF configuration are less attractive because of the high voltage loss and relative low transconductance of the VF buffers at high output currents.

7.4 GA-GA-VF Configuration

Most of the favorite classic bipolar operational amplifiers have been made in the GA-GA-VF configuration. These include the μA741 [11], and many others.

It was about this configuration that James E. Solomon wrote his famous tutorial study in 1974 [8]. It is still a good overview paper about this type of OpAmp in bipolar technology.

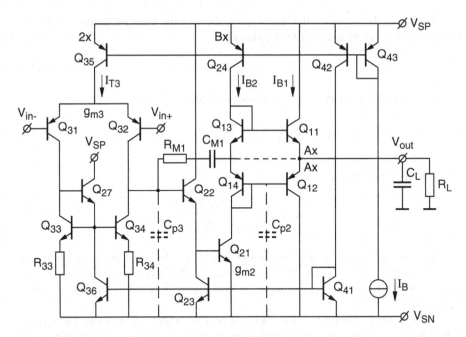

Fig. 7.20 General class-AB GA-GA-VF configuration in bipolar technology with Miller compensation (MC)

7.4.1 General Bipolar Class-AB Operational Amplifier with Miller Compensation

The general topology is shown in Fig. 7.20. It is composed of the GA-GA circuit of Fig. 7.13 and the basic VF output stage of Fig. 5.7a. Most of the HF behavior is already described with the GA-GA stage, except for the addition of the limiting pole frequency of the output stage, which is described with the GA-CF-VF circuit of Fig. 7.16.

In overviewing the HF behavior we have two limiting pole frequencies. One at the capacitive loaded VF stage:

$$f_1 = \frac{g_{m1}}{2\pi C_L} = \frac{I_{B1}}{\pi C_L V_T} \tag{7.32}$$

with: $g_{m1} = g_{m11} + g_{m12}$.

And the other limiting frequency at the Miller compensated output of the intermediate stage:

$$f_2 \approx \frac{g_{M2}}{2\pi (C_{p2} + C_{M1})} = \frac{I_{B2}}{2\pi (C_{p2} + C_{M1}) V_T} \tag{7.33}$$

with $g_{m2} = g_{m21}$.

When we choose to let the load capacitance set the limiting frequency, f_1, the other one f_2 must be made at least twice f_1:

$$f_2 \geq 2f_1 \tag{7.34}$$

The zero-dB frequency must be chosen half of f_1:

$$f_0' = \frac{1}{2}f_1 = \frac{I_{B1}}{2\pi C_L V_T} \tag{7.35}$$

And the value of C_{M1} becomes:

$$C_{M1} = \frac{g_{m3}}{2\pi f_0'} = \frac{I_{T3}/2}{I_{B1}}C_L \tag{7.36}$$

The parasitic pole in the Darlington pair Q_{21}, Q_{22} must be coped with, as described with the GA-GA stage. The value of R_{M1} must be chosen closely to $1/g_{m21}$ to eliminate the zero in the right half complex plane along with its extra phase shift.

An important drawback of the connection of the Miller capacitor to the input of the VF output stage is that the nonlinearity of the output-stage's voltage transfer is not suppressed by the Miller loop. For that reason a relatively high nonlinear signal component can be found at high frequencies where the external loop gain is low. This was also the case with the GA-CF-VF configuration. However, now we can try to incorporate the output buffer into the active Miller loop, by connecting C_{m1} to the final output V_{out} (see dotted line in Fig. 7.20). Stability of the Miller loop must now be ensured while the limiting pole frequency f_1 is incorporated into the Miller loop. The Miller loop becomes an GA-VF stage in itself, which must be compensated to:

$$f_{02}' = \frac{1}{2} f_1 \tag{7.37}$$

For this purpose the parasitic capacitor C_{p2} must be artificially enlarged to:

$$C_{p2} = \frac{2g_{m2}}{2\pi f_1} = \frac{I_{B2}}{I_{B1}}C_L \tag{7.38}$$

The bandwidth of the whole amplifier must now again be reduced a factor 2 for a phase margin of 60° to:

$$f_0' = \frac{1}{2} f_{02}' = \frac{1}{4}f_1 \tag{7.39}$$

This leads to a choice of C_{m1}:

$$C_{MI} = \frac{g_{m3}}{2\pi f_0'} = \frac{I_{T3}}{I_{BI}} C_L \qquad (7.40)$$

Though the external loop gain is reduced by a factor of 2, the loop gain around the output buffer is roughly increased by a factor C_{MI}/C_{p2} at high frequencies. This may improve the HF linearity by factor 2–10. It must be noted, however, that the slew-rate is reduced by a factor 2, according to the reduction of the bandwidth. So the nonlinearity due to the input stage increases at high frequencies (see also Sect. 6.4, and [12]).

An alternative solution to enlarging the parasitic capacitor C_{p2} is to split C_{ml} at its right hand, for instance, partly 2/3 to the final output, and partly 1/3 to the input of the output stage, like in forward nested Miller compensation (see Fig. 6.35). This solution does not reduce the bandwidth by a factor 2.

7.4.2 μA741 Operational Amplifier with Miller Compensation

An equivalent circuit of the μA741 operational amplifier of Fairchild [8, 11] is shown in Fig. 7.21.

The VF-CF connected input pair functions like an GA stage, only it is non inverting (NIGA). The NPN VF pair Q_1, Q_2 reduces the input bias current by its large current gain β to 80 nA at a collector bias current of 10 μA. The g_m is half that of a regular pair because we have four emitters in series, instead of two emitters. A

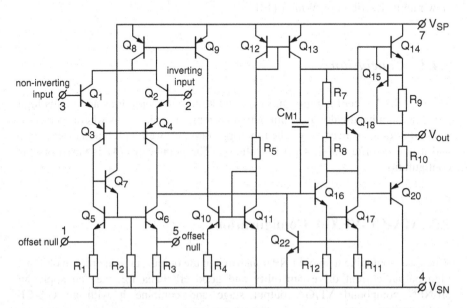

Fig. 7.21 Equivalent circuit of the μA741 GA-GA-VF operational amplifier with Miller compensation (MC)

common-mode feedback loop senses the sum of the collector currents of the NPN pair and compares it through a current mirror Q_8, Q_9 with the current of a current source Q_{10}. By this loop the common-base point of the PNP CF pair Q_3, Q_4 is controlled. In this way the voltage of the common base point follows the CM input voltage at the right bias.

The resistors R_1 and R_3 degenerate the current mirror load in order to lower its offset and noise contribution. The DC load current of the input stage is balanced by equal base currents through Q_7 and Q_{16} in order to avoid offset.

The biasing of the output transistors is performed by a current source Q_{13} and a transistor Q_{18} connected with R_7 and R_8 as a diode with about two V_{BE} voltages. The biasing current of the whole amplifier of 70 µA is derived from R_5.

A short-circuit current limiting transistor Q_{15} compares its V_{BE} with the voltage across a resistor R_9 which measures the upper output current. If the output current is larger than 25 mA, the driving of the upper output transistor Q_{14} is taken away.

The drive current of the lower output transistor Q_{20} is limited by transistor Q_{22} which similarly measures the current through R_{11}. The limited current gain β_{20} of the lower substrate output PNP Q_{20} further limits its current to 25 mA.

The intermediate Darlington stage Q_{16}, Q_{17} is Miller compensated by C_{m1} of 30 pF. The bandwidth f'_0 is slightly less than 1 MHz. The double voltage compliance $V_{in\ max}$ of the VF-CF input stage allows a two times larger slew rate S_r of 0.5 V/µs than Eq. 6.64 indicates. The double voltage compliance, on the other hand, increases any input offset voltage with a factor 2.

The two GA stages in cascade provide an abundance of DC gain A_0, which is in the order of 2.10^5.

A well-elaborated realization with a class-AB input stage that provides a large slew rate is described by Widlar [13].

7.4.3 Conclusion

The GA-GA-VF configuration has been made in many variations and yields high gain, medium bandwidth, medium output current, in applications where no rail-to-rail output range is necessary. The topology combines well with bipolar transistors and it is a favorite classic OpAmp solution. The popular µA741 is based on this configuration.

7.5 GA-CF-VF/GA Configuration

When we would like to avoid PNP transistors in the output stage in order to obtain a high output current drive capability and good HF behavior, we can apply an all-NPN compound VF/GA output stage and combine it with an GA-CF configuration.

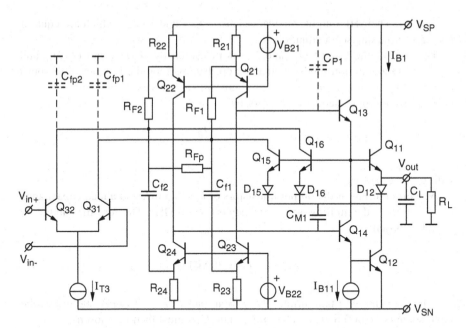

Fig. 7.22 High-frequency "All-NPN" GA-CF-VF/GA configuration with parallel compensation (PC) and Miller compensation (MC)

An in-between approach towards the GA-CF-VF/GA configuration would be to use the GA-CF-VF configuration of Fig. 7.16 in combination with one of the many semi-complementary compound output stages with feedforward biasing of Figs. 5.14–5.16, or with feedback biasing of Figs. 5.31 and 5.32.

However, the signal path through the PNP transistor, which is still present in these semi-complementary compound output stages impairs the HF behavior of the lower output transistor and lowers its limiting pole frequency.

7.5.1 High-Frequency All-NPN Operational Amplifier with Mixed PC and MC

A better approach is to optimally utilize the large current and HF capability of the NPN transistors in an "all-NPN" GA-CF-VF/GA as shown in Fig. 7.22.

Figure 7.22 shows the combination of the GA-CF-VF configuration of Fig. 7.16 combined with the VF/GA output stage of Fig. 5.33. The high-frequency path is made up of only NPN transistors [1].

It is interesting to see how similarly the upper VF output Darlington pair Q_{11}, Q_{13} functions in regard to the lower GA output Darlington pair Q_{12}, Q_{14} when both these pairs are driven by current sources from the outputs of the CF intermediate stages. Their HF transadmittances are $2\pi C_{P1}$ and $2\pi C_{M1}$, respectively, are equal if

$C_{P1} = C_{M1}$. Their HF output impedances are $1/g_{m11}$ and $1/g_{m12}$, which are equal if $I_{11} = I_{12}$ in the quiescent situation.

The sum of the V_{BE} of Q_{11} and D_{12} is measured by $Q_{15}D_{15}$ and $Q_{16}D_{16}$ and regulated in a common-mode feedback loop. By this loop the product of the push and pull output currents is regulated at a constant value, as explained with Fig. 5.33.

The limiting pole frequency f_1 is set by:

$$f_1 = \frac{g_{m1}}{2\pi C_L} = \frac{I_{B1}}{\pi C_L V_T} \tag{7.41}$$

with $g_{m1} = g_{m11} + g_{m12}$.

At a bias current of the output transistors of $I_B = 0.5$ mA and a load capacitance of $C_L = 100$ pF, the limiting pole frequency is $f_1 \approx 60$ MHz. The zero-dB bandwidth must be chosen:

$$f_0' = \frac{1}{2} f_1 \approx 30\,\text{MHz} \tag{7.42}$$

When we choose a tail current of the input pair of $I_{T3} = 100$ µA, the g_{m3} of the input stage is $1/1000$ S. The value of C_{P1} and C_{M1} must then be chosen:

$$C_{P1} = C_{M1} = \frac{g_{m3}}{2\pi f_0'} = 6\,\text{pF} \tag{7.43}$$

The currents must be chosen slightly higher than calculated. At the output, there is the extra series diode impedance of D_{12} which lowers the limiting pole frequency of the lower Miller-compensated output transistor. So the output bias current must be chosen about 30 % larger to obtain the right g_{m1}.

The input tail current must be chosen 20 % larger to overcome the leakage into the parasitic capacitors C_{fp1} and C_{fp2}. These are partly caused by the substrate sides of the feedforward capacitors C_{F1} and C_{F2}. At low frequencies the gain is equally reduced by a shunt resistor R_{FP}.

One of the problems of Darlington output transistors is their peaking, as we have shown with Fig. 7.17. This is particularly the case when we wish to deliver large output currents, i.e., 100 mA or higher. At high output currents, the output Darlingtons have a parasitic pole in between the two stages, caused by the emitter impedance of the first transistor and the diffusion capacitor C_{BE}, which becomes large at high currents.

To cope with this the bias current I_{B11} of the first transistors must be high, and the bulk base resistance of the first transistor low by a fingered layout.

A general cure is to bypass the first transistor of the Darlington pair for high frequencies, as is shown in Fig. 7.23.

The value of the upper turnover frequency $f_H = 1/2\pi\,R_F C_F$, and of the lower on $f_L = 1/2\pi\,\beta_1 R_F C_F$.

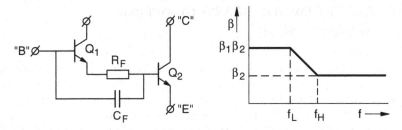

Fig. 7.23 HF bypassing of the first transistor of a Darlington pair to cure the HF peaking

The offset and noise of the operational amplifier is somewhat degraded by the offset and noise of the current source resistors R_{21} through R_{24} and the class-AB regular transistors Q_{15} and Q_{16}, which add with their current offset and noise to the equivalent input voltage offset and noise.

7.5.2 Conclusion

The "all-NPN" GA-CF-VF/GA configuration of Fig. 7.22 is an efficient circuit in regard to bandwidth/power ratio, disregarding the diode voltage losses at the output. It has a large programmable quiescent current range.

The amplifier works perfectly at very low quiescent currents in the order of 10 µA, while the bandwidth depends on the load capacitor. Even at high quiescent currents of the order of 10 mA, without Darlingtons in the output stage, the amplifier has been used up to 1 GHz in a 50 Ω source and load environment [14]. However, at a 50 Ω load and no Darlington transistors, the voltage gain is restricted to 40–50 dB. To provide a larger gain we have to insert an GA intermediate stage, as we will describe in the next chapter.

The input voltage noise of the amplifier is somewhat degraded by the extra noise of the current source resistors in the intermediate stage and the class-AB control transistors.

7.6 GA-GA-VF/GA Configuration

When we want more gain, lower voltage noise and a lower offset voltage than with the previous GA-VF-VF/GA configuration, it is better to shield the noise and offset sources of the output stage from the input stage by an GA intermediate gain stage. We then obtain the GA-GA-VF/GA configuration.

7.6.1 LM101 Class-AB All-NPN Operational Amplifier with MC

The design of the famous LM101 was described by Widlar in 1967 as a monolithic OpAmp with simplified frequency compensation [15]. Before that time the architecture of operational amplifiers resembled some of the limitations of the early tube versions. Widlar reduced the high-frequency compensation scheme to a simple Miller compensation. Though the output stage principally has a compound VF/GA configuration with all-NPN output transistors Q_{16} and Q_{11}, the stage has the overall function of a semi-complementary full voltage-follower VF stage. At the lower output side the voltage-follower VF function of Q_{12} dominates, while being boosted by the GA slave-connected transistor Q_{11}. At the upper output side we have the folded-Darlington VF pair Q_{13}, Q_{16}. The total compound stage has earlier been described with Fig. 5.16. The frequency compensation could be reduced to one Miller capacitor C_M which has been connected across the intermediate GA Darlington pair Q_9 and Q_{10}. Its frequency compensation has already been described with Figs. 6.14 and 7.20.

The input stage consists of a non-inverting general amplifier (NIGA). It uses NPN input transistors Q_1 and Q_2 with good β for a low input bias current. These input transistors are connected as voltage followers. The complementary PNP transistors Q_3 and Q_4 function as cascodes to the emitters of the input transistors. The whole functions like a differential PNP input pair with an input bias current of that of an NPN pair. Only, the offset voltage is larger than that of a single differential pair because the offset of the NPN and PNP transistor pairs are summed. The input stage is biased by the diode-connected secondary collectors of Q_3 and Q_4 in combination with a current source Q_{21}. These diode-connected secondary collectors and the primary collectors of Q_3 and Q_4 can be seen as a pair of current mirrors.

These mirrors are biased at the "tail" of the diode connected secondary collectors by the collector current source of Q_{21}. The three-transistor mirror Q_5 through Q_7 provides for a single-ended output of the input stage. This structure provides for equal early voltages for the output of the input transistors at Q_3 and Q_4 and the mirror load current sources Q_5 and Q_6. Also, the base current load of Q_9 is compensated by the base current load of Q_7.

The other transistors provide bias and protection functions. The junction FET Q_{18} functions as a biasing resistor to the bias generators $Q_{19}Q_{20}$, $Q_{21}Q_{22}$, and the multiple-collector transistor Q_{17}. The input transistors are protected against reverse base-emitter voltages by the reverse-connected diodes Q_{23} and Q_{24}.

Output current limitation is provided by the resistors R_{10} and R_{11} in combination with the translinear loop of $Q_{11}Q_{12}Q_{14}Q_{16}$ for positive output currents and with the base-emitter voltage of Q_{15} for negative going output currents. The voltage divider R_7, R_8 in combination with Q_8 limits the drive current of Q_{11}. This would otherwise become too large in regard to the drive current of the upper output transistor Q_{16}. The latter is limited by the bias current source provided by a collector of Q_{17}.

The LM101 has a bandwidth of about 1 MHz, and a straight 6 dB/oct roll-off from a DC gain of 100 dB (see Fig. 7.24).

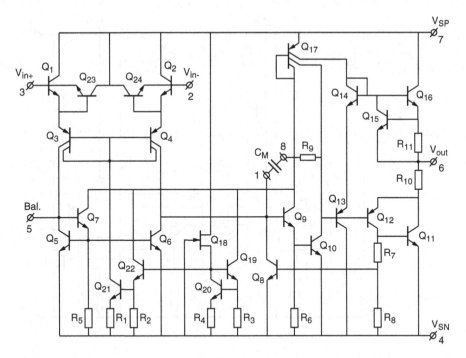

Fig. 7.24 Equivalent circuit of the LM101 Class-AB all-NPN operational amplifier with GA-GA-VF/GA configuration and Miller compensation (MC)

7.6.2 NE5534 Class-AB Operational Amplifier with Bypassed NMC

The classic NE5534 audio OpAmp with GA-GA-VF/GA configuration is depicted in Fig. 7.25 [16]. The input stage consists of Q_1 and Q_2. The intermediate stage has emitter-resistor degenerated transistors Q_8 and Q_{11} loaded with a mirror Q_{61}, Q_{62}, and Q_{10} at E_2. The output stage at the GA side consists of the Darlington pair Q_{13}, Q_{17}, while at the VF side Q_{21} is driven by Q_{17} through the class-AB biasing network D_4, Q_{15} and D_5, as explained with Fig. 5.30. A direct HF path for driving Q_{21} is provided by the collector connection of Q_{13} to the base of Q_{21}. Transistors Q_{19} and Q_{20} limit the positive and negative output currents respectively. The limits are set by the resistor values of R_{18} and R_{19} respectively. Q_{10} at E_1 and Q_{16} are clipping diodes, preventing the saturation of Q_{17} when overdriven.

Principally, the NE5534 is a GA-GA-GA configurated OpAmp with a VF upper output transistor functioning as a slave. So we can regard this circuit as a special case of the GA-GA-GA configuration.

The HF compensation scheme can best be denoted by a capacitive feedforward coupled nested Miller compensation. The latter has been described with Fig. 6.22.

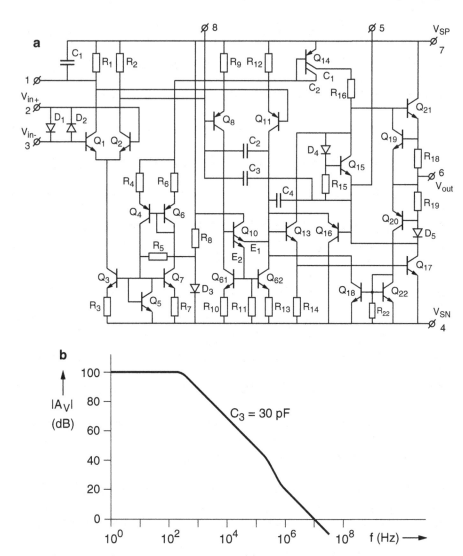

Fig. 7.25 (a) The GA-GA-VF/GA configuration of the classic NE5534 OpAmp and (b) Frequency characteristic of the classic NE5534 OpAmp with pole-zero doublet

The inner nest is compensated by C_4. The outer nest is compensated by C_3. Capacitor C_1 is a balancing capacitor for providing virtual ground to the reference input side of the differential intermediate stage to prevent a zero in the transfer of the Miller compensated intermediate stage far below its regular zero $f_z = g_{m2}/2\pi C_3$.

The nested Miller compensation would have been adequate in itself if the bandwidth of the intermediate stage with the lateral PNP transistors Q_8 and Q_{11} was not so low.

To overcome this problem a multipath according to Fig. 6.24 should have been used. But instead an HF feedforward capacitor C_2 has been inserted between the input and output of the intermediate stage so that this stage is overbridged at higher frequencies. The amplifier can now be compensated for a zero-dB bandwidth of 10 MHz. However, the overbridging of the intermediate stage changes the nested Miller compensation network at high frequencies into a single Miller compensated output stage, as C_3 and C_4 effectively become connected in parallel by C_2. This evokes a pole-zero doubled at 300 kHz where the gain is 30 dB in the frequency characteristic of the NE5534, as shown in Fig. 7.25b. This means that the amplifier has a low phase margin, if utilized at a feedback gain of about 30 dB where the roll-off of the amplitude characteristic is larger than 20 dB/dec. This makes the stability tricky in a tone-control circuit, where the feedback around the OpAmp is variable. Moreover, a slow settling component will distort the step response if used at a bandwidth above 300 kHz.

7.6.3 Precision All-NPN Class-AB Operational Amplifier with NMC

A better approach is to insert an NPN GA intermediate stage into the "all-NPN" GA-CF-VF/GA configuration of Fig. 7.22 and to use a pure Nested Miller Compensation scheme, as explained with Fig. 6.22. The result is presented in the circuit of Fig. 7.26 [17].

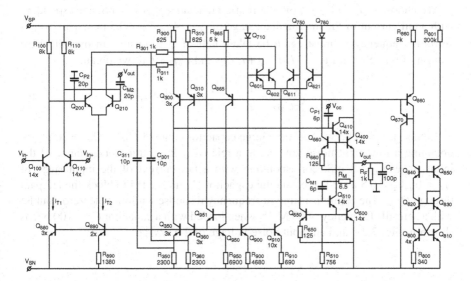

Fig. 7.26 Precision class-AB all-NPN operational amplifier with GA-GA-VF/GA configuration and nested Miller compensation

The input stage consists of Q_{100} and Q_{110}. The intermediate stage has Q_{200} and Q_{210}. The "all-NPN" level-shift stage consists of Q_{300}–Q_{360}, and was explained with Fig. 7.16. The all-NPN Darlington VF/GA output stage consists of Q_{400}, Q_{410} and Q_{500}, Q_{510}. The feedback-biasing class-AB regulating circuit consists of Q_{601}, Q_{602}, Q_{611}, Q_{621} and Q_{650}, Q_{660} together with the diode-connected transistors Q_{710}, Q_{750}, Q_{760}, and has been described with Fig. 5.34.

The Nested Miller Compensation has C_{M1} as an inner Miller capacitor on the GA output side, while C_{P1} balances C_{M1} for the VF output side. The outer nest consists of capacitor C_{M2}, while C_{P2} balances C_{M2} at the reference input of the intermediate stage. Without the capacitor C_{P2} a bypass for high frequency would exist, and a zero appears in the frequency characteristic. Starting out with a quiescent current of 2 mA through the output transistors Q_{400} and Q_{500}, both emitters have a resistance of $r_e = 1/g_m = V_T/I_E = 12.5\,\Omega$. As both push and pull stages are in parallel regarding the load capacitor $C_L = 100$ pF, we obtain a limiting pole frequency at the output of:

$$f_1 = \frac{g_{m1}}{2\pi C_L} \approx 250\,\text{MHz} \tag{7.44}$$

with $g_{m1} = g_{m400} + g_{m500}$.

The zero-dB frequency f_0' of the intermediate and Miller compensated output stage must be half f_1 to obtain 60° phase margin when the loop is closed by the second nest. So with $g_{m2} = 1/(r_{e200} + r_{e210})$ and $C_{M1} = C_{P1}$, we obtain:

$$f_0' = \frac{g_{m2}}{2\pi C_{M1}} = \frac{1}{2}f_1 \approx 125\,\text{MHz} \tag{7.45}$$

We choose $C_{M1} = C_{P1} = 6$ pF at a transconductance $g_{m2} \approx 5$ mS, corresponding with a tail current of the intermediate stage of $I_{T2} = 4g_{m2}V_T \approx 0.5$ mA. The overall zero-dB frequency f_0' must again be taken half of f_0' to ensure an overall phase margin of 60°. So with $g_{m3} = 1/(r_{e100} + r_{e110})$ and $C_{M2} = C_{P2}$ we obtain:

$$f_0'' = \frac{g_{m3}}{2\pi C_{M2}} = \frac{1}{2}f_0' = \frac{1}{4}f_1 \approx 60\,\text{MHz} \tag{7.46}$$

In this case we choose a relative large capacitor value of $C_{M2} = C_{P2} = 20$ pF for obtaining a low input noise voltage, at a relatively large transconductance of the input stage $g_{m3} \approx 8$ mS, corresponding with a tail current of the input stage of $I_{T1} = 4g_{m}V_T = 0.8$ mA. The balancing capacitor C_{P2} is needed to block the HF path through Q_{200}. The result is a straight frequency response without pole-zero doublet with a zero-dB frequency of 60 MHz, when loaded with a capacitor $C_L < 100$ pF, as shown in Fig. 7.27, and an input noise voltage of 2 nV/√Hz.

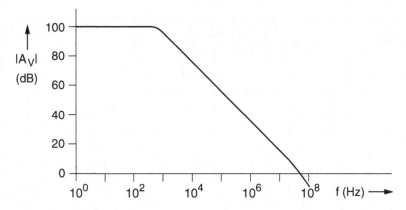

Fig. 7.27 Frequency response of the GA-GA-VF/GA configuration with nested Miller compensation of Fig. 7.26

7.6.4 Precision HF All-NPN Class-AB Operational Amplifier with MNMC

Comparing the Nested Miller Compensated GA-GA-VF/GA configuration of Fig. 7.26 with the simpler Miller compensated GA-CF-VF/GA configuration of Fig. 7.22, we see an increased low-frequency gain, a lower offset and noise, but half the maximum obtainable bandwidth under the condition that the quiescent output currents were chosen equal. The question arises: is it possible to have the high gain and low noise of the three-stage amplifier, while maintaining the high bandwidth of the two-stage amplifier at an equal quiescent current? A feedforward capacitor which bridges the intermediate stage, as in the NE 5534, results in a strong pole-zero doublet, as shown with Fig. 7.25. So then the Multi-path Nested Miller compensation technique (MNMC), as explained with Fig. 6.24, presents an adequate solution. The idea is to have a two-stage amplifier independently in parallel with a three stage amplifier. With this topology one can have the best of the two and three-stage amplifier without strong pole-zero doublets. The resulting circuit is shown in Fig. 7.28, together with its frequency characteristic in Fig. 7.29 [17].

As with the circuit of Fig. 7.26, starting with a quiescent current of 2 mA in the output transistors, resulting in $r_e = 1/g_m = V_T/I_e = 12.5\ \Omega$ and a load capacitance of 100 pF, we obtain a limiting pole frequency of:

$$f_1 = \frac{g_{ml}}{2\pi C_L} \approx 250\,\mathrm{MHz} \tag{7.47}$$

with $g_{ml} = g_{m400} + g_{m500}$

The zero-dB frequency f_0' of the two-stage amplifier through the second input stage Q_{105}, Q_{115} must be half f_1, so with $g_{m32} = 1/(r_{e105} + r_{e115})$ and $C_{M1} = C_{P1}$, we obtain:

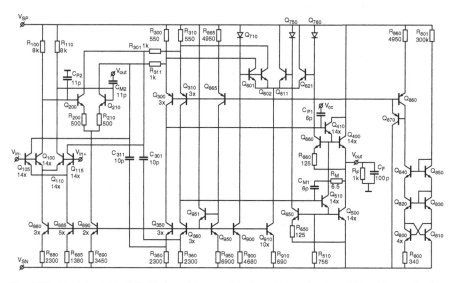

Fig. 7.28 Precision HF all-NPN class-AB operational amplifier with GA-GA-VF/GA configuration and multipath nested Miller compensation (MNMC)

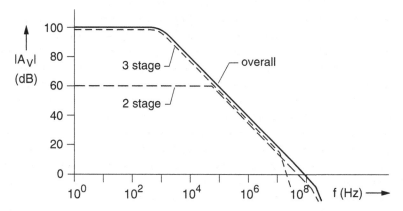

Fig. 7.29 Amplitude characteristic of the GA-GA-VF/GA configuration with a multipath nested Miller compensation

$$f_0' = \frac{g_{m32}}{2\pi C_{M1}} \frac{1}{2} f_1 \approx 125\,\text{MHz} \tag{7.48}$$

We have chosen $C_{M1} = C_{P1} = 6$ pF, and $I_{T31} = 4g_{m32}V_T \approx 0.6$ mA.

The overall amplitude characteristic of the three-stage amplifier must run along with that of the two-stage amplifier. So with $g_{m31} = 1/(r_{e100} + r_{e110})$ and $C_{M2} = C_{P2}$, we obtain likewise for the first input stage:

$$f_0' = \frac{g_{m31}}{2\pi C_{M2}} \frac{1}{2} f_1 \approx 125 \, \text{MHz} \tag{7.49}$$

We have chosen $C_{M2} = C_{P2} = 11$ pF, and $I_{T31} = 4g_{m31}V_T \approx 0.9$ mA.

Next we have to choose the g_{m2} of the intermediate stage with Q_{200} and Q_{210}. According to the reasoning with (Eq. 6.44), by which we argued that with a unity overall feedback the second input stage becomes connected in parallel with the unity gain Miller feedback intermediate stage, which could endanger the phase margin of the inner loop, we have to choose g_{m2} much smaller than g_{m32}. This leads to:

$$g_{m2} < 1/3 \, g_{m32} \tag{7.50}$$

For this reason we degenerated the intermediate stage with emitter resistor R_{200} and R_{210} of 500 Ω each, making $g_{m2} \approx 1$ mS, which is much lower than $g_{m32} \approx 6$ mS.

7.6.5   1 GHz, All-NPN Class-AB Operational Amplifier with MNMC

One of the disadvantages of the Darlington all-NPN output stage is that the upper output voltage range is limited to at least the voltage of two diodes plus a saturation voltage plus a current-source resistor voltage.

The three-stage GA-GA-VF/GA configuration has enough gain to leave out the Darlington VF driver transistors in the output stage. An example is shown in Fig. 7.30. It is a 1 GHz OpAmp powered at 5 V single supply and loaded with a 50 Ω resistive load at a voltage gain of 76 dB. The circuit is almost equal to that of

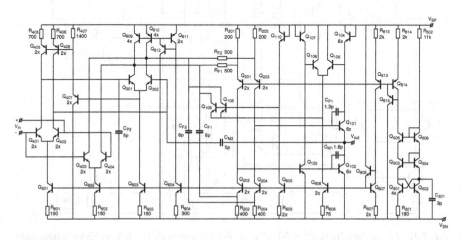

Fig. 7.30 1 GHz 5 V class-AB operational amplifier with all-NPN GA-GA-VF/GA configuration and a multipath nested Miller compensation

Fig. 7.28, except for the single output transistors instead of Darlington ones. Importantly the class-AB feedback is now connected in front of the all-pass feedforward network R_FC_F to accommodate class-AB control up to 1 GHz. The circuit is fully described in Ref. [14].

The 1 GHz amplifier has a supply-current consumption of 20 mA. But the bandwidth and supply current can easily be scaled down, while the resistor values are scaled up. At a bandwidth of 1 MHz, the supply current need only be 20 μA.

7.6.6 2 V Power-Efficient All-NPN Class-AB Operational Amplifier with MDNMC

When we do need more gain without losing the extra diode voltage of a Darlington transistor in the output stage of Fig. 7.26, we can take another GA driver while adding another nest to the nested HF compensation structure. This is shown in the power-efficient "all-NPN" OpAmp of Fig. 7.31.

In the circuit of Fig. 7.31 we easily distinguish, going from left to right, an GA input stage, an GA intermediate stage, an GA-VF/GA output stage, and a feedback biasing class-AB regulator. The GA-VF/GA output stage includes a Multipath Nested Miller compensation scheme. This multipath scheme allows the Darlington VF transistor to be replaced by an GA boost transistor with a current mirror, Q_{102}, Q_{106}, Q_{105} in the upper VF half, and Q_{104}, Q_{106}, Q_{107} in the lower half to drive the output transistors Q_{101} and Q_{103}, respectively. The primary Miller compensation capacitors are C_{M1} and C_{P1}. The secondary nested capacitors are C_{M2} and C_{P2}. The tertiary nested capacitors are C_{M3} and C_{P3}. Strictly, the topology is not a three-stage amplifier but a four-stage one GA-GA-(GA-VF)/(GA-GA). However, the

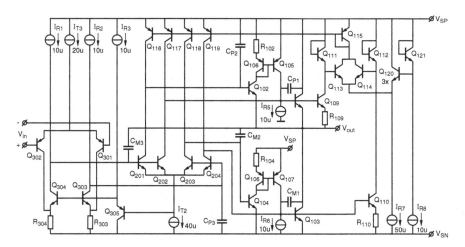

Fig. 7.31 2 V power-efficient "all-NPN" class-AB OpAmp with GA-GA-VF/GA configuration and a multipath double nested Miller compensation (MDNMC)

resemblance with a three-stage amplifier is so strong that we like to place it in this chapter. Only the output transistors and the drivers have a multipath topology generated in the intermediate stage. The quiescent current of the whole OpAmp is 0.35 mA. The peak load current is 100 mA. The bandwidth is 1 MHz. The output-voltage range reaches from 0.2 V above the negative supply rail up to 1.0 V below the positive supply rail. The supply voltage must be larger than 2 V.

7.6.7   Conclusion

We have shown that the GA-GA-VF/GA configuration is ideally suited to high-quality bipolar OpAmps with an all-NPN topology.

The classic NE5534 with a capacitive feedforward bridged intermediate stage has a pole-zero doublet in the frequency characteristic.

The Nested Miller compensation provides a straight frequency characteristic combined with a large voltage gain. The Multipath Nested Miller compensation allows a two-times larger bandwidth-over-power ratio with a frequency characteristic having a straight 20 dB/dec roll-off.

The bandwidth can be taken up to the GHz range when we take out the Darlington VF driver transistors. Taking out the Darlington transistors also increases the effective usable output voltage range. When we cannot miss the current gain of the Darlington transistors, nor allow a reduction of the output voltage range, the Darlington VF transistor can be replaced by another nested GA transistor driver stage. This results in a power-efficient "all-NPN" output stage.

7.7   GA-CF-GA Configuration

The demand for lower supply voltages such as 3.0, 2.0 and eventually 1.0 V by very dense VLSI processes with sub-micron gate lengths necessitates an efficient use of the supply voltage range by the output voltage range. This can only be accomplished if we use a rail-to-rail.

(R-R) GA-output stage, because only then we avoid the loss of one or more diode voltages. The GA-CF-GA configuration is the simplest way to obtain this.

7.7.1   Compact 1.2 V R-R-Out CMOS Class-A OpAmp with MC

When we do not care about a current-efficient class-AB output stage but instead can live with a class-A version, the simple class-A compact 1.2 V CMOS OpAmp with an GA-CF-GA configuration of Fig. 7.32 results. Its behavior is comparable with

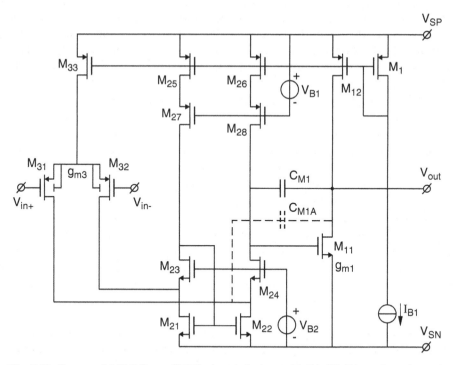

Fig. 7.32 Compact 1.2 V R-R-out CMOS class-A OpAmp with GA-CF-GA configuration and Miller compensation. Compact 1.0 V OpAmp without the upper cascodes M_{27}, M_{28}

the GA-GA circuit of Fig. 7.15, except that now we are using a folded cascode stage in between the two GA stages.

This enables the input CM voltage to include the negative rail. Further, we can choose between a direct (with C_{M1}) or active (through the cascode, explained with Fig. 6.18) (with C_{M1A} dotted) connection of the Miller capacitor, depending on circumstances, or a combination of both.

P-channel transistors have been chosen for the differential input pair because the back-gate can be bootstrapped with the common-source voltage. This has the advantage that the input transistors are purely driven by the differential input voltage and not that much by a common-mode input voltage through the back-gates as would be the case with N-channel transistors. This results in the highest obtainable CMRR (see Sect. 4.3).

The folded-cascode intermediate stage functions in addition to a current mirror as a summing circuit for the differential signals from the input stage.

N-channel transistors are chosen for the folded cascode mirror connection because they have the highest f_T.

To keep the offset and noise contribution of the folded cascodes low, in regard to those of the input stage, the g_m or W/L ratio of the current-determining transistors in the folded cascodes must be lower than the g_m or W/L ratio of the input-stage transistors. Moreover, the input transistors as well as the current-determining transistors in the folded cascodes may have a cross-coupled quad layout to further reduce the offset. The upper bias current sources M_{25}, M_{26} of the folded cascode branches must be chosen equal to the tail current of the differential input pair to ensure that the full differential output current of the input pair is available to drive the Miller capacitors C_{MI} and C_{M1A}. This is needed for a maximal slew-rate.

In respect to noise and offset the GA-CF-GA configuration of Fig. 7.32 has a disadvantage over the GA-GA configuration of Fig. 7.15. The intermediate summing and mirror CF stage adds its noise and offset to that of the input stage. This addition is more than the noise and offset of the "telescopic" current mirror load of the input stage of Fig. 7.15.

We may alternatively choose for a lower noise and offset as well as a lower power consumption instead of choosing a high-frequency behavior for the intermediate cascode and summing circuit. This can be realized by taking the mirror connection at the upper P-channel side instead of at the lower N-channel side. In that case, the lower current-source transistors M_{21} and M_{22} need to provide a current equal to the tail current of the differential input pair.

As a result, the two branches of the summing circuit would carry only a quiescent current equal to half of the tail current. Yet the output of the summing circuit would symmetrically deliver the maximum of plus or minus the full tail current of the input pair into the compensation capacitor. Thus proving full slew rate. This has been further explained with Fig. 7.2b.

The limiting pole frequency, if loaded with a load capacitance C_L, is roughly:

$$f_1' = \frac{g_{m1}}{2\pi C_L} \tag{7.51}$$

with $g_{m1} = g_{m11}$.

The zero-dB bandwidth must be chosen half of f_1':

$$f_0' = \frac{g_{m3}}{2\pi C_{M1}} = \frac{1}{2} f_1' = \frac{1}{2} \frac{g_{m1}}{2\pi C_L} \tag{7.52}$$

Hence the Miller capacitor has to be chosen roughly:

$$C_{M1} = 2C_L \frac{g_{m3}}{g_{m1}} \tag{7.53}$$

The bandwidth can be increased by roughly a factor 2 if we make use of the active-Miller connection (dotted) with C_{M1A}, as explained with Fig. 6.18. The maximum low-frequency voltage gain obtained without R_L is roughly:

$$A_{VOM} \approx g_{m3} r_{ds22} \mu_{24} \mu_{11} \tag{7.54}$$

with $\mu_{24} = g_{m24} r_{ds24}$, $\mu_{11} = g_{m11} r_{ds11}$, and disregarding the output resistance of the upper cascoded current source.

A practical value is 80 dB.

If loaded with a load resistor R_L the gain A_{V0} is lowered by a load factor $R_L/(R_L + r_{ds11})$.

The dominating pole frequency without R_L is:

$$f_2' = f_0'/A_{VOM} \tag{7.55}$$

The OpAmp of Fig. 7.32 has a minimum supply voltage of about 1.2 V as a result of the addition: $V_{S\ min} = V_{diode\ 11} + V_{sat\ 28} + V_{sat\ 26}$.

When we further simplify the circuit and leave out the upper cascode transistors M_{27}, M_{28}, a very simple GA-CF-GA OpAmp remains that may be powered at the absolute minimum supply voltage of 1 V, made up by one diode voltage of M_{11} and one saturation voltage of M_{26}. However, the maximum voltage gain will drop to:

$$A_{VOM} \approx g_{m3}r_{ds26}\mu_{11} \tag{7.56}$$

A practical value then is in the order of 60 dB.

7.7.2 Compact 2 V R-R-Out CMOS Class-AB OpAmp with MC

When we need a higher output current and still want a low quiescent current we need to incorporate a current efficient class-AB biasing of the output stage. A compact and robust OpAmp in an GA-CF-GA configuration [18] is shown in Fig. 7.33.

The GA output stage M_{11}, M_{12} is feedforward biased in class-AB by a mesh of head-to-tail connected transistors M_{13}, M_{14}. The output stage has been described with Fig. 5.27. At first sight, the bias connections to the sources of M_{13} and M_{14} seem to lower the impedance at the gates of the output transistors M_{11} and M_{12} and thus lower the gain. However, the drain connections of M_{13} and M_{14} cancel the low source impedances by a positive feedback loop for CM driving voltages. Meanwhile, the bias impedances for DM driving voltages are strongly fixed to accurately bias the output transistors an a class-AB characteristic, as has been described by Eq. 5.10.

The class-AB mesh with M_{13} and M_{14} has been incorporated into the folded cascode with M_{21} through M_{24}. This has the important advantage that no additional bias currents have to be used for class-AB biasing. The offset and noise of these extra bias currents would otherwise have been added to the offset and noise of the input stage. To further reduce the noise and offset contribution of the folded-cascode stage, one could have chosen for the mirror connection at the upper side instead as of at the lower side. By so doing, the currents in the folded cascodes can be reduced by a factor two. This has been extensively explained with Figs. 7.2b and 7.32.

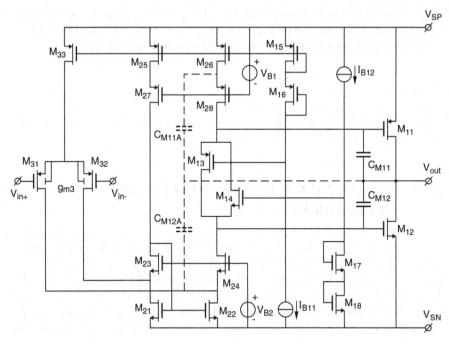

Fig. 7.33 Compact 2 V R-R-out CMOS class-AB OpAmp with GA-CF-GA configuration having a R-R output stage with Miller compensation

The limiting pole frequency, if loaded with a load capacitance C_L, is roughly:

$$f_1' = \frac{g_{ml}}{2\pi C_L} \tag{7.57}$$

with $g_{ml} = g_{ml1} + g_{ml2}$.

The zero-dB bandwidth must be chosen half of f_1':

$$f_0' = \frac{g_{m3}}{2\pi C_{MI}} = \frac{1}{2} f_1' = \frac{1}{2} \frac{g_{ml}}{2\pi C_L} \tag{7.58}$$

with $C_{MI} = C_{M11} + C_{M12}$

Hence, the Miller capacitors have to be chosen roughly:

$$C_{MI} = 2C_L \frac{g_{m3}}{g_{ml}} \approx C_L \frac{g_{m3}}{g_{ml1,12}} \tag{7.59}$$

with $g_{m3} = g_{m31} = g_{m32}$ due to the mirror in the folded cascode and $g_{ml1,12} \approx g_{ml1} \approx g_{ml2}$.

The bandwidth can increase by roughly a factor 2, if we make use of the active-Miller connection (dotted) with C_{M11A} and C_{M12A}, as explained with Fig. 6.18.

The maximum low-frequency voltage gain obtained without R_L is:

$$A_{VOM} \approx g_{m3}r_{ds22}\mu_{24}\mu_{11,12} \tag{7.60}$$

with: $\mu_{24} = g_{m24}r_{ds24}$, $\mu_{11,12} \approx \mu_{11} \approx \mu_{12}$.

If loaded with a load resistor R_L, the gain is lowered by a load factor $R_L/(R_L + r_{ds11}\|r_{ds12})$.

7.7.3 Compact 2 V R-R-In/Out CMOS Class-AB OpAmp with MC

The minimum supply voltage of the circuit of Fig. 7.33 is set by two-diode voltages and a saturation voltage through the class-AB mesh M_{13}, M_{14}, which is of the order of 2 V. This allows for the addition of a rail-to-rail (R-R) input stage. As a first approach, we could extend the circuit of Fig. 7.33 with a R-R input stage as shown in Fig. 7.34. When the commode-mode (CM) input voltage is at the negative supply rail voltage V_{SN}, the circuit behaves exact like the circuit of Fig. 7.33, because the

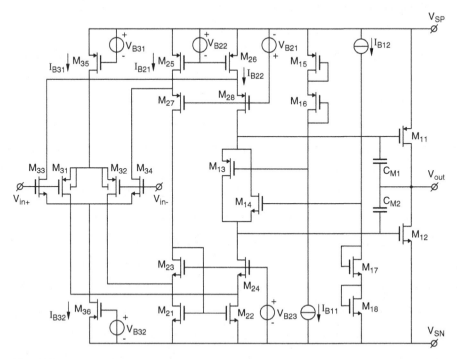

Fig. 7.34 Compact 2 V R-R-in/out CMOS class-AB OpAmp with GA-CF-GA configuration and Miller compensation (first approach)

N-channel pair transistors M_{33}, M_{34} do not have enough gate-source voltage to conduct, and hence the current source M_{36} is being cut off.

When the input CM voltages rise above the threshold voltage V_{TH} of M_{33}, M_{34}, these transistors start conducting a current, and so does the tail-current source transistor M_{36}. When M_{36} conducts the full tail current, both input pairs function normally, as explained with Figs. 4.19 and 4.20. Now the total g_m of the R-R input stage is twice that of a single rail input stage. We suppose that $I_{B32} = I_{B31}$, and that the W/L ratios of both input pairs have been chosen such that they compensate the ratio of about three in the mobilities between the N and P-channel type, so that both input pairs have the same g_m.

Now two problems arise: firstly, the g_m of the input stage changes a factor 2 when the input CM voltage swings from rail-to-rail, and we cannot optimize the frequency compensation. We will address this problem later. Secondly, the changing of the currents through the N-channel input pair from 0 to half the tail current I_{B32} also changes the currents through the upper folded cascodes M_{27} and M_{28}. And this changes the quiescent current of the class-AB output stage, which is undesirable.

The changing of the quiescent current as a function of the CM input voltage can be stopped by using current mirror connections both in the upper and lower folded cascodes. This is depicted in the second approach of Fig. 7.35. Both current mirrors

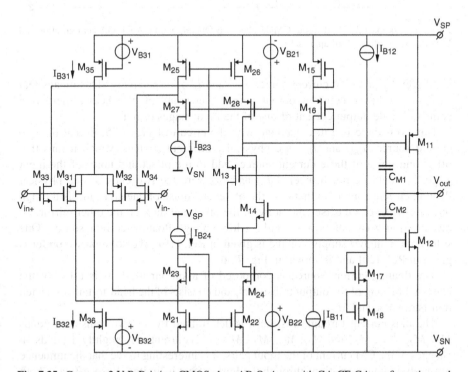

Fig. 7.35 Compact 2 V R-R-in/out CMOS class-AB OpAmp with GA-CF-GA configuration and Miller compensation (second approach)

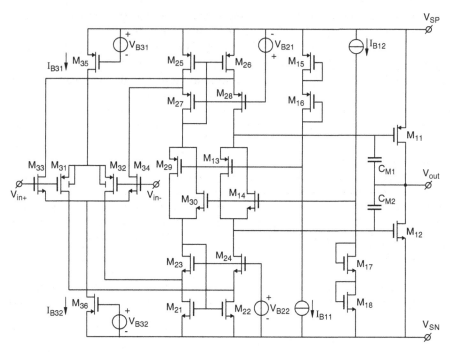

Fig. 7.36 Compact 2 V R-R-in/out CMOS class-AB OpAmp with GA-CF-GA configuration and Miller compensation (third approach)

are driven by an equal current source I_{B23} and I_{B24}, respectively. This means that each mirror has a constant output current, independent of whether there is a common-mode output current of one of the input stages or not.

However, there is a disadvantage with the circuit of Fig. 7.35, in that the two current sources I_{B23} and I_{B24} are physically different sources, which means that offset and noise of these current sources add to the offset and noise of the input stage. It would be much nicer if I_{B3} and I_{B4} were the two ends of one physical floating current source. In that case the offset and noise of one end cancels through one mirror the equal offset and noise of the other end through the other mirror. So the question arises of how to build a low-voltage floating current source. One solution at a higher supply voltage is given in Ref. [19]. The solution we prefer is given in Ref. [20] and is shown in Fig. 7.36.

The floating-current source is composed of a similar mesh as is used for the class-AB biasing of the output transistors, and consists of the head-to-tail connected transistors M_{29} and M_{30}.

The bias current through this mesh is determined by the two translinear loops M_{25}, M_{29}, M_{15}, M_{16} and M_{21}, M_{30}, M_{17}, M_{18}. This current only slightly depends on changes of the CM current of the input pairs. It is interesting to see that the influence of supply voltage variations through the Early effect is limited. Early effect in both

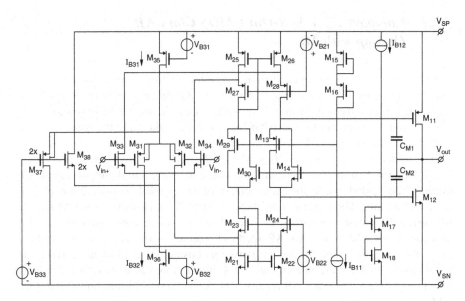

Fig. 7.37 Compact 2 V R-R-in/out CMOS class-AB OpAmp with GA-CF-GA configuration and Miller compensation (final approach)

meshes cancel each other in their influence on the biasing of the output transistors [20].

For the final circuit, we have still to solve the problem of the changing g_m of the R-R input stage. For this purpose we choose one of the solutions of Sect. 4.4. For example we choose the solution with the current switches of Fig. 4.25, because of its simplicity [21, 22].

The final approach is shown in Fig. 7.37. The result is a compact circuit with excellent specifications and which is robust against device parameter variations. An example of specifications is a bandwidth of 2 MHz at a load capacitance of 20 pF and a quiescent current of 200 μA. The DC gain is 85 dB at a resistive load of 10 kΩ, and the CMRR is 80 dB at a CM input voltage around the negative supply rail voltage, but 43 dB when just crossing the CM voltage level of V_{B33}. Most specifications can easily be programed by a factor 1/10 or a factor 10 by choosing other W/L ratios and other voltages and currents.

The bandwidth may be enlarged by roughly a factor 2 without changing the bias currents nor the load capacitor by connecting the Miller capacitors, not directly to the gates of the output transistors, but indirectly through a connection at the sources of the folded cascode transistors M_{28} and M_{24}. This has been described in Sect. 6.2 with Fig. 6.18 [20]. At very low bias currents the transistors function in weak inversion. For a constant g_m the spill-over switches must be chosen two times wider than the inpt stages.

7.7.4   Compact 1.2 V R-R-Out CMOS Class-AB
OpAmp with MC

When we want to further reduce the supply voltage while maintaining a class-AB bias for the output stage, the feedforward class-AB biasing mesh cannot be used anymore, because its two stacked diode voltages can no longer be provided.

To avoid this problem the mesh has been folded, as shown in the compact 1.2 V CMOS OpAmp of Fig. 7.38 [23].

The folded mesh consists of M_{202}, M_{2021}, M_{203}, M_{2031}. Somewhat similar to the non-folded mesh, at first it does not seem that it has such a high output impedance because the folded mesh does not consist of two folded cascodes, but of two differential pairs. Their output impedance is only $2r_{DS}$. However, for CM signals that drive the output transistors, the folded mesh may be considered as a folded cascode. Even when the class-AB feedback control at the gates of M_{203}, M_{2031} keeps the gate voltage of one output transistor constant at its minimum current, the regulation is such that the gate voltage of the other output transistor is fully driven at a high impedance. The class-AB control acts like a cascode 2-times boost circuit for CM driving voltages.

The DM driving is strongly controlled by the feedback class-AB measuring circuit. In this example, the minimum selector of Sect. 5.4 is used and is explained with Fig. 5.40 [23, 24].

The circuit needs one diode voltage and two collector-source saturation voltages, which allows a minimum supply voltage of about 1.2 V over temperature variations.

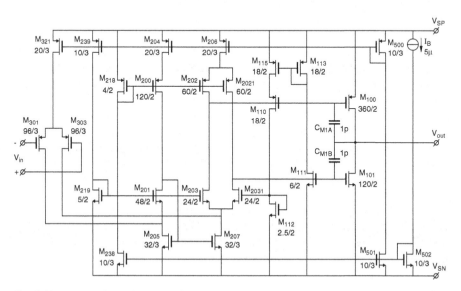

Fig. 7.38 Compact 1.2 V R-R-out CMOS class-AB OpAmp with GA-CF-GA configuration and Miller compensation

The circuit can be extended with a rail-to-rail input stage at 2 V supply voltage. The easiest way is with an input stage having constant g_m using spilling-over switches and adding the currents from the switches to obtain a constant output current [23]. A quiescent current of 200 µA was needed for bandwidth of 4 MHz at a capacitive load of 5 pF, and a DC gain of 85 dB at a resistive load of 10 kΩ.

7.7.5   Conclusion

Robust and high-quality compact low-voltage CMOS VLSI OpAmp cells can be designed with the GA-CF-GA configuration. If the simple GA-CF configuration of Sect. 7.1 does not provide adequate LF gain, the GA-CF-GA configuration is the next best choice. The output may nearly swing from rail-to-rail at a minimum supply voltage as low as one diode and two saturation voltages, which is in the order of 1.2 V, or even 1 V. The CM input voltage may include the negative supply voltage rail, and at a supply voltage of 2 V even from rail-to-rail. The bandwidth can roughly be designed up to $f_T/4$. The circuits stay relatively simple.

7.8   GA-GA-GA Configuration

The previous GA-CF-GA configuration may not always deliver enough LF gain. This may particularly be the case when the supply voltage has to go down to the absolute minimum value of 1 V of one diode voltage plus one saturation voltage, or if we want to have a bipolar output stage. At such low voltages we can no longer use CMOS cascode or bipolar Darlington transistors. In those cases we have to take resource to another GA general amplifying stage instead of a CF intermediate stage, resulting in the GA-GA-GA configuration. In this section we will discuss a CMOS, a BiCMOS and a full bipolar design.

7.8.1   1 V R-R-Out CMOS Class-AB OpAmp with MNMC

When we have to drive relatively heavy output loads in CMOS at a low supply voltage, the GA-CF-GA configuration with a folded cascode intermediate stage no longer meets the needs. The folded cascode with its two saturation voltages in series with a higher gate-source voltage at high currents may require a too high supply voltage. Moreover, we may need more power gain. To solve these two problems, a third GA stage may be needed instead of the CF intermediate stage. A design example is given in Fig. 7.39 [25].

The 1.0 V CMOS OpAmp with GA-GA-GA configuration of Fig. 7.39 has an input stage M_{301}, M_{302} and a folded-cascode mirror circuit M_{211} through M_{216}. The second stage M_{201}, M_{202}, drives through mirrors M_{203}, M_{205} and M_{204}, M_{206} the

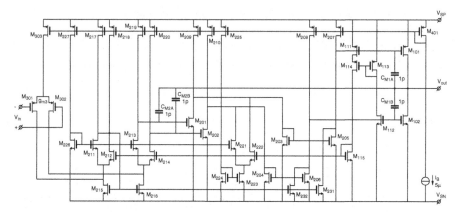

Fig. 7.39 1.0 V CMOS R-R-out class-AB OpAmp with GA-GA-GA configuration with multipath nested Miller compensation

output push–pull stage M_{101}, M_{102}. The driver mirrors are needed to provide an equal phase to the intermediate and output stage in order to allow a nested Miller frequency compensation. The driver mirrors also separate the direct class-AB control amplifier M_{221}, M_{222} from the gates of the output transistors in order to allow these gates to be driven from rail-to-rail. For the same purpose we avoided folded cascodes to drive the output transistors. This whole architecture does not have any branch between the supply lines that contain more than one gate-source diode voltage plus one-drain-source saturation voltage. This makes the architecture of Fig. 7.39 suitable for operation at a supply voltage of 1.0 V, or over wide temperature and process variations 1.2 V.

The high-frequency compensation scheme has been multipath nested. The first nest across the output transistors is shaped by C_{M1A} and C_{M1B}. The second nest runs from the output to the input of the intermediate stage M_{201}, M_{202} through C_{M2A} and C_{M2B}. The gain path goes through the intermediate stage. The feedforward path has been tapped single-sidedly from the folded-cascode mirror connection and duplicated directly into the gates of the output transistors by M_{231} and M_{232}, by passing the intermediate stage.

The limiting-pole frequency is set by the output stage at its quiescent current by:

$$f_1' \approx \frac{g_{m1}}{2\pi C_L} \tag{7.61}$$

with $g_{m1} = g_{m101} + g_{m102}$.

The zero-dB frequency of the two-stage direct amplifier path must be half of the limiting-pole frequency, so:

$$f_0' = \frac{1}{2} \, f_1' \approx \frac{1}{2} \frac{g_{m1}}{2\pi C_L} \tag{7.62}$$

The Miller capacitors of the first nest must be chosen:

$$C_{MI} = \frac{g_{m3}}{2\pi f_0'} = \frac{g_{m3}}{g_{ml}} C_L \qquad (7.63)$$

with $C_{MI} = C_{M1A} + C_{M2B}$.

under the assumption that the W/L ratios of $M_{215}, M_{216}, M_{231}, M_{232}$ are equal. On one hand we have half the g_{m3} of the input stage because the mirror only senses one input half, but on the other hand we double the current in two transistors M_{231}, M_{232}. The zero-dB frequency of the whole three-stage amplifier may be equal to that of the two-stage one because of the multipath nested Miller compensation structure, hence:

$$f_0'' = f_0' = \frac{1}{2} f_1 \qquad (7.64)$$

which leads to the choice of the Miller capacitors of the second nest:

$$C_{M2} = \frac{g_{m3}}{2\pi f_0''} = \frac{g_{m3}}{g_{ml}} C_L \qquad (7.65)$$

with $C_{M2} = C_{M2A} + C_{M2B}$.

The two-stage amplifier must dominate the high-frequency behavior at the amplifier's bandwidth, therefore the transconductance through the intermediate stage must be smaller than the transconductance through the feedforward path. This causes the requirement:

$$g_{m2} < \frac{1}{3}\frac{1}{2} g_{m3} \qquad (7.66)$$

with $g_{m2} = g_{m201} + g_{m202}$.

The factor 1/3 stems from the explanation with Fig. 6.25. The factor 1/2 comes from the fact that only half of the input stage contributes through its mirror connection to the transconductance of the feedforward path through M_{231}, M_{232}.

For measuring the class-AB relation between the push and pull currents we have chosen a variation to the Seevinck version, described with Fig. 5.40. Note that the functions of M_{111} and M_{114} have been interchanged in regard to Fig. 5.40.

The direct class-AB control loop is realized by the control amplifier M_{221}, M_{222} which differentially controls the biasing of the output transistors via the mirrors M_{203}, M_{205} and M_{204}, M_{206}. The control amplifier has been referenced on the left side by M_{226}. The bandwidth of this first or direct class-AB loop is determined by the product of the g_m of the control amplifier and the impedance of a parallel compensation capacitor, which consists of the series connection of C_{M1A} and C_{M1B} in between the output of the mirrors. The bandwidth of the class-AB control loop must be stable and of the same order as the bandwidth of the whole amplifier.

The intermediate stage with the grounded transistors M_{201}, M_{202} may provide such large driving currents that the direct class-AB control may be overruled. To avoid this, the class-AB control has been given a second, or gain path, through the intermediate stage in a multipath compensation topology. To this end, the folded cascode circuit between the input and intermediate stage has also been given a differential amplifier function by splitting the cascode transistors. In this way the intermediate stage is driven in parallel for input signals, and differentially for class-AB control signals. The transconductance of this extra gain path for class-AB control signals is determined by the product of the g_m of the differential amplifier M_{213}, M_{214}, the impedance of a parallel compensation capacitor, which consists of the series connection of C_{M2A} and C_{M2B}, and the g_m of the intermediate stage M_{201}, M_{202}. This transconductance must be smaller than that of the direct path at the bandwidth of the direct path, otherwise the direct path does not dominate the stability of the class-AB loop at high frequencies.

A realized example of such a three-stage CMOS 1.0-voltage OpAmp has a bandwidth of 5 MHz at a load of 5 pF and 10 kΩ, at a quiescent current of only 200 μA. The voltage gain is 80 dB. Over wide temperature and process variations the minimum supply voltage is 1.2 V [25, 26].

7.8.2 Compact 1.2 V R-R-Out BiCMOS Class-AB OpAmp with MNMC

At low supply voltages we can no longer use Darlington transistors to drive bipolar output transistors. An excellent solution for driving bipolar output transistors at low supply voltages is to use the low-voltage GA-CF-GA configuration of Fig. 7.37 as a class-AB driver. The result is shown in the 1.2 V BiCMOS OpAmp of Fig. 7.40 with an GA-GA-GA configuration [27].

The nested Miller compensation requires an equal phase across the first and second nested Miller capacitors. This forces us to place mirrors M_{203}, M_{204}, and Q_{205}, Q_{206} between the CMOS driver intermediate transistors M_{201} and M_{202} and the output transistors Q_{101} and Q_{102}. These mirrors allow the transfer of class-AB driving currents. This was not needed in the CMOS OpAmp of Fig. 7.39, in which case we could choose an all N-channel intermediate stage and N-channel mirror driver for best HF performance.

The complementary CMOS intermediate stage and mirror drivers may generate such large class-AB dominating currents, that a local class-AB control differential amplifier M_{207}, M_{208} is not sufficient. Therefore the class-AB regulation is extended with a multipath control amplifier M_{213}, M_{214} as to also control the intermediate stage. The control amplifier is made by splitting the cascode transistors of the folded cascode and mirror following the input stage.

The push and pull currents are sensed by the bipolar transistors Q_{111} and Q_{112}. To limit their currents, emitter resistors may be inserted. The minimum selector M_{116}, M_{115} senses the lowest of one of the two push or pull output currents across

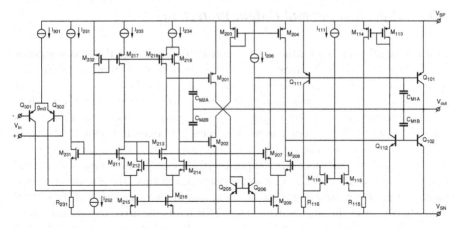

Fig. 7.40 Compact 1.2 V BiCMOS class-AB OpAmp with GA-GA-GA configuration and multipath nested Miller compensation

the measurement resistors R_{116} and R_{115}. The class-AB control amplifiers M_{207}, M_{208} and M_{213}, M_{214} are referenced at the left hand by the diode M_{231} and resistor R_{231}. The result is a robust class-AB regulation that keeps the lowest of the push or pull output currents above the minimum value.

The driving of the complementary intermediate stage by a folded cascode mesh M_{213}, M_{214}, M_{218}, M_{219}, as explained with Fig. 7.37, allows a minimum supply voltage of one diode voltage and two saturation voltages. The result is a minimum supply voltage of about 1.2 V, or over a wide range of temperature and process variations 1.4 V. The three transistor combination M_{207}, M_{208}, M_{209} can be regarded as a third folded cascode, to directly drive the output transistors. This allows the combination of the direct class-AB control path through M_{207}, M_{208} and the feedforward path through M_{209} from the input stage to the bases of the output transistors.

The multipath nested Miller frequency compensation is similar to that of Fig. 7.39 and has been described by the Eqs. 7.61–7.66. In the last formula $g_{m231} + g_{m232}$ must be replaced for g_{m209}. The main differences to the circuit of Fig. 7.39 are that the output and input transistors are bipolar and that the second stage is driven by cascodes at the bottom as well as at the upper side. This means that a larger bandwidth can be obtained, and that the DC gain is larger than those of the circuit of Fig. 7.39. Moreover, the input offset voltage is lower.

By example, a 30 MHz bandwidth was obtained at a supply current of 800 μA at a load of 5 pF and a supply voltage of 1.2 V, or 1.4 V over wide temperature and process variations. The DC gain is in the order of 100 dB at a load resistor of 10 kΩ [27].

7.8.3 Bipolar Input and Output Protection

When the amplifier is used at the input or output of a VLSI chip, the bipolar input or output must be protected. Firstly, ESD diodes must be used from each input and output to both power rails. Secondly, two antiparallel diodes across the input terminals are needed to protect the input transistors against large reverse bias voltages. Thirdly, to avoid HF ringing when the output transistors are drivers into saturation, saturation detection transistors must be connected across the output transistors, as was explained earlier with Fig. 5.42. These transistors must be of the same type as the concerned output transistors and with a base-to-base and collector-to-collector connection with the output transistors. The emitter in reverse mode is used as a collector and can be connected to the related gate of the CMOS driver transistor of the intermediate stage. This limits the drive current and keeps the output transistors from saturation. Fourthly, the output current must be limited to protect the output transistors from being destroyed. To this end the current-limiting topology of Fig. 5.45 can be used with a direct path to the base of the output transistor and a multipath to the gate of the related driver transistor of the intermediate stage [26]. These protection issues must be dealt with in all described circuits.

7.8.4 1.8 V R-R-In/Out Bipolar Class-AB OpAmp (NE5234) with NMC

If we want to obtain a high output current capability in pure bipolar technology, Darlington output transistors offer the most simple solution. However, a Darlington output stage needs a supply voltage of minimally two diode voltages and one saturation voltage, which is 1.8 V over a wide range of temperature and process variations. An example circuit of a fully bipolar OpAmp with an GA-GA-GA configuration and Darlington output transistors, the NE5234, is shown in Fig. 7.41 [28, 29].

The output stage consists of the complementary Darlington transistors Q_{11}, Q_{12} on the NPN side and the folded Darlington transistors Q_{13}, Q_{14} Q_{15} on the PNP side. The PNP side is current-boosted (see Sect. 6.1) by Q_{140}, Q_{141}, Q_{130}.

The class-AB control amplifier consists of a differential pair Q_{16}, Q_{17} with a reference voltage across Q_{20}, Q_{21} at the left-hand side and a minimum selector Q_{200}, Q_{201} at the right-hand side, which gets its push and pull current information directly from the NPN output transistor Q_{11} and indirectly from the PNP output transistor Q_{13} through Q_{18} and Q_{19} (see Sect. 5.4.3).

The differential pair senses the differential voltage between the voltage reference and the minimum selector and regulates the difference in the ideal case at zero. When driving the current of one output transistor high, the current of the other output transistor will be regulated at half its quiescent current (see Sect. 5.4.3). If this happens, the driving current for the transistor with the lowest output current is

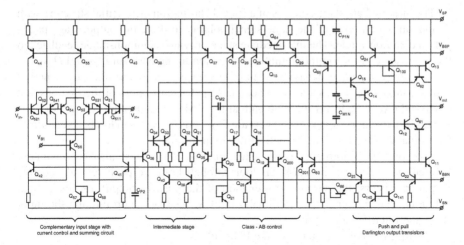

Fig. 7.41 1.8 V bipolar class-AB operational amplifier NE5234 with R-R input and output ranges in GA-GA-GA configuration and nested Miller compensation

channeled away from this transistor and fed to the other output transistor through the differential pair, thus doubling the drive current of the output transistor with the highest current. The result is that both currents from the intermediate stage are either used to drive both output transistors in parallel in the quiescent situation, or to jointly drive the output transistor that needs to drive the largest current, while the other is idle at half the quiescent current.

Bipolar output transistors must be protected against excess current, otherwise they can be destroyed, and against saturation, otherwise the bandwidth may drastically decrease and the circuit may start oscillating. Moreover, saturation of NPNs creates large amounts of charge carriers in the substrate, which may create unwanted signal paths. Current limiting is provided by the sense transistors Q_{63}, Q_{64} for the NPN output side, and Q_{65}, Q_{66} for the PNP output side. Saturation protection is provided by the sense transistors Q_{61} and Q_{62} for the NPN and PNP sides, respectively. These transistors are used in reverse mode. They match their base–collector diodes with those of the output transistors over process variations.

The R-R input stage has the two complementary differential input stages Q_{51}, Q_{52} and Q_{53}, Q_{54}. The current switch Q_{56} and mirror Q_{57}, Q_{58} keeps the sum of the tail currents constant at the collector current of Q_{55}. The folded cascodes Q_{41}, Q_{42}, Q_{43}, Q_{44} feed the output currents of the input long-tail-pairs to the second stage while these currents are being summed.

Bipolar input stages have the nasty property that their signal transfer is reversed when the transistors are driven into saturation. This is caused by conductance of the reverse base–collector diodes. Particularly for low-voltage amplifiers, input voltages can easily surpass the supply voltage. To protect the input stage from signal reversion each input transistor is paralleled with a two-times larger transistor, which is connected as a collector–base diode, while its collector is cross-coupled to the other transistor of the belonging input pair (see Sect. 4.4).

The intermediate stage consists of a folded Darlington differential structure with double output currents of the same signal polarity to drive the output stage push–pull in common-mode. The intermediate stage provides common-mode feedback to the input stage at its common-emitter connection to the bases of Q_{41} and Q_{42}.

The high-frequency behavior of the OpAmp is well determined by a nested Miller compensation structure (see Sect. 6.2.3). The capacitors of the first nest are C_{MIN} and C_{MIP}. To lower the loop gain at the PNP side a parallel capacitor C_{PI} is used. The capacitors of the second nest are C_{M2} and C_{P2}. Ground has been chosen as the virtual reference for the second Miller loop through C_{P2}.

Note that the stabilizing of the other internal loops must also be separately secured, as there are: the class-AB loop through both the positive and negative output transistors; the current booster loop of the positive output transistor; both output-current limiter loops; and finally the CM feedback loop through the input stage and intermediate stage.

The limiting-pole frequency at the PNP side is:

$$f_1' = \frac{g_{m1}}{2\pi C_L (1 + C_{PI}/C_{MIP})} \tag{7.67}$$

with load capacitor C_L and transconductance of the output stage g_{m1}.

The zero-dB frequency of output and intermediate stages is:

$$f_0' = \frac{f_1'}{2} = \frac{g_{m2}}{2\pi C_{MI}} \tag{7.68}$$

with g_{m2} as the transconductance of half the second stage, and C_{MI} the average of C_{MIN} and C_{MIP}.

The zero-dB bandwidth of the whole amplifier becomes:

$$f_0'' = \frac{f_0'}{2} = \frac{f_1'}{4} = \frac{g_{m3}}{2\pi C_{M2}} \tag{7.69}$$

with g_{m3} as the transconductance of the second stage, and $C_{M2} = C_{P2}$ the second Miller-loop capacitors.

In the example of the NE5234, the overall bandwidth is 2 MHz at a total supply current of 700 µA.

It is interesting to see how the intermediate stage functions like a mirror in adding the voltage across C_{P2} (in series) with the voltage across C_{M2}. This provides the factor 2 in gain comparable to that of a mirror in the output of an input stage in subtracting the two differential output currents, thus adding the absolute values into a single output.

The overall low-frequency gain is:

$$A_0 = g_{m1} R_L g_{m2} r_{p2} g_{m3} r_{p3} \tag{7.70}$$

in which r_{p2} is the parasitic parallel resistance at the output of the intermediate stage, and r_{p3} that at the output of the input stage. In the example of the NE5234, the DC gain is in the order of 3×10^5 at a 10 kΩ load resistance.

The minimum supply voltage is set by the Darlington output transistors and the minimum selector, which take at least two diodes and a saturation voltage. Over the full temperature range the minimum single supply voltage is 1.8 V.

When we have to work at supply voltages down to 1 V, we can fold the Darlington output transistors and boost the bias currents as in the LM10 [30].

However, these stages have a poor HF behavior. Therefore, it is better to replace the VF-GA Darlington output stage for an GA-GA multipath nested output stage [31]. In fact, we then obtain a four-stage GA-GA-GA-GA configuration, which will be discussed in Sect. 7.9.

7.8.5 Conclusion

We have seen that the GA-GA-GA configuration may provide an abundance of gain. Therefore, it may be used for the design of precision low-voltage operational amplifiers and for heavy loads. Three important design examples have been evaluated.

7.9 GA-GA-GA-GA Configuration

It is a general rule that if we cannot improve the gain of a single GA stage by a bipolar Darlington or a CMOS cascode connection because of lack of supply-voltage room, we have to cascade more GA stages. In this paragraph we will present two four-stage GA-GA-GA-GA OpAmps, the first in full bipolar technology and the second in CMOS technology.

7.9.1 1 V R-R-In/Out Bipolar Class-AB OpAmp with MNMC

If we need to work at supply voltages as low as 1 V, we can no longer use bipolar Darlington output transistor combinations with stacked diode voltages. Nor would we like to use the folded Darlington output stage with a current boost [30] because of its poor HF behavior. Hence, the way out is to replace the VF-GA Darlington output stage by an GA-GA output stage. This leads to a bipolar OpAmp design with the GA-GA-GA-GA configuration of Fig. 7.42 [32] with a simplified schematic, and Fig. 7.43a, b with a full schematic.

Starting at the output, we have the complementary output transistors Q_{110} and Q_{120}. They have first inner-nested Miller capacitors C_{M11} and C_{M12}, respectively. The output transistors are driven by complementary driver transistors and mirrors,

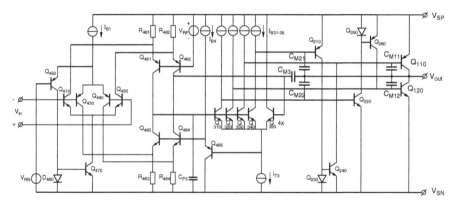

Fig. 7.42 Simplified schematic of a 1 V bipolar class-AB OpAmp with R-R (in)/out, GA-GA-GA-GA configuration, and multi-path nested Miller compensation.

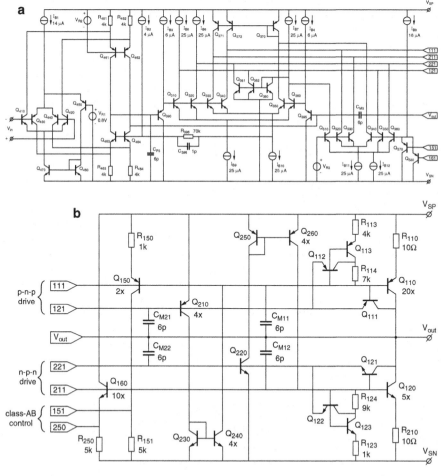

Fig. 7.43 (a) GA input stage, GA intermediate stage, and class-AB current control of a 1 V R-R-(in)/out OpAmp with GA-GA-GA-GA configuration. (b) GA output stage with multipath GA booster of a 1-V R-R-in/out OpAmp with GA-GA-GA-GA configuration

respectively: Q_{210} through mirror Q_{230}, Q_{240} and Q_{220} through mirror Q_{250}, Q_{260}. These drivers and Cascaded mirrors may provide the output transistors with an abundance of current, particularly because we can give the mirrors an extra gain.

The drivers combined with the mirrors have an non-inverting gain, so we may simply connect a second nest of Miller capacitors C_{M21} and C_{M22} to the inputs of the two drivers respectively.

We decided to apply a multipath to the nest so in order to regain a factor 2 in bandwidth. To this effect, we designed an intermediate stage that provides two output currents for each of the N and P output sides. The intermediate stage for the N output side Q_{310}, Q_{320}, Q_{360} has been terminated by two folded N-type mirrors Q_{380}, Q_{381}, Q_{382}. The folding has been done to give the mirror input side the same voltage level as the mirror output sides to balance out the effect of variations in the supply voltage through the Early effect. The intermediate stage for the P output side Q_{330}, Q_{340}, Q_{360} has been terminated by two P-type mirrors. This gives the mirror-input side the same voltage-level as the mirror-output side, which also balances the effect of supply voltage changes through the Early effect at the P-side. The current gain of the intermediate stage is further enhanced by folded Darlington transistors Q_{490}, Q_{495}. The differential intermediate stage is HF compensated by a third nest C_{M3} and C_{P3}. The capacitor C_{P3} is needed to provide the intermediate stage with an HF ground terminal. Without C_{P3} the left input of the intermediate stage will not be a virtual ground, but terminated by R_{590}. This would provide an HF sneaky path, that keeps the frequency compensation from roll-off at 6 dB/oct above $f_z = 1/2\pi R_{590}C_{P3}$. C_{P3} must be equal to C_{M3} for balancing purposes. The level-shift resistor R_{596} has been inserted to make it possible for the whole amplifier to function at a supply voltage of 1 V between the positive and negative supply rails.

The input stage Q_{410}, Q_{420}, Q_{430}, Q_{440} has been connected as a rail-to-rail input stage (see Sect. 4.4). The CM range is from rail to rail at a supply voltage of 1.8 V or higher, but at 1 V the circuit already functions with CM ranges around both rails. The switching between the PNP side and NPN side occurs at a voltage V_{RI}. The summing circuit Q_{461}, Q_{462}, Q_{563}, Q_{464} functions as a folded cascode and has a differential output. The summing circuit receives its common-mode feedback at the bottom by the intermediate stages through the level shift resistor R_{396}.

The class-AB regulator amplifier Q_{510} through Q_{560} has been connected in parallel to the intermediate stage. The class-AB amplifier also provides a multipath output to directly drive the output transistors. The output currents are measured in parallel with the output transistors by Q_{150} and Q_{160} which have series emitter resistors. The lowest of the output current is sensed by the minimum selector Q_{570}, Q_{580} and transferred to the class-AB regulator amplifier on its right-hand side. A reference voltage is offered on the left-hand side.

The limiting frequency f_1' is at the output transistors:

$$f_1' = \frac{g_{m1}}{2\pi C_L} \qquad (7.71)$$

with $g_{m1} = g_{m110} + g_{m120}$.

The g_m is taken at its quiescent current level. The zero-dB frequency of the feedforward path of the intermediate stage and output stage is:

$$f_0' = \frac{1}{2} f_I' = \frac{1}{2} \frac{g_{ml}}{2\pi C_L} \tag{7.72}$$

So the choice of the first nest of Miller capacitors of the feedforward path should be:

$$C_{MI} = \frac{g_{m3}}{2\pi f_0'} = 2 \frac{g_{m3}}{g_{ml}} C_L \tag{7.73}$$

with $C_{MI} = C_{MII} + C_{MI2}$ and $g_{m3} = g_{m330}$.

The zero-dB frequency of the gain path of the intermediate stage, driver stage, and output stage is:

$$f_0'' = f_0' = \frac{1}{2} f_I' \tag{7.74}$$

This results in a choice for the second nest of Miller capacitors:

$$C_{M2} = \frac{g_{m3}}{2\pi f_0''} = 2 \frac{g_{m3}}{g_{ml}} C_L \tag{7.75}$$

with $C_{M2} = C_{M21} + C_{M22}$ and $g_{m3} = g_{m340}$.

The zero-dB frequency of the whole amplifier is:

$$f_0''' = \frac{1}{2} f_0'' = \frac{1}{2} f_0' = \frac{1}{4} f_I' \tag{7.76}$$

The factor ½ results from the third nest, in which we have not used a second multipath to make the amplifier less complicated. The bandwidth of the four-stage amplifier with multipath output stage and driver stage is thus comparable with a three-stage amplifier without a multipath. The third nest of Miller capacitors is:

$$C_{M3} = \frac{g_{m4}}{2\pi f_0'''} = 4 \frac{g_{m4}}{g_{ml}} C_L \tag{7.77}$$

with $g_{m4} = g_{m410}$.

The transconductance g_{m4} is that of the R-R input stage with mirrored (through tail of intermediate stage) output of the summing circuit.

Care must be taken that the transconductance of the driver stage and current mirror is lower (see explanation with Fig. 6.25) than the transconductance of the feedforward path from the intermediate stage. Because at high frequencies the direct HF path must dominate the LF gain path. This results in:

$$4g_{m2} \; < \; g_{m3} \qquad\qquad (7.78)$$

with $g_{m2} = g_{m210} + g_{m220}$ and $g_{m3} = g_{m340}$.

At high output currents, this can become difficult to comply with. A solution can be to degenerate the emitters of the driver transistors Q_{210} and Q_{220}.

Bipolar output transistors have to be protected against saturation, otherwise their HF behavior drastically reduces and ringing may occur. They must also be protected against overdriving, otherwise these transistors can easily be destroyed. For this purpose, the saturation detectors Q_{111} and Q_{121} have been placed in parallel with the output transistors with their collectors and bases. If the output transistors saturate then the saturation detectors also saturate and feed the saturation current in reverse mode through their emitters to the bases of the regarding drivers. Current limitation is effected by a combination of small emitter-resistors in the output transistors, and sense transistors Q_{113} and Q_{123} with emitter resistors R_{113} and R_{123} with their bases in parallel with the output transistors. The collectors of the sense transistors have been connected through collector resistors R_{114} and R_{124} respectively, to the concerning bases. If the currents of these sense transistors become so large that they saturate, this saturation current is sensed by additional sense detectors Q_{112} and Q_{122}. Their emitter currents limit the current at the base of the driver transistors.

The result is a 1 V Bipolar OpAmp with a bandwidth of 3 MHz at a load of 100 pF and a total quiescent supply current of 700 μA, a DC gain of 100 dB at a load of 10 kΩ, and a maximum current of 10 mA. These values can be scaled up or down easily by a factor 10 [32].

7.9.2 1.2 V R-R-Out CMOS Class-AB OpAmp with MHNMC

As a last example of the use of more than three GA-stages in cascade, a 1.2 V CMOS OpAmp with an GA-GA-GA-GA configuration will be given. It has a multipath hybrid nested Miller compensation. The circuit has an abundance of gain and can drive low-impedance loads. Nowhere are more than one diode and two saturation voltages stacked. The output transistors are driven by transistors which need only one saturation voltage. This results in a supply voltage as low as 1.2 V over temperature and process variations. The circuit is shown in Fig. 7.44a [33] with a simplified schematic, and Fig. 7.44b with a full schematic.

The output stage M_{110} and M_{120} has a first Miller nest of capacitors C_{M11} and C_{M12}. The output stage is driven by a driver stage M_{210}, M_{220}. The intermediate stage M_{310} and M_{320} has a second Miller nest of capacitors C_{M31} and C_{M32}. Across the intermediate, driver, and output stages the outer nest is made up by C_{M2}. The input stage M_{410} and M_{420} is followed by a folded cascode M_{431}, M_{432} and loaded with a mirror M_{433}, M_{434}. The mirror also produces two feedforward paths through M_{435} and M_{436}, which directly drive the output transistors in a mulitpath way. The mirror and intermediate stage-transistors are chosen of equal type, so that the mirror transistors have about equal drain-source voltages and do not contribute to supply-voltage dependent offset.

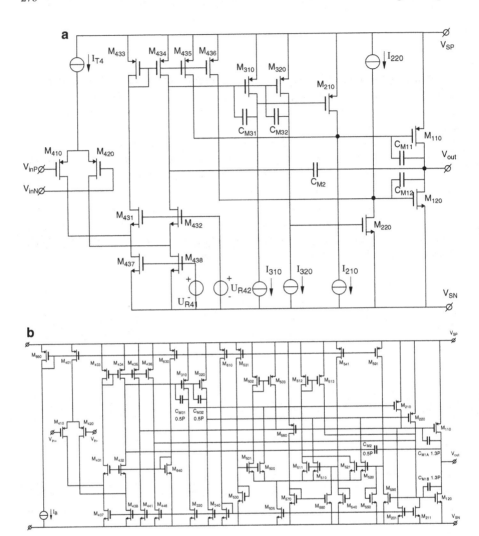

Fig. 7.44 (a) Simplified schematic of a 1.2 V R-R-out CMOS class-AB OpAmp with GA-GA-GA-GA configuration, and multi-path hybrid nested Miller compensation. (b) 1.2 V R-R-out CMOS class-AB OpAmp with GA-GA-GA-GA configuration with multipath hybrid nested Miller compensation

The class-AB regulator has output current sensors M_{560} with mirror M_{570}, M_{580} and M_{590}. The sense currents are first subtracted from current sources M_{541} and M_{591}, and then fed into the measuring diodes M_{540} and M_{550}, respectively. By this action a minimum current in one of the output transistors is translated into a maximum current in one of the measuring diodes. A class-AB control amplifier M_{500} through M_{521} has been connected to the measuring diodes as an and-gate or "maximum selector" at its right-hand side. The four outputs of the control amplifier bias the driver-stage and output transistors in a multipath nested manner. This ensures a robust and stable class-AB biasing of the output transistors [33]. It has further been explained with Fig. 5.39.

The frequency compensation has been explained with Fig. 6.28 [12].

The limiting-pole frequency is:

$$f_1' = \frac{g_{ml}}{2\pi C_L} \tag{7.79}$$

in which C_L is the output load capacitor, and $g_{ml} = g_{ml10}$.

The zero-dB frequency f_0' of the driver and output stage combination must be half that of the limiting-pole frequency, so:

$$f_0' = \frac{1}{2} f_1' \tag{7.80}$$

The gain around the driver and output stage is determined by the inverting intermediate stage M_{310}, M_{320} with a gain of C_{M2}/C_{M3}, which leads to the choice of:

$$\frac{g_{m2}}{2\pi C_{M1}} = \frac{1}{2} \frac{C_{M2}}{C_{M3}} \frac{g_{ml}}{2\pi C_L} \tag{7.81}$$

with $g_{m2} = g_{m210} + g_{m220}$ and $C_{M1} = C_{M11} + C_{M12}$ and $C_{M3} = C_{M31} + C_{M32}$.

When a multipath connection is made directly from the input stage to drive the output transistors, the overall bandwidth f_1'' may be taken equal to f_0' or half f_1', so:

$$f_0'' = f_0' = \frac{1}{2} f_1' \tag{7.82}$$

which leads to a choice for C_{M2}:

$$\frac{g_{m4}}{2\pi C_{M2}} = \frac{1}{2} \frac{g_{ml}}{2\pi C_L} \tag{7.83}$$

in which g_{m4} is the gain path of the transconductance of the input stage M_{410}, M_{420} provided through the mirror M_{433}, M_{434} to the overall nest C_{M2}, while the direct path of the transconductance $\frac{1}{2} g_{m4}$ is provided through M_{435} and M_{436} to the first nest C_{M1A} and C_{M1B} respectively.

It must be prevented that the gain-path dominates the HF path at high frequencies. This leads to the choice of a lower transconductance of the driver stage.

Hence:

$$\frac{C_{M2}}{C_{M3}} \frac{g_{m2}}{2\pi C_{M1}} \leq \frac{1}{3} \frac{\frac{1}{2} g_{m4}}{2\pi C_{M1}} \tag{7.84}$$

As the gain path does not contribute much at high frequencies, the constraint for the bandwidth f_3' of the intermediate stage is also relaxed to $f_3' \geq \frac{1}{3}f_1'$. This leads to a choice:

$$\frac{g_{m3}}{2\pi C_{M2}} \leq \frac{1}{3}\frac{g_{m1}}{2\pi C_L} \tag{7.85}$$

with $g_{m3} = g_{m310} + g_{m320}$.

Finally, we must match the 6 dB roll-off of the gain path and the multipath, to avoid a pole-zero doublet.

This leads to the choice:

$$\frac{g_{m4}}{2\pi C_{M2}} = \frac{\frac{1}{2}g_{m4}}{2\pi C_{M1}} \tag{7.86}$$

7.9.3 Conclusion

The conclusion can be drawn that the hybrid nested Miller compensation technique leads to robust GA-GA-GA-GA amplifiers with an abundance of gain and, when provided with a multipath, a bandwidth comparable to that of a regular Miller compensated OpAmps. As an example two CMOS OpAmp were made, one with and one without a multipath. At a bias current I_B of 10 μA, the total quiescent current was 300 μA with a supply voltage of 1.5 V, a bandwidth was obtained of 2 and 6 MHz, respectively for the non-multipath and the multipath version, at a load capacitance of 10 pF.

The open-loop voltage gain was 120 dB at a load resistor of 10 kΩ. The OpAmp does not use resistors and can be easily scaled up or down. At a bias current I_B of 1 μA, a total supply current of only 15 μA, a supply voltage of 1.2 V, and a bandwidth was obtained of 0.2 and 0.6 MHz, respectively [33].

7.10 Problems and Simulation Exercises

7.10.1 Problem 7.1

Figure 7.2b shows a folded-cascode operational amplifier with class-A output stage. Considering the MOS devices sized $W/L_{15} = W/L_{16} = 3.5$ $W/L_{13} = 3.5 W/L_{14} = 35\,\mu/1\,\mu$, $W/L_{17} = W/L_{18} = 100\,\mu/3\,\mu$, $W/L_{19} = 60\,\mu/3\,\mu$ and $W/L_{20} = W/L_{21} = 20\,\mu/3\,\mu$ biased with the current source $I_{B1} = 20\,\mu$A at a supply voltage of $V_{SP} = -V_{SN} = 1.35$ V, design the input pair and the current sources M_3, M_{11} and M_{12} so the gain of the amplifier will be 60 dB with capacitive load $C_L = 1$ pF and resistive load $R_L = 5$ MΩ, and under these conditions the slew-rate will be $S_r = 40$ V/μs. Transistor parameters are $V_{THN} = 0.5$ V, $V_{THP} = -0.6$ V, $K_N = 56\,\mu$A/V^2, $K_P = 16\,\mu$A/V^2, $\lambda_N = \lambda_P = 0.1$ V^{-1}. Calculate the 0-dB frequency and the biasing voltages V_{B1}, V_{B2}.

7.10.1.1 Solution 7.1

The tail current of the input stage limits the slew rate of the whole amplifier under given capacitive load, the second stage being only a current follower

$$S_r = \frac{I_{D3}}{C_L} = \frac{2I_B}{C_L} \tag{7.87}$$

which for the given slew-rate value requires an I_B current of

$$I_B = \frac{S_r C_L}{2} = 20\,\mu\text{A} \tag{7.88}$$

The corresponding sizes for the current source transistors are determined taking care to keep the same channel length for the current mirrors using M_{19} as reference

$$\frac{W}{L_3} = \frac{120\,\mu}{3\,\mu} \tag{7.89}$$

Transistors M_{11} and M_{12} are sized knowing their drain current equal to $2I_B$

$$\frac{W}{L_{11}} = \frac{W}{L_{12}} = \frac{2I_B}{I_{B1}}\frac{W}{L_{21}} = \frac{40\,\mu}{3\,\mu} \tag{7.90}$$

Without load resistance, the DC voltage gain of the amplifier is

$$A_{V0} = g_{m1}\left(\mu_{14}r_{ds12} \,\|\, \mu_{16}r_{ds18} \,\|\, R_L\right) \tag{7.91}$$

For a saturated MOS transistor, the voltage gain μ and r_{ds} above can be calculated based on device sizes and drain current

$$\mu = \frac{1}{\lambda\sqrt{\dfrac{2I_D}{K\dfrac{W}{L}}}}$$
$$r_{ds} = \frac{1}{\lambda I_D} \tag{7.92}$$

Numerically, using device sizes and the biasing given by Eq. 7.92, this allows calculating the amplifier output impedance

$$Z_0 = 31.2\,M\Omega \tag{7.93}$$

which is much larger than the load resistance

$$Z_0 \gg R_L \tag{7.94}$$

Thus, g_{ml} can be calculated as

$$g_{ml} = \frac{A_{V0}}{R_L} = 200\,\mu S \qquad (7.95)$$

Using the value above in Eq. 7.8 the sizes of input pair transistors result

$$\frac{W}{L_{1,2}} = \frac{g_{ml,2}^2}{4K_P I_B} = 32 \qquad (7.96)$$

For low input offset, the channel length of the input pair transistors is usually non-minimal, similar to the current mirrors

$$\frac{W}{L_{1,2}} = \frac{100\,\mu}{3\,\mu} \qquad (7.97)$$

Because the input transistors are actually close to weak inversion, their size is usually large in real circuits. The 0-dB frequency of the amplifier designed above will be

$$f_0 = \frac{g_{ml}}{2\pi C_L} = 32\,MHz \qquad (7.98)$$

The biasing sources V_{B1}, V_{B2} should be designed to keep $M_{17,18}$ and $M_{11,12}$ in saturation even when the input pair is slewing, sourcing all available current through one of the input transistors

$$\begin{aligned} V_{B1} &= (V_{GS16} + V_{sat18})|_{I_{D16}=2I_B} = 0.977 + 0.380 = 1.36\,V \\ V_{B2} &= (V_{GS14} + V_{sat12})|_{I_{D12}=2I_B} = 0.077 + 0.330 = 1.10\,V \end{aligned} \qquad (7.99)$$

7.10.2 Problem 7.2

Figure 7.16 shows a rail-to-rail output two-stage amplifier compensated by a Miller configuration which does not introduce a positive zero in its AC behavior. Considering the MOS devices sized as $W/L_{28} = 2\,W/L_{14} = 10\,W/L_{32} = 80\,\mu/2\,\mu$, $W/L_{27} = 2\,W/L_{13} = 10\,W/L_{31} = 100\,\mu/2\,\mu$, $W/L_{21} = W/L_{22} = 60\,\mu/2\,\mu$, $W/L_{23} = W/L_{24} = 20\,\mu/2\,\mu$, $W/L_{25} = W/L_{26} = 10\,\mu/2\,\mu$, $W/L_{12} = 40\,\mu/1\,\mu$ and $W/L_{11} = 80\,\mu/1\,\mu$, the biasing source $I_B = 10\,\mu A$ and the amplifier loaded with only the capacitor $C_L = 10$ pF, calculate the D-C voltage gain A_{V0} and the Miller compensation capacitor C_M for a phase margin of the compensated amplifier $\varphi_m = 60°$. Transistor parameters are $V_{THN} = 0.5$ V, $V_{THP} = -0.6$ V, $K_N = 56\,\mu A/V_2$, $K_P = 16\,\mu A/V^2$, $\lambda_N = \lambda_P = 0.1$ V^{-1}.

7.10.2.1 Solution 7.2

The biasing currents for M_{14} and M_{28} can be calculated knowing the reference current through M_{32}.

$$I_{D14} = I_{D32}\frac{W\ L}{L_{14}W_{32}} = 50\,\mu A$$

$$I_{D28} = I_{D32}\frac{W\ L}{L_{28}W_{32}} = 100\,\mu A \tag{7.100}$$

The DC voltage gain with no resistive load is given by Eq. 7.26 as

$$A_{V0} = \mu_2\mu_1\mu_3 \tag{7.101}$$

each μ voltage gain corresponding to M_{21}–M_{22}, M_{11} and M_{12} respectively

$$A_{V0} = \frac{1}{\lambda_P\sqrt{\dfrac{2I_{D22}}{K_P\dfrac{W}{L_{22}}}}\lambda_N}\frac{1}{\sqrt{\dfrac{2I_{D11}}{K_N\dfrac{W}{L_{11}}}}\lambda_N}\frac{1}{\sqrt{\dfrac{2I_{D12}}{K_N\dfrac{W}{L_{12}}}}} = 96\,dB \tag{7.102}$$

There is a huge gain which can be obtained with this amplifier structure. The price is the reduced input and output common-mode ranges.

The Miller compensation is calculated forcing the unity gain bandwidth to be half the value of the nondominant pole introduced by the capacitive load

$$f_0 = \frac{1}{2}\frac{g_{m11}}{2\pi C_L} = 10.6\,MHz \tag{7.103}$$

As this frequency should be set by the transconductance of the input stage and the Miller capacitor, the value of C_M results

$$C_M = \frac{g_{m22}}{2\pi f_0} = 6.6\,pF \tag{7.104}$$

7.10.3 Problem 7.3

The class-AB amplifier in Fig. 7.33 is designed with transistors sized $W/L_{12} = 7.5\,\mu/2\,\mu$, $W/L_{17} = W/L_{18} = 2\,\mu/2\,\mu$, $W/L_{11} = 26\,\mu/2\,\mu$, $W/L_{15} = W/L_{16} = 7\,\mu/2\,\mu$, $W/L_{25} = W/L_{26} = W/L_{33} = 42\,\mu/2\,\mu$, $W/L_{31} = W/L_{32} = 10\,\mu/2\,\mu$, $W/L_{13} = 14\,\mu/2\,\mu$, $W/L_{14} = 4\,\mu/2\,\mu$, $W/L_{21} = W/L_{22} = 4\,\mu/2\,\mu$, $W/L_{23} = W/L_{24} = 6\,\mu/2\,\mu$, $W/L_{27} = W/L_{28} = 21\,\mu/2\,\mu$ and biased with $I_{B11} = I_{B12} = 0.5\,\mu A$. For a capacitive load only, $C_L = 10\,pF$, calculate the Miller capacitors C_{M11} and C_{M12} so the phase margin becomes $\varphi_m = 60°$. Calculate the unity gain bandwidth of the compensated amplifier. What is the maximum current which can be supplied by the class-AB output stage? Transistor parameters are $V_{THN} = 0.5$ V, $V_{THP} = -0.6$ V, $K_N = 56$ μA/V^2, $K_P = 16$ μA/V^2, $\lambda_N = \lambda_P = 0.1$ V^{-1}.

7.10.3.1 Solution 7.3

The drain currents for M_{33}, M_{25} and M_{26} are all equal because of equal sizing of these transistors

$$I_{D33} = I_{D25} = I_{D26} = I_{D15} \frac{W}{L_{33}} \frac{L}{W_{15}} = 3\,\mu A \tag{7.105}$$

The quiescent current of output transistors can be calculated using one of the translinear loops, for example V_{GS11}-V_{GS13}-V_{GS16}-V_{GS15}

$$I_{D11} = W/L_{11} \left(\sqrt{\frac{I_{D15}}{W/L_{15}}} + \sqrt{\frac{I_{D16}}{W/L_{16}}} - \sqrt{\frac{I_{D13}}{W/L_{13}}} \right) = 1.1\,\mu A \tag{7.106}$$

This value is equal with I_{D12} at quiescent operating point. Knowing the drain current and the sizes of the transistors, the DC gain voltage can be calculated

$$A_{V0} = g_{m3} r_{ds22} \mu_{24} \mu_{12} = \frac{\sqrt{2K_P \dfrac{W}{L_{32}} I_{D32}}}{\lambda_N^3 I_{D22} \sqrt{\dfrac{2I_{D24}}{K_N \dfrac{W}{L_{24}}}} \sqrt{\dfrac{2I_{D12}}{K_N \dfrac{W}{L_{12}}}}} \tag{7.107}$$

The Miller capacitors result from Eq. 7.59

$$C_{M11} = C_{M12} = C_L \frac{g_{m3}}{g_{m12}} = 7.6\,pF \tag{7.108}$$

Unity gain bandwidth of the compensated amplifier can now be calculated as

$$f_0' = \frac{g_{m3}}{2\pi C_L} = 246\ kHz \tag{7.109}$$

The maximum output push or pull current can be calculated if all the available current from M_{28} flows either through M_{13} or M_{14}. Using (Eq. 7.106) under the assumption that $I_{D13} = 0$, the push current limit results

$$I_{D11} = W/L_{11} \left(\sqrt{\frac{I_{D15}}{W/L_{15}}} + \sqrt{\frac{I_{D16}}{W/L_{16}}} \right)^2 = 7.4\,\mu A \tag{7.110}$$

7.10.4   Problem 7.4

The compact amplifier shown in Fig. 7.38 can be tuned to operate at 1.2 V supply voltage and 10 μA supply current. Considering the transistors sized as shown, adjust the bias current to the required value to operate the amplifier at the aforementioned supply current, then calculate the Miller compensation capacitors C_{M1A}, C_{M2A} and the 0-dB bandwidth for a phase margin $\varphi_m = 60°$ when the amplifier is loaded with a capacitor $C_L = 10$ pF. The PMOS transistors have $K_P = 16$ μA/V$_2$.

7.10.4.1   Solution 7.4

All the biasing currents are obtained from I_B in this circuit so the sum of all these currents can be obtained as a function of I_B.

$$I_{D321} = I_{D204} = I_{D206} = 2I_B$$

$$I_{D239} = I_{D238} = I_{D501} = I_B$$

$$I_{D112} = I_{D113} = \frac{1}{2}I_{D219} = \frac{1}{2}I_B \qquad (7.111)$$

$$I_{D101} = \frac{W}{L_{101}}\frac{L}{W_{111}}\frac{1}{2}I_B = 10I_B$$

Summing all the currents calculated above except for the biasing current itself, the total supply current is

$$I_{supply} = 20I_B \qquad (7.112)$$

which requires a bias current

$$I_B = \frac{10\,\mu A}{20} = 0.50\,\mu A \qquad (7.113)$$

In order to calculate the Miller capacitors and the bandwidth of the compensated amplifier, two g_m values must be calculated

$$g_{m3} = g_{m301} = g_{m302} = \sqrt{2K_P\frac{W}{L_{301}}I_{D301}} = 22.6\,\mu A/V \qquad (7.114)$$

The above value seems to be too large, considering the level of drain current for the input transistors. The maximum g_m for a MOS transistor operated at a drain current I_D is attained when the transistor works in weak inversion and is typically limited to $25I_D$ at room temperature, similar to a bipolar transistor. Therefor, the value of g_{m3} will be limited to

$$g_{m3} = 25I_{D301} = 12.5\,\mu A/V \qquad (7.115)$$

In the same manner, the output g_m can be calculated

$$g_{m1} = 2g_{m100} = 50I_{D100} = 250\,\mu A/V \qquad (7.116)$$

The Miller capacitors are compensating one output transistor each, so their value is

$$C_{M1A} = C_{M2A} = \frac{g_{m3}}{g_{m1}}C_L = 0.5\,pF \qquad (7.117)$$

and the corresponding 0-dB frequency of the compensated amplifier becomes

$$f'_0\frac{g_{m3}}{2\pi(C_{M1A} + C_{M2A})} = 2\,MHz \qquad (7.118)$$

7.10.5   Simulation Exercise 7.1

Use the AC simulation circuits shown in Fig. 2.13 to plot the gain and phase characteristics of the folded-cascode amplifier in Fig. 7.45. Note the single pole characteristic and change the unity gain bandwidth to double its value by adjusting the load capacitor and/or the input stage g_m. Also run AC simulation with and without a load resistor and note the major change in DC gain. Using the original and the modified amplifier, run transient simulations with the amplifier in a non-inverting configuration and using a pulsed voltage for input signal. Estimate the slew-rate of

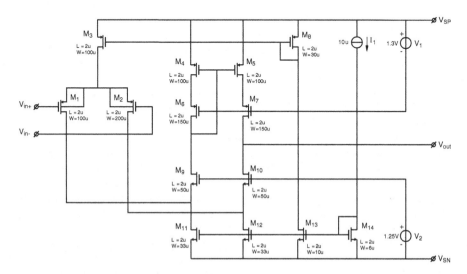

Fig. 7.45 Folded cascode amplifier

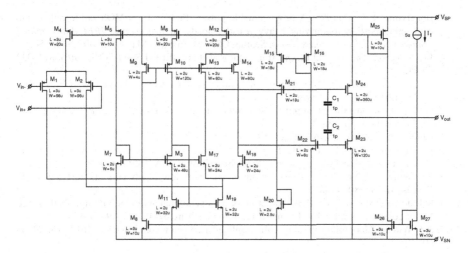

Fig. 7.46 Compact low voltage amplifier

both versions. By changing the input signal to a variable amplitude sinusoidal voltage and running transient and Fourier analyses, a maximum limit can be obtained for the output voltage by imposing a limit for the accepted distortion THD $= 0.1\%$.

7.10.6 Simulation Exercise 7.2

The class-AB compact amplifier shown in Fig. 7.46 can be operated at a supply voltage as low as 1.2 V. Run AC simulation for gain and phase characteristic, then transient simulations for slew-rate and maximum output range in a manner similar to the one described in the previous exercise. During the slew-rate measurement, plot the drain currents of both the output transistors and observe the minimum current which is drawn by the inactive transistor.

References

1. J.H. Huijsing, F. Tol, Monolithic operational, amplifier design with improved HF behavior. IEEE J Solid-St Circ **SC-11**, 323–328 (1976)
2. M. Steyaert, W. Sansen, Opamp design towards maximum gain-bandwidth, in *Analog Circuit Design*, ed. by J.H. Huijsing et al. (Kluwer Academic Publishers, Boston, MA, 1993), pp. 63–85
3. R. Caprio, Precision differential voltage-current converter. Electron Lett **9**, 147–148 (1973)
4. B.W. Lee, B.J. Shen, A high-speed CMOS amplifier with dynamic frequency compensation. J Semicustom ICs **8**(3), 42–46 (1991)
5. L.G.A. Callewaert, W. Sansen, Class AB CMOS amplifiers with high efficiency. IEEE J Solid-St Circ **25**(3), 684–691 (1990)
6. K. Bult, G.J.G.M. Geelen, A fast settling CMOS op amp for SC-circuits with 90-dBDC gain. IEEE J Solid-St Circ **25**(3), 1379–1383 (1990)

7. Hogervorst R et al. (1995) A programmable power-efficient 3-V CMOS rail-to-rail opamp with gain boosting for driving heavy resistive loads, in *Proceedings IEEE International Symposium on Circuits and Systems*, Seattle, USA, 30 Apr to 3 May 1995, pp. 1544–1547

8. J.E. Solomon, The monolithic Op Amp: a tutorial study. IEEE J Solid-St Circ **9**(6), 314–332 (1974)

9. P.R. Gray, R.G. Meijer, MOS operational amplifier design – a tutorial overview. IEEE J Solid-St Circ **17**(6), 969–982 (1982)

10. G.M. Cotreau operational amplifiers and voltage regulators. ISSCC 85, in *Proceedings*, vol. 28, THAM 11.3, pp. 138–139

11. Fairchild Data Sheet of μA 741 opamp

12. R.G.H. Eschauzier, J.H. Huijsing, *Frequency Compensation Techniques for Low-Power Operational Amplifiers* (Kluwer Academic Publishers, Boston, MA, 1995), p. 245

13. R.J. Widlar, M. Yamatake, A fast-settling op amp with low supply current. IEEE J Solid-St Circ **24**(3), 796–802 (1989)

14. K.J. de Langen et al., A 1-GHz bipolar class-AB operational amplifier with multipath nested Miller compensation for 76-dB gain. IEEE J Solid-St Circ **32**(4), 488–498 (1997)

15. R.J. Widlar, Monolithic op amp with simplified frequency compensation. IEEE **15**, 58–63 (1967)

16. Signetics/Philips Data Sheet NE 5534

17. R.G.H. Eschauzier et al., A 100-MHz 100-dB operational amplifier with multipath nested Miller compensation structure. IEEE J Solid-St Circ **27**(12), 1710–1717 (1992)

18. D.M. Monticelli, A quad CMOS single-supply op amp with rail-to-rail output swing. IEEE J Solid-St Circ **21**(6), 1026–1033 (1986)

19. W.C.S. Wu et al., Digital-compatible high-performance operational amplifier with rail-to-rail input and output ranges. IEEE J Solid-St Circ **29**(1), 63–66 (1994)

20. R. Hogervorst et al., A compact power-efficient 3 V CMOS rail-to-rail input/output operational amplifier for VLSI cell libraries. IEEE J Solid-St Circ **29**(12), 1505–1513 (1994)

21. R. Hogervorst, J.H. Huijsing, *Design of Low-Voltage Low-Power Operational Amplifier Cells* (Kluwer Academic Publishers, Boston, MA, 1996), pp. 35–63, 147–203, 207

22. K.J. de Langen et al., Translinear circuits low-voltage operational amplifiers, in *Analog Circuit Design*, ed. by W. Sansen et al. (Kluwer Academic Publishers, Boston, MA, 1996), pp. 357–386

23. K.J. de Langen, J.H. Huijsing, Compact low-voltage power efficient operational amplifier cells for VLSI. IEEE J Solid-St Circ **33**(10), 1482–1496 (1998)

24. E. Seevinck et al., A low-distortion output stage with improved stability for monolithic power amplifiers. IEEE J Solid-St Circ **23**(3), 794–801 (1988)

25. K.J. de Langen, J.H. Huijsing, *Ultimate Low-Voltage Compact Three-Stage Operational Amplifiers Using Nested Miller and Mirrored Nested Miller Compensation* (Kluwer Academic Publishers, Boston, MA, 1999), p. 249

26. K.J. de Langen, J.H. Huijsing, *Compact Low-Voltage and High-Speed CMOS, BICMOS, and Bipolar Operational Amplifiers* (Kluwer Academic publishers, Boston, MA, 1999), p. 249

27. K.J. de Langen, J.H. Huijsing, *Compact Low-Voltage Three Stage BiCMOS Operational Amplifier Cell* (Kluwer Academic Publishers, Boston, MA, 1999), p. 249

28. J.H. Huijsing, D. Linebarger, Low-voltage operational amplifier with rail-to-rail input and output ranges. IEEE Solid-St Circ **20**(6), 1144–1150 (1985)

29. M.J. Fonderie, J.H. Huijsing, *Design of Low-Voltage Bipolar Operational Amplifiers* (Kluwer Academic Publishers, Boston, MA, 1993), p. 193

30. R.J. Widlar, Low-voltage techniques. IEEE J Solid-St Circ **13**(6), 838–846 (1978)

31. M.J. Fonderie et al., I-V operational amplifier with rail-to-rail input and output ranges. IEEE J Solid-St Circ **24**(6), 1551–1559 (1989)

32. M.J. Fonderie, J.H. Huijsing, Operational amplifier with 1-V rail-to-rail multipath driven output stage. IEEE J Solid-St Circ **26**(12), 1817–1824 (1991)

33. R.G.H. Eschauzier et al., A programmable 1.5 V Class-AB operational amplifier with hybrid nested Miller compensation for 120 dB gain and 6 MHz UGF. IEEE J Solid-St Circ **29**(12), 1497–1504 (1994)

Chapter 8
Fully Differential Operational Amplifiers

Abstract As the supply and signal voltages go down to lower values from 30, 12, 5, 3, 2, and finally 1 V, the signal-to-noise-and-interference ratio becomes increasingly worse. An important way to cope with this problem is to use fully differential signal paths. The differential peak-to-peak signal then becomes maximally twice the total supply voltage $V_S = V_{SP} - V_{SN}$. But even more important will be that the influence of substrate interference on the two balanced signals will largely cancel one another. All kinds of amplifiers, filters, sigma-delta converters, and other circuits using fully differential OpAmps may thus be designed in a fully balanced or differential way. In this chapter several practical design examples of fully differential operational amplifiers are presented.

As the supply and signal voltages go down to lower values from 30, 12, 5, 3, 2, and finally 1 V, the signal-to-noise-and-interference ratio becomes increasingly worse. An important way to cope with this problem is to use fully differential signal paths. The differential peak-to-peak signal then becomes maximally twice the total supply voltage $V_S = V_{SP} - V_{SN}$. But even more important will be that the influence of substrate interference on the two balanced signals will largely cancel one another. All kinds of amplifiers, filters, sigma-delta converters, and other circuits using fully differential OpAmps may thus be designed in a fully balanced or differential way.

For this purpose, we have to design fully differential OpAmps which have two outputs of which the voltages accurately move opposite to each other in regard to a constant common output reference voltage.

The main additional problem is to design a common-mode output control circuit that is accurate and that can handle voltages that move close to the supply-rail voltages, so that the amplifier can function at low supply voltages. This will be elaborated in this chapter for the GA-CF, GA-CF-GA, and the GA-GA-GA-GA configuration with several different solutions.

The control of the differential output voltages of this chapter is in contrast with the OFA approach (see Chap. 9) in which the relation between the output currents is controlled in an accurate differential way.

© Springer International Publishing Switzerland 2017

J. Huijsing, *Operational Amplifiers*, DOI 10.1007/978-3-319-28127-8_8

8.1   Fully Differential GA-CF Configuration

The simplest OpAmp is the GA-CF configuration. It has been shown in Fig. 7.2b. with a single-ended output. To properly connect the differential input to the single output a mirror connection was needed. If we want a differential output this mirror connection must be removed and a common-mode output voltage control must be added.

8.1.1   Fully Differential CMOS OpAmp with Linear-Mode CM-Out Control

A very simple CM output control circuit arises when we use CMOS transistors in their linear mode as CM sensors. An example of a simple circuit is shown in Fig. 8.1a [1, 2].

The common mode output control consists of two transistors M_9 and M_{10} biased in their linear resistive region and connected at their drains with an equal drain-source voltage $V_{DS9} = V_{DS10} = V_{DS9,10}$ of only 100 mV, while the effective gate-source voltage $V_{GT} = V_{GS} - V_{TH}$ is larger.

$$V_{GT} = V_{GS} - V_{TH} > V_{DS} \tag{8.1}$$

Under this condition, the channel is not pinched-off anywhere, and we can write for the drain current

$$I_D = \mu C_{ox} \frac{W}{L} V_{DS} \left(V_{GT} - \frac{1}{2} V_{DS} \right) \tag{8.2}$$

This relation is linearly dependent on V_{GT} or V_G. So we connect one gate V_{G9} to one output V_{out-} and the other gate V_{G10} to the other output V_{out+}, and we keep the sum of the two drain currents $I_{D9} + I_{D10} = I_C$ constant, and the drain voltages V_{DS9}, V_{DS10} equal: $V_{DS9} = V_{DS10}$. The average gate voltage $(V_{G9} + V_{G10})/2$ will now regulate itself at the common-mode level V_{CM} of the gate voltage V_{G4} of a model transistor M_4 that is driven with the same current density as M_9 and M_{10}.

This follows from:

When we choose $I_C = 2I_{D4}$, $W_9/L_9 = W_{10}/L_{10} = W_4/L_4$, and

$$I_C = I_{D9} + I_{D10} = \mu C_{OX} \frac{W_{9,10}}{L_{9,10}} V_{DS9,10} (V_{GT9} + V_{GT10} - V_{DS9,10}) \tag{8.3}$$

$$2I_{D4} = \mu C_{OX} \frac{W_4}{L_4} V_{DS4} (2V_{GT4} - V_{DS4}) \tag{8.4}$$

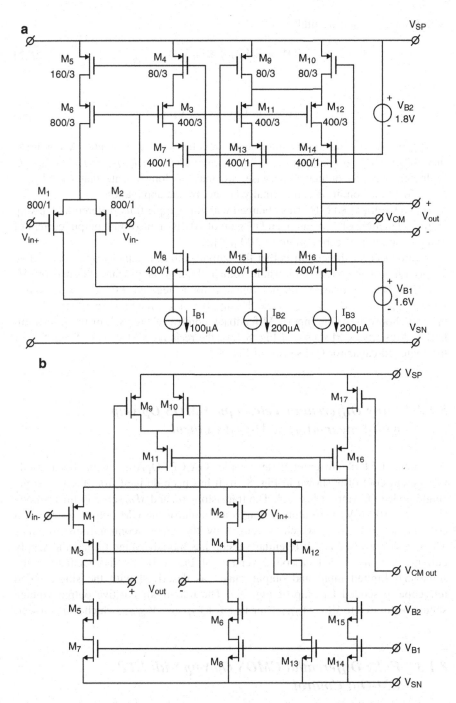

Fig. 8.1 (a) Fully differential operational amplifier with an GA-CF configuration having common-mode output-voltage control with CMOS transistors in their linear mode. (b) Fully differential telescopic operational amplifier with GA-CF configuration with common-mode output-voltage control with CMOS transistors in their linear mode

with $V_{DS9,10} = V_{DS4}$, we find:

$$(V_{G9} + V_{G10})/2 = V_{G4} \tag{8.5}$$

and hence:

$$(V_{out-} + V_{out+})/2 = V_{CM} \tag{8.6}$$

The result is a robust common-mode regulation of the output. A condition therefore is that no common-mode output signal current $I_{outCM} = (I_{out-} + I_{out+})/2$ is drawn from the output. Otherwise, we may no longer state that $I_{D9} + I_{D10} = I_C = 2I_{D4}$. This condition can normally be met by the application.

A bandwidth of 80 MHz was obtained with this simple fully differential OpAmp at load capacitors of 1 pF, an open DC gain of 70 dB, a maximum output current of 100 μA, and a quiescent current of 500 μA [2].

A drawback of the circuit is that the output voltages cannot reach the positive supply-rail voltage within a threshold voltage $|V_{TH}|$ of the p-channel transistors M_9 and M_{10}. These transistors do not function for a negative effective gate-source voltage $|V_{GT}| = |V_{GS}| - |V_{TH}|$. To use the supply voltage range more effectively, we have to choose from the input CM feedback control of Fig. 8.3, or the rail-to-rail buffered resistive CM sensor of Fig. 8.4, or the resistive CM sensor of Fig. 8.5, or the switched-capacitor CM sensor of Fig. 8.7.

8.1.2 Fully Differential Telescopic CMOS OpAmp with Linear-Mode CM-Out Control

A simple CMOS differential telescopic GA-CF OpAmp with linear-mode CM-output control is shown in Fig. 8.1b. It has been derived from the telescopic single-ended OpAmp of Fig. 7.3. The transistors M_9 and M_{10} sense in linear-mode ($V_{DS} \approx$ 50–100 mV) the output voltages and control the CM output voltage at a constant level $V_{CM\ out}$ which is sensed by the model control transistor M_{17}. The telescopic cascoded differential version has the advantage of half the supply current over the folded cascoded version of Fig. 8.1a. A disadvantage is the somewhat limited input and output range, as described with the single-ended telescopic cascoded OpAmp of Fig. 7.3. The maximum positive output voltage should be limited to $V_{SP} - V_{TH9,10}$, or $V_{in\ CM} + V_{TH1,2} - V_{sat\ 3,4}$, whichever is lower.

8.1.3 Fully Differential CMOS OpAmp with LTP CM-Out Control

A set of two long-tail pairs (LTP) can also sense the common mode output voltage. An example of a simple circuit is shown in Fig. 8.2 [3, 4].

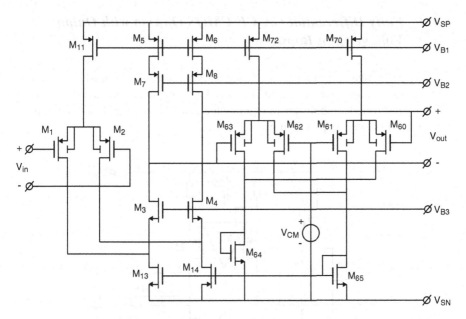

Fig. 8.2 Fully differential operational amplifier with an GA-CF configuration having common-mode output voltage control with a set of two long-tail pairs

Two long-tail pairs M_{60}, M_{61} and M_{62}, M_{63} are connected with their gates between the desired CM voltage level V_{CM} and each of the output terminals V_{out+} and V_{out-}.

Each long-tailed pair has a nonlinear transconductance. But if the nonlinearities of both pairs is identical, we can still accurately measure differences in the input voltage of each pair. The sum of the output currents of the transistors M_{61} and M_{62}, whose gates are connected to V_{CM} is used to control the common-mode output voltages CM through M_{65}, M_{13}, M_{14}. This sum current is constant if the gate voltage of M_{60} moves opposite to that of the gate voltage of M_{63}. This results in:

$$V_{G60} - V_{G61} = V_{G62} - V_{G63} \qquad (8.7)$$

or

$$(V_{G60} + V_{G63})/2 = (V_{G61} + V_{G62})/2 = V_{CM} \qquad (8.8)$$

Note that we certainly used P-channel pairs instead of N-channel ones to avoid differences in g_m by signal dependent back-gate modulation.

The CM regulation of the circuit naturally has the same bandwidth as the differential signal bandwidth.

The circuit of Fig. 8.2 with long-tail pair CM-output-voltage control works well in a certain output voltage range. The same drawback, or even a little worse, is present as with control by CMOS in the linear mode. The long-tail pair sensors do not work when one of the gate voltages is higher than one saturation voltage plus one threshold voltage below V_{SP}. So the circuit cannot effectively use the full supply-voltage range at its output.

8.1.4 Fully Differential GA-CF CMOS OpAmp with Output Voltage Gain Boosters

In many cases, like in the first stage of a sigma-delta converter or passive integrator, a very large voltage gain is needed at a very large output impedance. This can be made by inserting gain boosters to regulate the gates of the output cascode transistors. The situation is depicted in Fig. 8.3 [5].

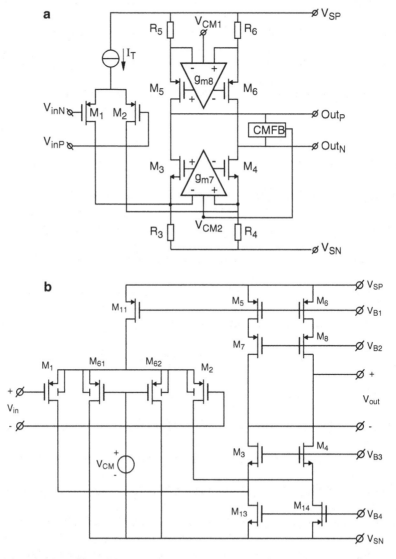

Fig. 8.3 (a) Fully differential operational amplifier with an GA-CF configuration and output voltage gain boosters. (b) Fully differential operational amplifier with an GA-CF configuration and output CM control by input CM feedback

The differential voltages between the pairs of sources of the output transistors are measured by the boost amplifiers G_{m7} and G_{m8} and regulated to zero. Therefore, the output voltages have no longer influence on the current. This results in a high output impedance and hence a high voltage gain. The CM level has still to be measured and regulated to a constant reference voltage. This is symbolized in Fig. 8.3a by the regulation of V_{CM2} of G_{m7}. The boost amplifiers can be easily made as shown in Fig. 8.3b.

8.1.5 Fully Differential GA-CF CMOS OpAmp with Input-CM Feedback CM-Out Control

If the application, such as a filter, includes overall positive DC CM feedback, the CM sensor can be combined with the input stage in a very simple way, see Fig. 8.3b [5].

The input stage has been provided with two extra transistors in a common-source connection, having their gates connected to a desired CM voltage V_{CM} at the input, and their drains connected to the negative rail or ground. If a positive DC feedback between the output and input exists, the CM level at the output is regulated so that the CM level at the input is equal to V_{CM} at the gates of M_{61} and M_{62}.

If the CM level at the input rises, the CM current in M_1 and M_2 is lowered and taken away by M_{61} and M_{62} to ground. The result is that the CM current in M_3 and M_4 increases. This pulls the CM level at the output back down.

A very simple CM control can be made by regulating the CM level of the input stage if the application has positive overall DC CM feedback. The advantage of this solution is that the CM range at the output is not restricted by a regulation circuit, and can approach an R-R behavior very closely.

8.1.6 Fully Differential CMOS OpAmp with R-R Buffered Resistive CM-Out Control

If we could connect two equal resistors, one to each of the output voltages, and control the common mode level by sensing the center connection voltage between the two resistors, a nearly rail-to-rail (R-R) output signal range would result. However, the connection of the resistors between both outputs will ruin the DC gain. In order to avoid this we can drive the resistors not directly by the outputs, but indirectly through R-R buffers. This is shown in Fig. 8.4a.

The voltage on the center point in Fig. 8.4a of the CM resistors R_{CM1} and R_{CM2} is sensed by the differential pair M_{31}, M_{32} and compared with the desired CM voltage level V_{CM}. The transistor M_{32} is connected as a diode.

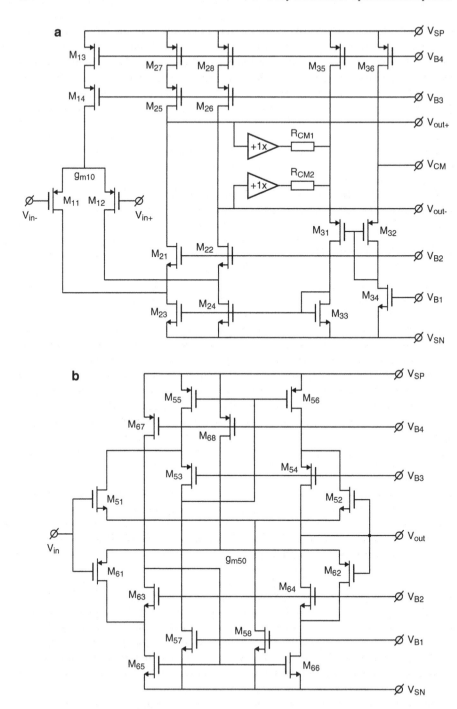

Fig. 8.4 (a) Fully differential operational amplifier with an GA-CF configuration having CM output voltage control by an R-R buffered resistive sensor. (b) Rail-to-rail (R-R) voltage follower

A circuit example of an R-R voltage follower is shown in Fig. 8.4b, see also Chap. 4.

Two complementary differential, mirrored, and folded-cascoded transconductance amplifiers M_{51}–M_{58} and M_{61}–M_{68} have been connected together at their outputs and connected as a unity-gain voltage follower. If the total supply voltage is larger than two saturation voltages and two diode voltages, which is normally the case at 2 V, at least one of the differential pairs is functioning, and the output will follow the input voltage nearly from rail-to-rail.

The transconductances g_{m50} and g_{m60} are supposed to be equal. To this end the W/L ratios of the P-channel transistors have been corrected for their mobilities in regard to the N-channel transistors.

The transconductance of the circuit $(g_{m50} + g_{m60})$ is a factor two larger, when the input voltage is in the middle between the supply voltages and both pairs are functioning, than when the input voltage is closer to one of the rail voltages and one pair g_{m50} or g_{m60} is cut off.

This results in a gain error change ΔE_T

$$\Delta E_T \approx -1/R_{CM1.2}(g_{m50} + g_{m60}) \tag{8.9}$$

This means that the product $R_{CM1.2}(g_{m50} + g_{m60})$ must be chosen much larger than 1. However, if V_{CM} is chosen in the middle, the error of one buffer compensates that of the other, because both buffers switch from 2 to $1g_m$ at the same time, so the requirements are relaxed.

For HF stability of the CM feedback loop, it may be necessary that R_{CM1} and R_{CM2} each are bridged by a small capacitor C_{CM1} and C_{CM2}. In general, the bandwidth of the CM loop must be chosen as large as the bandwidth for differential signals. This means that the inverse CM parallel resistance $1/R_{CM1} + 1/R_{CM2}$ must have the same value as the transconductance g_{10} of the input stage.

If the HF loads at the outputs are balanced, the HF output signals will stay balanced even if the CM loop is functioning only at lower frequencies than the differential bandwidth. In that case the CM parallel conductance $1/R_{CM1} + 1/R_{CM2}$ may be much smaller than g_{10}.

The CM control by an R-R buffered resistive sensor works fine for a total supply voltage larger than two saturation voltages and two threshold voltages, which is about 2 V. A drawback is the extra circuiting and extra supply current needed to do the job.

8.2 Fully Differential GA-CF-GA Configuration

When we need an amplifier with a higher transconductance than we can get with the GA-CF configuration, the GA-CF-GA configuration is a good choice at low supply voltages. It also has an R-R output range except for one saturation voltage at each rail. The high transconductance allows us to load the output directly with common-mode feedback resistors without losing too much gain.

8.2.1 Fully Differential CMOS OpAmp with R-R Resistive CM-Out Control

An example of a fully differential GA-CF-GA configuration with rail-to-rail resistive CM-out control is given in Fig. 8.5 [2].

The circuit is based on the compact CMOS OpAmp of Fig. 7.2. The two outputs V_{out+} and V_{out-} are sensed by a resistor string R_{CM1} and R_{CM2} and paralleled by two small capacitors C_{CM1} and C_{CM2} for HF stability of the CM loop. The voltage of the central sensing point is compared with a desired common-mode voltage V_{CM} by the transistors M_{51}, M_{52} and M_{53}. The latter is connected as a diode. The CM output voltage is translated into a current by R_{CM1} and R_{CM2} and fed through cascodes M_{51} and M_{52} into the folded cascodes M_{45} and M_{46}. The bandwidth of the CM loop should be taken as large as that of the whole OpAmp. The latter one is:

$$f_o' = g_{m30}/2\pi(C_{MI1} + C_{MI2} + C_{M2I} + C_{M22}) \tag{8.10}$$

The transconductance of the input stage g_{m30} has the same value as in Fig. 7.2, but now it drives four Miller capacitances instead of two. The transconductance of the CM loop circuit should therefore be taken equal to $\frac{1}{2}g_{m30}$ to obtain the same bandwidth, so:

$$\frac{1}{2}g_{m30} = 1/(1/(g_{m51} + g_{m52}) + 1/(1/R_{CM1} + 1/R_{CM2})) \tag{8.11}$$

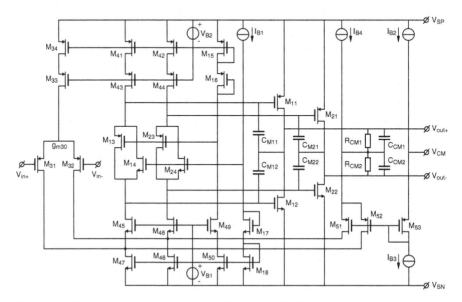

Fig. 8.5 Fully differential operational amplifier with GA-CF-GA configuration and with a resistive CM output control

An example of the above circuit had a bandwidth of 7.5 MHz at a capacitive load of 5 pF, a DC gain of 86 dB at a load resistor of 10 kΩ, and a maximum output current of 5 mA at a quiescent current of 450 μA [2].

The circuit of Fig. 8.5 needs a supply voltage of two diode voltages and two saturation voltages, which is in the order of 2 V. If we would have taken the circuit of Fig. 7.6 with the folded cascode summing circuit as a basis for the differential OpAmp, a supply voltage of one diode and two saturation voltages would have been sufficient, which is of the order of 1.2 V. In that case, the CM output control circuit in the right-hand part of Fig. 8.5 has to be replaced by a folded CM output control circuit which can function at 1.2 V An example of such a circuit is given in Fig. 8.6. The resistors R_{CM1} and R_{CM2} may now carry a strong DC voltage component, because the central point may only be one saturation voltage away from the positive rail. This is compensated by the same voltage on the left-hand side of the compensation resistor R_{CM3}. At its right-hand side the set point of the CM voltage V_{CM} can be chosen. The DC currents through the three resistors are generated by M_{54} through M_{56} in excess of the bias current I_{B3}.

The use of current followers (cascodes) M_{51} and M_{52} for sensing the error current at the CM point of the measuring resistors R_{CM1} and R_{CM2} has the advantage of a better HF behavior of the CM feedback loop (current feedback) than if we would have used a differential amplifier to sense the error voltage at the CM point of the resistors [6].

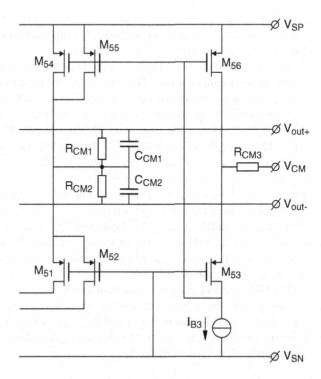

Fig. 8.6 1.2 V CM output control circuit with folded cascode

8.2.2 Conclusion

The resistive CM output control fits well with low-voltage OpAmps with two GA stages or more in cascade where the transconductance is large enough to directly drive the resistive CM control resistors. It can be used with the GA-CF-GA, GA-GA-GA, and even for the GA-GA-GA-GA configuration.

8.3 Fully Differential GA-GA-GA-GA Configuration

When more gain is needed, for example to linearize a heavy duty loudspeaker output stage a fully differential OpAmp with an GA-GA-GA-GA configuration can be used. Apart from CM output control with resistors, switched-capacitors can also be used, as shown next.

8.3.1 Fully Differential CMOS OpAmp with Switched-Capacitor CM-Out Control

As a final example, a circuit is presented of a fully differential OpAmp in an GA-GA-GA-GA configuration that has CM output control by switched-capacitors. The circuit is shown in Fig. 8.7 [7].

The purpose of the circuit is to provide low-distortion at a relatively high output power for audio applications. The large gain of four cascaded GA stages and the double nesting around the output stage provide a high loop gain around the output stage up to relatively high frequencies. The input stage M_1, M_2 is connected to a differential folded cascode stage with M_3 through M_9. The two balancing second stages M_{10} through M_{13} with indices A for the left-hand side and B for the right-hand side have a mirror output to bring the phase back to positive. The third stages M_{14} through M_{19} with indices A and B respectively also have a current mirror in the output. They drive the upper transistors. The lower output transistors are driven in parallel with the third stages. This provides for a multipath. This also provides a kind of very nonlinear class-AB behavior. The gain of the three or four-stage amplifier is so high though, that the strong nonlinear behavior of the output stage is sufficiently linearized. The Miller capacitors C_3, C_2 and C_1 shape the double-nested HF compensation structure.

The CM feedback circuit is shown in Fig. 8.7b.

The resistors have been replaced by capacitors C_1 and C_2 to control the CM level. The switched-capacitors C_{S1} and C_{S2} control the DC drift on the central CM point by taking samples of a CM voltage V_{CM} and frequently correct the control capacitors with the right DC level.

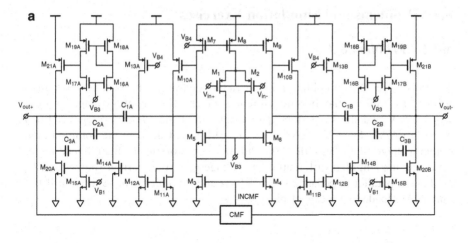

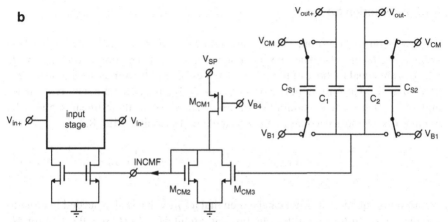

Fig. 8.7 (a) Fully differential operational amplifier with GA-GA-GA-GA configuration and CM output feedback control. (b) CM output feedback control by switched-capacitors

8.3.2 Conclusion

The fully differential power OpAmp has a bandwidth of 2 MHz at a load of 200 pF and 50 Ω. It has a maximum output current of 70 mA at a quiescent current of 2 mA and at a supply voltage of 5 V. The total harmonic distortion at 1 kHz is −86 dB. The circuit functions well at a supply voltage of 3 V.

The circuit could also work at voltages as low as 1.2 V if the CM feedback had been devised with the 1.2 V CM output control circuit of the previous circuit Fig. 8.6.

8.4   Problems and Simulation Exercises

8.4.1   Problem 8.1

The fully differential amplifier in Fig. 8.1a has the transistors sized as shown in the picture, as well as the biasing current sources. The supply voltage is $V_{SP} = V_{SN} = 1.5$ V and the transistors' parameters are $V_{THN} = 0.8$ V, $V_{THP} = -1.0$ V, $K_N = 75$ μA/V^2, $K_P = 22$ μA/V^2, $\lambda_N = \lambda_P = 0.1$ V^{-1}. Calculate the common-mode output voltage and check that M_9 and M_{10} are working in their linear region. Calculate the amplifier differential DC voltage gain if all current sources I_{B1}, I_{B2}, I_{B3} have a parallel impedance $r_{cs} = 200$ kΩ and the unity gain bandwidth if both outputs are loaded with $C_L = 1$ pF.

8.4.1.1   Solution 8.1

The common-mode output voltage, according to Eq. 8.5, equals the source-gate voltage of M_9 and M_{10} when no input signal is present. The drain currents for M_9 and M_{10} are equal to those of the current sources I_{B2} and I_{B3} minus the drain currents of M_1 and M_2. The current mirror M_7-M_3-M_4-M_5-M_6 makes the drain currents of M_1 and M_2 100 μA each. Assuming M_9 and M_{10} work in their linear region, the common-mode output voltage can be calculated by applying Eq. 8.2

$$V_{GS9} = V_{GS10} = \frac{I_{D9}}{K_P \frac{W}{L_9} V_{DS9}} - V_{THP} + \frac{1}{2} V_{DS9} \qquad (8.12)$$

From the equation above it can be seen that a V_{DS} value is also needed in order to calculate V_{GS}. In this amplifier, the biasing circuit M_7-M_3-M_4 sets the V_{DS} for M_9 and M_{10} to

$$V_{DS9} = V_{GS9} - V_{GS11} \qquad (8.13)$$

These two equations can be used to write a second order equation with only V_{GS9} as unknown parameter

$$V_{GS9} = \frac{I_{D9}}{K_P \frac{W}{L_9} (V_{GS9} - V_{GS11})} - V_{THP} + \frac{1}{2} (V_{GS9} - V_{GS11}) \qquad (8.14)$$

V_{GS11} can be calculated considering M_{11} saturated and knowing its drain current

$$V_{GS11} = \sqrt{\frac{2 I_{D11}}{K_P \frac{W}{L_{11}}}} = 1.26 V \qquad (8.15)$$

Solving Eq. 8.14 and picking the positive solution

$$
\begin{aligned}
V_{GS9} &= 1.37\,\text{V} \\
V_{DS9} &= V_{GS9} - V_{GS11} = 0.11\,\text{V} < V_{GS9} + V_{THP} = 0.36\,\text{V}
\end{aligned}
\tag{8.16}
$$

The results show that both M_9 and M_{10} are working in the linear region and that the common-mode output voltage is close to the middle of the supply voltage. The differential DC voltage gain can be calculated taking into account that half the input signal is amplified to one of the outputs using half of the differential circuit, so

$$
A_{V0} = g_{m1} r^{\mu}_{CS16} = \frac{\sqrt{2K_P \frac{W}{L_1} I_{D1}}\, r_{CS}}{\lambda_N \sqrt{\dfrac{2I_{D16}}{K_N \frac{W}{L_{16}}}}} = 93\,\text{dB}
\tag{8.17}
$$

The 0-dB frequency is given by the input g_{m1} and the load capacitor, for each side of the differential output

$$
f_0 = \frac{g_{m1}}{2\pi C_L} = 300\ \text{MHz}
\tag{8.18}
$$

8.4.2   Problem 8.2

For the fully differential depicted in Fig. 8.1b, the biasing voltages V_{B1} and V_{B2} together with transistors M_5 through M_8 and M_{13} through M_{15} are sized for drain currents $I_{D1} = I_{D2} = 100\ \mu\text{A}$, $I_{D13} = 50\ \mu\text{A}$ and $I_{D14} = 50\ \mu\text{A}$. Design the common-mode control loop M_9 through M_{11} and M_{16}, M_{17} such that the output common-mode level is placed 1.5 V lower than V_{SP}. $V_{THp} = -1.0\ \text{V}$, $K_p = 22\ \mu\text{A/V}^2$.

8.4.2.1   Solution 8.2

The output common-mode is set by transistors M_9, M_{10} which should be forced to work in the linear region. For transistor M_{17}, which is the model transistor for common-mode control, the equation voltage-current is

$$
V_{GS17} = \frac{I_{D17}}{K_P \frac{W}{L_{17}} V_{DS17}} - V_{THP} + \frac{1}{2} V_{DS17}
\tag{8.19}
$$

As this transistor should stay in its linear region even when noise or power voltages spikes are added, a safe margin for the difference between gate-source effective voltage and drain-source voltage is

$$V_{GS17} + V_{THP} = V_{DS17} + 350 \text{ mV} \qquad (8.20)$$

Replacing the equality above in Eq. 8.19 and considering the requested common-mode voltage which is identical to V_{GS17}, the aspect ratio of transistor M_{17} can be calculated

$$\frac{W}{L_{17}} = \frac{I_{D17}}{\left(V_{GS17} + V_{THP} - \frac{1}{2}V_{DS17}\right)K_P V_{DS17}} = 35.6 \qquad (8.21)$$

Sizing of M_9 and M_{10} is made considering that in the linear region the drain current is proportional to W/L for the same V_{DS}

$$\frac{W}{L_9} = \frac{W}{L_{10}} = \frac{I_{D9}}{I_{D17}}\frac{W}{L_{17}} = 178 \qquad (8.22)$$

This will produce equal V_{GS} and V_{DS} values for all transistors working in their linear region, keeping the source voltage for the saturated current sources M_{16} and M_{11} at the same value, so sizing for these transistors must be in the ratio

$$\frac{W/L_{11}}{W/L_{16}} = \frac{I_{D11}}{I_{D16}} = 5 \qquad (8.23)$$

The actual size for these transistors is calculated based on the needed common-mode input range, which provides a minimum size for M_{11}.

8.4.3 Simulation Exercise 8.1

The differential output amplifier shown in Fig. 8.8 can be simulated for AC analysis using the circuit in Fig. 8.9. Run AC simulation for gain and phase using the test circuit. An important factor in the functioning of the circuit is the reference common-mode voltage V_1 which has to be in a certain range to allow correct biasing of all devices. Use a stepped voltage for V_1 to find the limits of this range by observing the biasing of all MOS transistors. Change the input AC signal from differential to common-mode and measure the common-mode to differential crosstalk. Change the test circuit to one suited for transient simulation, using two feedback resistors and two resistors as voltage-to-current converters at the input of the amplifier and simulate the resulted inverting amplifier using a sinus differential voltage input to analyze the output differential voltage range if a total harmonic distortion of 0.1 % is allowed.

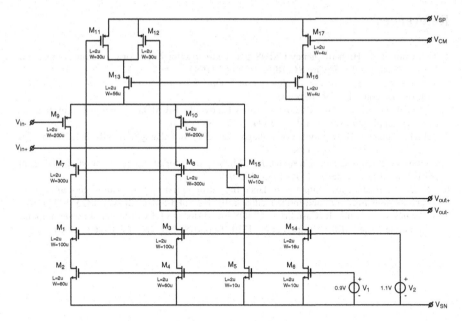

Fig. 8.8 Telescopic differential amplifier

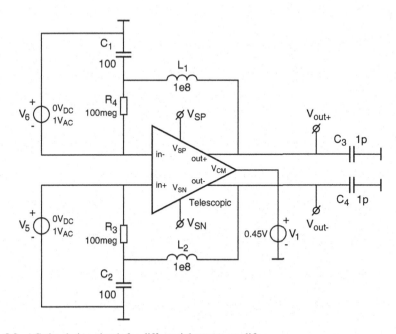

Fig. 8.9 AC simulation circuit for differential output amplifier

References

1. T.C. Choi et al., High-frequency CMOS switched-capacitor filters for communications applications. IEEE J Solid-St Circ **SC-18**(6), 652–664 (1983)
2. R. Hogervorst, J.H. Huijsing, *Design of Low-Voltage Low-Power Operational Amplifier Cells* (Kluwer, Boston, MA, 1996), pp. 195–203
3. G. Nebel, U. Kleine, H.J. Pfleiderer, Large bandwidth BiCMOS operational amplifiers for SC-video-applications. Proc ISCAS **1994**, 5.85–5.88 (1994)
4. J.H. Huijsing et al., Low-power low-voltage VLSI operational amplifier cells. IEEE Trans Circ Sys I Fundam Theory Appl **42**(11), 841–852 (1995)
5. K. Bult, G.J.G.M. Geelen, A fast-settling CMOS op amp for SC circuits with 90-dB DC gain. IEEE J Solid-St Circ **25**(6), 1379–1383 (1990)
6. J.N. Babanezhad, A low-output-impedance fully differential op amp with large output swing and continuous-time common-mode feedback. IEEE J Solid-St Circ **26**(12), 1825–1833 (1991)
7. S. Pernici, G. Nicolini, R. Castello, A CMOS low-distortion fully differential power amplifier with double nested Miller compensation. IEEE J Solid-St Circ **28**(7), 758–763 (1993)

Chapter 9
Instrumentation Amplifiers and Operational Floating Amplifiers

Abstract With the definition of universal active devices in Chap. 1 we have seen that the operational floating amplifier (OFA) is the most universal active device, even more universal than the operational voltage amplifier (OVA), abbreviated to OA or OpAmp, because of its most wide usage. The OpAmp provides us with accurate output voltage control. Additionally, the OFA provides us with accurate control of an output current, independently of the output voltage. So with the OFA we are able to create controlled current sources. These can be used for the transmission of current signals independent of ground or reference voltage differences and for instrumentation amplifier applications mentioned in Sect. 3.4 (Huijsing, Analog Integr. Circ. Signal Process 4(2):115–129, 1993).

It may be as important to open up our mind to the more simple system approach which the OFA offers over the OpAmp approach in many cases (Tellegen, IEEE Trans. Circ. Theory CT-13:466–468, 1966). Only after finding that the simple OFA system solution cannot be implemented because of lack of good OFA realizations, a more complicated system design with more OpAmps can be justified.

9.1 Introduction

The symbol for an OFA is shown again in Fig. 9.1.

Ideally, the OFA should obey the nullor requirements, as explained in Chap. 1:

$$V_{i1} - V_{i2} = V_{idiff} \approx 0, \quad \text{or} \quad V_{i2} \approx V_{i1} \tag{9.1}$$

$$I_{i1} + I_{i2} = 2I_{ibias} \approx 0 \tag{9.2}$$

$$I_{o1} + I_{o2} = 2I_{obias} + 2I_{o1}/H_o \approx 0, \quad \text{or} \quad I_{o2} \approx -I_{o1} \tag{9.3}$$

The equality of $I_{o2} = -I_{o1}$ makes it possible to establish accurate external current relations, in the same way as that the equality of $V_{i2} = V_{i1}$ makes it possible to establish accurate external voltage relations.

© Springer International Publishing Switzerland 2017   307
J. Huijsing, *Operational Amplifiers*, DOI 10.1007/978-3-319-28127-8_9

Fig. 9.1 Symbol for
an operational floating
amplifier (OFA)

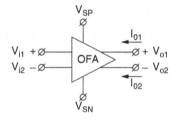

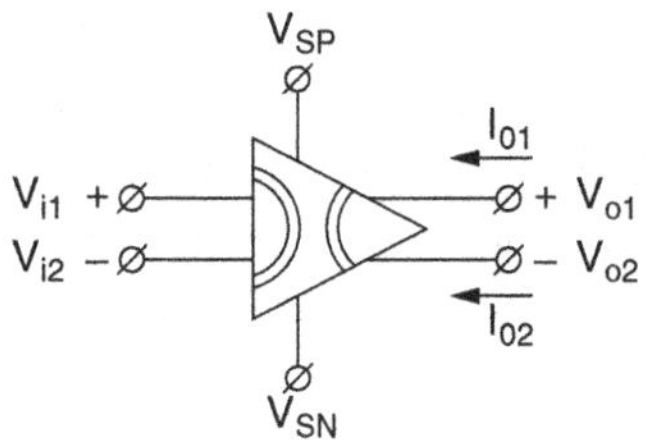

Fig. 9.2 Isolation barriers
in an OFA

Practically, we can approach this by the following four requirements: high gain,
low input offset voltage and current, low input bias current, and a low bias current at
the output. We have seen in Chap. 2 that we could obtain the dynamic input
requirements by isolation of the input stage, while we need not depend on accurate
internal components. Similarly, we will see that the dynamic output requirements
too can be met by isolation of the output stage and that we need not depend on
accurate internal components.

Isolation of the input and output stages is symbolized in Fig. 9.2. Current-source
isolation is the way to go at the input stage as is shown in Chap. 4.

However, at the output stage current-source isolation is more complicated than
at the input stage. This is because of the need to supply larger output currents. This
is the topic of this chapter. In the first five sections, Sects. 9.1–9.5, we will explore
ways to obtain accurate current signals without having to resource to a fully
universal OFA. In Sects. 9.6–9.8 the design of fully universal class-A and class-B
OFAs will be discussed.

Section 9.2 explains how simple unipolar three-terminal OFAs can be used to
realize unipolar voltage-to-current converters. Section 9.3 shows the design of
differential voltage-to-current converters using unipolar three-terminal OFAs.
Section 9.4 discusses how differential voltage-to-current converters can be used to
realize high-quality instrumentation amplifiers, without requiring universal OFAs.

Section 9.5 shows how an accurate universal voltage-to-current converter can be
realized without an OFA by using an instrumentation amplifier. In Sect. 9.6 we see
that current-source isolation can be used to realize class-A biased output stages for
high-quality universal OFAs. Section 9.7 discusses how the problem of designing a
class-AB output stage, that is isolated from ground, can be shifted to the use of a

floating power supply. Finally, Sect. 9.8 shows how difficult it is to design class-AB biased output stages for universal OFAs. Only the use of an instrumentation amplifier provides a high-quality solution.

9.2   Unipolar Voltage-to-Current Converter

In many cases we use the OFA as a three-terminal element. In those cases, the input nullator is connected with the output norator [1] to create a voltage-to-current converter as shown in Figs. 9.3 and 9.4.

The circuit functions as follows: at the input of the OFA the input voltage V_{IN} is carried over at no loss, or followed by the voltage V_G across a conductance G. The current I_G through G is carried over at no loss, or followed by the current at the output I_{out}. So:

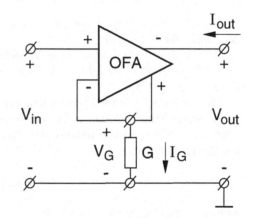

Fig. 9.3 OFA connected as a three-terminal element used as a voltage-to-current converter

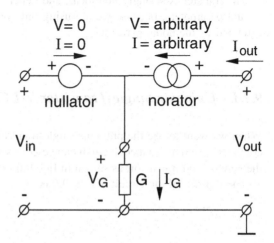

Fig. 9.4 OFA dissected in its basic parts of a nullator and a norator as a three-terminal element used as a voltage-to-current converter

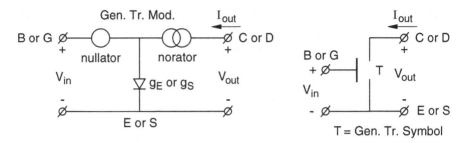

Fig. 9.5 A three-terminal nullator/norator combination closely resembles the Ebers–Moll model for a forward-biased bipolar transistor, but may also be used to model a CMOS transistor in a generalized transistor model

$$V_G = V_{in}$$
$$I_G = V_{in} \cdot G \qquad\qquad (9.4)$$
$$I_{out} = I_G = V_{in} \cdot G$$

The result is a voltage-to-current converter in which only one passive element G determines the V-I transfer.

The simplest realization of the previous idea is a single transistor. The Ebers–Moll [2] model of a forward biased bipolar transistor resembles the previous idea closely, see Fig. 9.5.

The conductance g_e for a bipolar transistor or g_s for a CMOS transistor models the forward transconductance. At one supply-voltage polarity, the current is restricted to the regarding polarity. For this reason the transfer is unipolar. The idealized transfer is:

$$I_{out} = g_m V_{in} \qquad\qquad (9.5)$$

with: $g_m = g_e$, or $g_m = gs$.

The transfer is strongly nonlinear, and depends on many process and environmental parameters, because we are using only one transistor. The model may be extended by the nodal parasitics.

9.2.1 Unipolar Single-Transistor V-I Converter

When we want to use the previous single-transistor for a linear voltage-to-current converter, we may connect a conductance G in series with the emitter or source of the generalized transistor, as shown in Fig. 9.6.

The transfer of the circuit of Fig. 9.6 is:

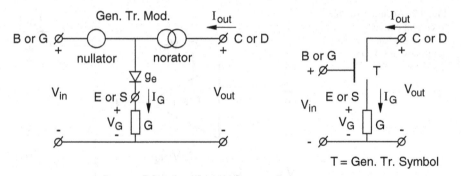

Fig. 9.6 A unipolar voltage-to-current converter composed of a single generalized transistor and a transconductance of G

$$I_{out} = G\left(\frac{1}{1 + G/g_m}\right)V_{in} \qquad (9.6)$$

If $g_m \gg G$, we may write:

$$I_{out} \approx GV_{in} \qquad (9.7)$$

The transfer is unipolar. If the output voltage is positive, the output current can only be positive.

9.2.2   Unipolar OpAmp-Gain-Boosted Accurate V-I Converter

For a higher accuracy we need to increase the g_m of the transistors and lower their parasitic effects. This can be obtained by applying more internal gain in a composite transistor circuit. The first approach is to use an OpAmp in combination with a CMOS transistor, as shown in Fig. 9.7.

The elegance of the unipolar voltage-to-current converter in Fig. 9.7 is its clear accuracy. The input voltage V_{in} is accurately followed by the voltage V_G because of the high gain of the OpAmp OA_1. The current I_G is accurately followed by the output current I_{out} because of the good channel-to-gate isolation and a presumably low input current of the OpAmp. The g_m of the composite transistor is equal to the g_m of the CMOS transistor multiplied by the gain A of the OpAmp. So the V-I converter has an accurate transfer according to Eq. 9.8:

$$I_{out} = G\left(\frac{1}{1 + G/A_v g_m}\right)V_{in} \approx GV_{in} \qquad (9.8)$$

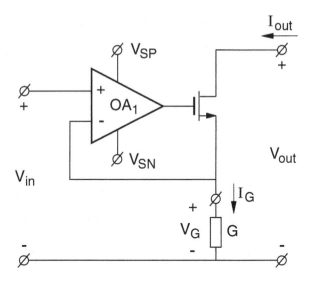

Fig. 9.7 Accurate voltage-to-current converter consisting of an OpAmp gain-boosted CMOS transistor

If the input of the OpAmp is able to reach the negative rail the voltage and current through G may function down to zero. However, the polarity cannot be reversed. The transfer remains unipolar.

9.2.3 Unipolar CMOS Accurate V-I Converter

The OpAmp used may be simple. When we use the GA-CF CMOS OpAmp of Fig. 7.2, the voltage-to-current converter with a GA-CF-VF configuration of Fig. 9.8 arises.

The example of Fig. 9.8 has a high accuracy. The error in the voltage-follower function is low. It is equal to the reciprocal gain of the OpAmp, which is of the order of 3×10^3. The error in the current-follower function is very low, because the channel of the output transistor is isolated from the rest of the circuit by gate oxide. The OpAmp will probably be sufficiently MF compensated by the parasitic stray capacitance at the gate of M_1.

A unique attribute of the above voltage-to-current converter is that it fully functions down to the voltage of negative rail. This allows voltages and currents to be processed from zero.

9.2.4 Unipolar Bipolar Accurate V-I Converter

The bipolar counterpart can be derived from the GA-VF voltage follower of Fig. 7.19 by eliminating the lower half of the output stage as shown in Fig. 9.9.

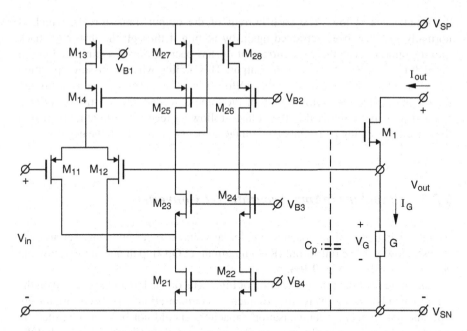

Fig. 9.8 Accurate CMOS unipolar voltage-to-current converter design with GA-CF-VF configuration

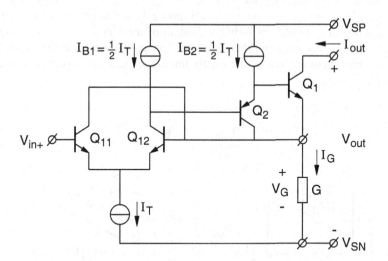

Fig. 9.9 Accurate unipolar bipolar voltage-to-current converter with GA-VF configuration

The voltage-to-current converter of Fig. 9.9 has several interesting features. The collector current of the input transistor Q_{11} has been used as bias current for the output driver Q_2. At the same time the collector of Q_{11} has been nicely bootstrapped at the same voltage of the collector of Q_{12}, so that the offset by the Early effect of the input stage is low. The base current of the output transistor Q_1, which is normally lost, has been retrieved again by adding it through the driver Q_2 back into the output. Even the base current of Q_2 has been retrieved again by adding it through Q_{12} and Q_{11} back into the output. This adding-what-is-missing operation takes care of a highly accurate current-follower function of the current in G through the output. So the high current gain of the three-cascaded bipolar transistors Q_{11}, Q_2, and Q_1 is not only used in the voltage-following feedforward path, but also in the current-following path. Only the voltage across G cannot reach the negative rail.

9.2.5 Unipolar OpAmp Accurate V-I Converter

An interesting general approach to a unipolar voltage-to-current converter arises if we use a low-voltage rail-to-rail (R-R) input/output OpAmp in one of the following connections [3], see Fig. 9.10a, b.

The choice between the connection of Fig. 9.10a on the left and Fig. 9.10b on the right depends on the way the frequency compensation has been organized. The internal frequency compensation capacitors should not be short-circuited by the external connections of input, supply, and output. So if the virtual ground of the frequency compensation is sitting on the bottom (normally V_{SN} connection), then we should choose the right-hand circuit of Fig. 9.10b. If it is sitting at the top (normally V_{SP} connection) than we should choose the left-hand circuit of Fig. 9.10a. The functionality should not forbid a shortcut of the output to one of the supply rails. If so, we can avoid a full shortcut by the connection of a diode, for example, in between the output and one of the supply lines. See the application note of the NE 5230 [3].

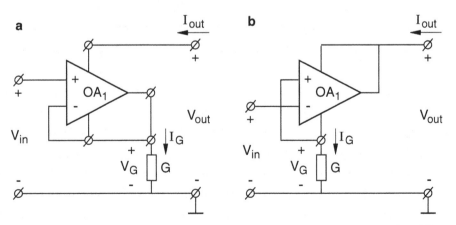

Fig. 9.10 (a, b) Accurate unipolar R-R-in/out OpAmp voltage-to-current converter using different OpAmps connections

Regarding the accuracy aspect, it is clear that the only errors are: the input offset voltage V_{ioffs} for the voltage-follower function, and the bias current I_{ibias} for the current-follower function. These errors may be very low, so that Eq. 9.8 is accurately true.

The output current is down-limited by the quiescent current of the OpAmp, which also flows through G. If this is a hindrance, one has to go back to the basic topology of Fig. 9.7.

9.2.6   Conclusion

The design of several accurate unipolar voltage-to-current converters have been shown. Basically they consist of an OpAmp boosted transistor. CMOS and bipolar realizations have been shown as well as a single OpAmp realization. The restriction of these solutions is that the V-I conversion can only be performed on physical unipolar signals. Physical bipolar current signals may be processed by differential V-I converters, see Sect. 9.3, and by using an instrumentation amplifier, see Sects. 9.4 and 9.5.

9.3   Differential Voltage-to-Current Converters

When we need to provide bipolar current signals instead of unipolar signals, we may apply the unipolar voltage-to-current converters of the previous Sect. 9.2 in a differential way.

9.3.1   Differential Simple V-I Converter

The simplest unipolar voltage-to-current converter can be made with the single transistor as a three-terminal OFA, as we have seen with Fig. 9.6.

If we balance the voltage-to-current converter, and bias it with current sources so that the transconductance G may be used floating or isolated from ground, the differential voltage-to-current converter of Fig. 9.11 arises.

If g_{m1} and $g_{m2} \gg G$, the transfer is roughly:

$$I_{od} = G\left(\frac{1}{1 + 2G/g_m}\right)V_{id} \approx GV_{id} \tag{9.9}$$

$$V_{od} \approx -G(R_3 + R_4)V_{id} \tag{9.10}$$

with $I_{od} = I_{o1} - I_{o2}$, $V_{id} = V_{i1} - V_{i2}$.

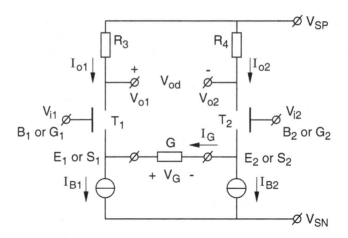

Fig. 9.11 A simple differential voltage-to-current converter with two transistors as two three-terminal unipolar OFAs

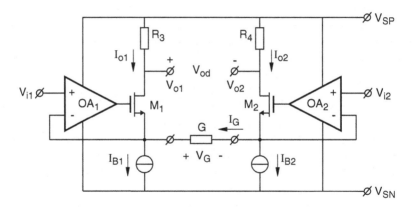

Fig. 9.12 Accurate differential voltage-to-current converter with two OpAmp-boosted unipolar voltage-and-current followers

9.3.2 Differential Accurate V-I Converter

For a high accuracy we have to increase the g_m of the transistors by artificial internal voltage gain. This can be realized by an OpAmp as shown in Fig. 9.12.

The circuit is closely related to the principle of Fig. 3.8.

The transfer is:

$$I_{od} = G\left(\frac{1}{1 + 2G/g_m A_v}\right)V_{id} \approx G\,V_{id} \qquad (9.11)$$

$$V_{od} = G\left(\frac{1}{1 + 2G/g_m A_v}\right)(R_3 + R_4)V_{id} \approx -G(R_3 + R_4)V_{id} \qquad (9.12)$$

With a relative error of:

$$E_{rel.} = \left(\frac{1}{1 + 2G/g_m A_v}\right) \tag{9.13}$$

The common-mode crosstalk ratio CMCR = 1/H may be high. It is the product of the isolation factor and the balancing factor (see Sects. 3.4 and 4.3). The overall CMCR is:

$$\frac{1}{H} = \left|\frac{G_B}{G} \times \frac{\Delta G_B}{G_B}\right| + \left|\frac{1}{H_1}\right| + \left|\frac{1}{H_2}\right| \tag{9.14}$$

with:

G_B = average in the parasitic conductances of I_{B1} and I_{B2}.
ΔG_B = difference in the parasitic conductances of I_{B1} and I_{B2}.
$H_{1,2}$ = common-mode rejection ratio of OpAmp 1 and 2, respectively.

9.3.3   Differential CMOS Accurate V-I Converter

The differential voltage-to-current converter of Fig. 9.12 can be realized using the OpAmp circuit as shown for the unipolar voltage-to-current converter in Fig. 9.8. This is shown in Fig. 9.13. However, insertion of the bias current sources I_{B1} and I_{B2} prevent the common-mode input voltages from reaching the negative rail.

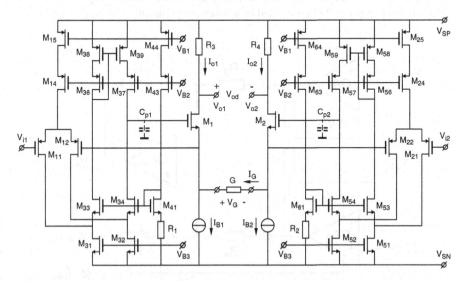

Fig. 9.13 Accurate differential CMOS V-I converter based on two accurate unipolar voltage-to-current followers [4]

This is not an obstacle, for instance if we want to read out a sensor like a bridge circuit, which has its common-mode voltage level between the negative and positive rail. But if we need to read out a sensor like a thermocouple which is grounded, then we need to adapt the circuit.

For instance, a level-shift can be built-in on both sides by inserting a diode M_{13} and M_{23} in series with the source of M_{11} and M_{21}, respectively [4]. This allows differential voltage sensing around the negative rail voltage.

The circuit functions at a minimum supply voltage of 2.5 V, has a bandwidth of 3 MHz, and a CMRR > 90 dB [4].

9.4 Instrumentation Amplifiers

The design of high-quality instrumentation amplifiers requires the basic function of an OFA, as we have seen in Sect. 3.4. However, interestingly, we do not need a general-purpose OFA with physical bipolar voltages and currents: a differential voltage-to-current converter (see Sect. 9.3) is sufficient to meet out needs. But firstly we will see what we can do with just operational amplifiers.

9.4.1 Instrumentation Amplifier (Semi) with Three OpAmps

When we need to design an instrumentation amplifier for the readout of small differential signals with a large common-mode voltage, we would probably first think of using the three OpAmp semi-instrumentation amplifiers of Fig. 9.14. It has a bridge output amplifier OA_3 preceded by a balanced preamplifier with OA_1 and OA_2.

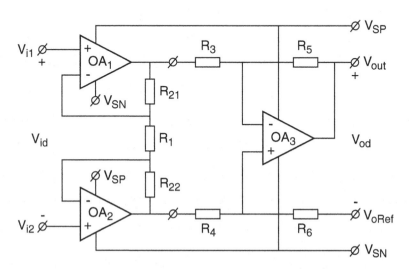

Fig. 9.14 Semi-instrumentation amplifier with a bridge amplifier OA_3 preceded by a balanced preamplifier with two amplifiers OA_1 and OA_2

The three-OpAmp semi-instrumentation amplifier has a voltage gain:

$$A_V = A_{VI,2} A_{V3} \qquad (9.15)$$

with: $A_{VI,2} = (R_{21} + R_{22} + R_1)/R_1$, and $A_{V3} = R_5/R_3$.

The common-mode crosstalk $1/H$ depends on matching of the bridge resisters and on the voltage gain:

$$1/H = -\Delta R_B/R_B(-A_{V3} + 1)A_{VI,2} \qquad (9.16)$$

with: $\Delta RB/R_B = 1 - R_5R_4/R_6R_3$.

For the explanation see ref. [4] in Sect. 3.2.

The semi-instrumentation amplifier will probably work fine together with a bridge-type sensor circuit. If the preamplifier stage $OA_{1,2}$ has enough gain $A_{VI,2}$, the common-mode crosstalk $1/H$ of the bridge amplifier OA_3 will probably be negligible in regard to that of the sensor bridge itself. But if we do not need a high gain, and we do not want to trim the bridge resistors, we have to find other solutions for designing high-quality instrumentation amplifiers. The OpAmps themselves may have a high CMRR because of the current-source isolation applied to the input stage (see Sect. 4.3). But OpAmps do not have an accurate fixed gain. For this reason the three OpAmp instrumentation amplifier has feedback resistors around its OpAmps. These feedback resistors are crossing the current-source isolation barrier. And the value of the CMRR falls back on the accuracy of matching (see Sect. 4.3).

If we want to design real instrumentation amplifiers without having to cross the isolation barrier by feedback resistors, we have to make use of voltage-to-current converters which have a built-in current-source isolation barrier at their output. This will be described next.

9.4.2 Instrumentation Amplifier with a Differential V-I Converter for Input Sensing

In the family of differential voltage-to-current converters of Fig. 9.12, a high input CMRR was achieved by the current-source character of the output, while maintaining an accurate signal transfer without the use of accurate matching. Hence, it is obvious to use the differential V-I converter of Fig. 9.12 as an input stage for an instrumentation amplifier design. The final problem to be solved is to shift the output common-mode level down to an arbitrary reference voltage V_R somewhere in between the positive and negative rail voltages.

As a first solution, we may again use the bridge OpAmp for an output level shift stage. This gives rise to an overall structure for an instrumentation amplifier as depicted in Fig. 9.15.

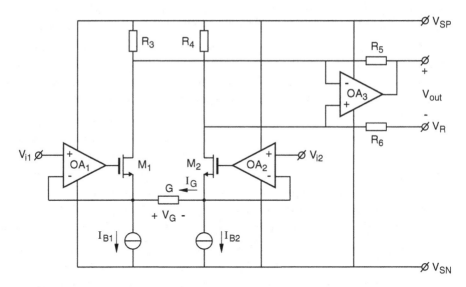

Fig. 9.15 Instrumentation amplifier with a differential voltage-to-current converter at the input and a bridge-type amplifier for output-voltage level shifting

The overall voltage gain is:

$$A_V = V_{od}/V_{id} = G(R_5 + R_6) \qquad (9.17)$$

The power supply rejection ratio (PSRR) still depends on the balancing of the resistive bridge R_3, R_4, R_5, and R_6 in the same way as calculated in formula 9.12. Calculated back to the input we find:

$$1/PSRR = -\Delta R_B/R_B(A_{VPS} + 1)A_V \qquad (9.18)$$

with: $\Delta R_B/R_B = 1 - R_5R_4/R_6R_3$, and $A_{VPS} = R_5R_3$.

A clear disadvantage of this first solution is that the PSRR depends on the matching of resistors.

9.4.3 Instrumentation Amplifier with Differential V-I Converters for Input and Output Sensing

In a second solution we largely avoid the dependency of PSRR on resistor matching, but instead use the current-source isolation principle with two differential V-I converters to sense the input voltage as well as the feedback output voltage, and amplify the current difference to drive the output.

The current difference can be obtained by connecting the two differential V-I converters in cascode on top of each other [5], or in parallel of each other. An

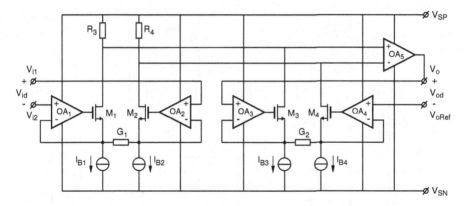

Fig. 9.16 Instrumentation Amplifier with accurate differential V-I converters for input and output sensing

advantage of the cascode connection is that we use the same bias current for both V-I converters, which results in a lower noise than if we would have connected the V-I converters in parallel. On the other hand, the cascode connection needs more supply-voltage room. The latter is more important today with even lower supply voltages than it was in earlier days. Therefore, we have chosen for the parallel connection: an instrumentation amplifier based on the parallel connection of two V-I converters, one for input sensing and one feedback output sensing is shown in Fig. 9.16.

The instrumentation amplifier of Fig. 9.16 uses the absolute minimum number of passive elements to determine the gain, viz. the ratio of two conductances g_1 and g_2.

The overall voltage gain is:

$$A_V = G_1/G_2 \tag{9.19}$$

The common-mode crosstalk ratio CMCR has been described by Eq. 9.14. This formula is equally valid to describe the CMCR at the output. The instrumentation amplifier of Fig. 9.16 is the most precise one, but also the most complex instrumentation amplifier.

The instrumentation amplifier of Fig. 9.16 has an interesting virtue, viz. that residual gain and nonlinearity errors in the input V-I converter largely cancel equal errors in the output V-I converter.

9.4.4 Instrumentation Amplifier with Simple Differential V-I Converters for Input and Output Sensing

It is a special attribute of the above instrumentation amplifier namely that gain and nonlinearity errors of one V-I-C cancel those of the other V-I-C. We may exploit

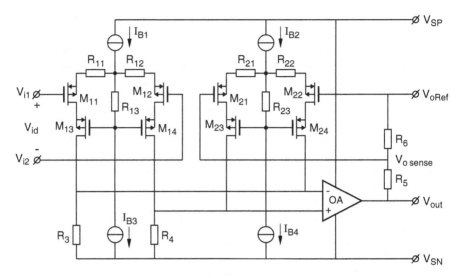

Fig. 9.17 Instrumentation amplifier with simple differential V-I converters for input and output sensing

this special attribute to simplify the instrumentation amplifier design by using simple degenerated differential transistor pairs as V-I converters and set the gain by an overall feedback resistor ratio.

This principle is shown in Fig. 9.17. We use P-channel CMOS transistors to eliminate back-gate modulation of their channels by bootstrapping the back-gates. A residual gain modulation by a g_m modulation of the CMOS transistors, due to differences in common-mode voltage levels between the input and output, has to be eliminated by using bootstrapped cascodes for the input transistors. The voltage gain A_V is:

$$A_V = \frac{R_5 + R_6}{R_6} \times \frac{R_2}{R_1} \tag{9.20}$$

with: $R_1 = R_{11} + R_{12}, R_2 = R_{21} + R_{22}$.

The voltage gain of the simple instrumentation amplifier of Fig. 9.17 is:

$$A_V = \frac{R_5 + R_6}{R_6}, \quad \text{at } R_1 = R_2 \tag{9.21}$$

This principle has been elaborated in the two following instrumentation amplifier realizations with a common-mode range that includes the negative rail voltage.

9.4.5 Instrumentation Amplifier Bipolar with Common-Mode Voltage Range Including Negative Rail Voltage

When we have to amplify differential voltage signals from sensors which have one terminal connected to the negative rail, we can no longer use the three-OpAmp semi-instrumentation amplifier design shown in Fig. 9.14, even though the input OpAmps might have an input CM range including the negative rail voltage. The feedback around the preamplifier stages prevents the common-mode range from including the negative rail voltage.

A possible way out would be to use PNP, or P-channel emitter follower, or source follower level-shift stages at the input. But these stages add additional noise and offset. Besides, the three-OpAmp semi-instrumentation amplifier has other drawbacks, as we have seen with Fig. 9.14.

Therefore, it is better to use PNPs or P-channel transistors directly as the input transistors of simple degenerated V-I converters followed by a folded cascode to realize a topology similar to that of Fig. 9.17. A bipolar design [6] is shown in Fig. 9.18.

The bipolar input transistors have a high output impedance and do not show much g_m modulation as a function of their common mode level. Hence, cascoding of these transistors is not needed. Moreover, there is not much voltage headroom for cascodes. If we set the voltage at R_{31} and R_{32} at about 200 mV, the bases of the input transistors may reach about -200 mV below the negative rail voltage. When we specify a maximum differential input voltage V_{id} of ± 100 mV we have to degenerate the input transistors by a factor 4, which means that $\frac{1}{2}R_1$ and $\frac{1}{2}R_2$ have a four times larger resistance value than the emitter resistance r_e of the transistors Q_{11}, Q_{12}, Q_{21}, Q_{22}.

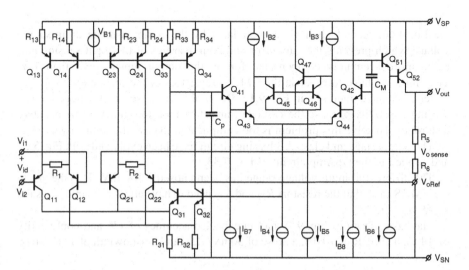

Fig. 9.18 Bipolar instrumentation amplifier with input common-mode range which includes the negative rail voltage [6]

The intermediate stage has Darlington transistors for a high current gain and a high voltage gain because the output impedance at the collector of Q_{44} is compensated for by the negative output impedance at the collector of Q_{43} through the action of a bootstrapped current mirror by Q_{47}. The circuit has been described in [6].

The output stage has Darlington transistors as an emitter follower. The output can reach the negative rail voltage because the gain setting resistor chain R_5, R_6 is connected to the negative rail. The output is also referenced to ground by the connection to ground of the output-sensing differential pair Q_{21} and Q_{22}.

The result is an instrumentation amplifier with an input bias current of 0.2 μA, an input offset voltage of 0.3 mV, an input noise voltage of 30 nV/√Hz, a CMRR of 90 dB, a common-mode voltage range up from −200 mV below the negative rail voltage, and an input voltage range of ±100 mV. The gain can be set between 1 and 1000 by R_5 and R_6. The gain error is typically 0.1 % and nonlinearity of typical 0.1 %. The bandwidth is 1 MHz. The minimum supply voltage is 2.5 V [6].

9.4.6 Instrumentation Amplifier CMOS with Common-Mode Voltage Range Including Negative Rail Voltage

A basic circuit is shown in Fig. 9.19.

The differential V-I converters in CMOS will have a lower accuracy than in the bipolar circuit. There are two reasons: first, their maximum voltage gain μ is so low that the CMRR will be too low, in the order of 70 dB

$$1/H = \frac{1}{\mu} \times \frac{\Delta\mu}{\mu}, \qquad (9.22)$$

see Eq. 4.16; second, their g_m is modulated by the common-mode drain-source voltage which prevents the gain error and nonlinearities of the input sensing pair being more accurately compensated for by the output sensing pair than 1 %. Therefore, we have to cascode the CMOS transistors of the differential V-I converters. We have done this in Fig. 9.17. But in that case, we lost the negative-rail sensing capability because the cascode transistors took too much of the negative headroom. To solve this problem is too complicated to describe in this overview.

The output stage and class-AB biasing circuit have been explained with Fig. 5.27 and applied in the OpAmp circuit of Fig. 7.33.

The differential input voltage range has been chosen ±100 mV. This means, in the CMOS case, that the resistors R_{11} and R_{12} need to have about the same g_m as the g_m of M_{11} and M_{12}.

The basic circuit of Fig. 9.19 has an expected accuracy of 1 % and nonlinearity of 0.1 %, a CMRR of 70 dB, a noise of 30 nV/√Hz, and a bandwidth of 1–10 MHz.

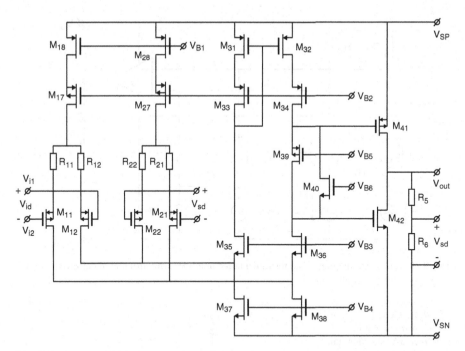

Fig. 9.19 Basic CMOS instrumentation amplifier with input common-mode range including the negative rail voltage

9.4.7 Instrumentation Amplifier Simplified Diagram and General Symbol

There is a need to come up with a simplified schematic diagram and a general symbol for an instrumentation amplifier. The first is given in Fig. 9.20, the latter in Fig. 9.21.

9.4.8 Conclusion

In this section we have seen how we can realize highly accurate instrumentation amplifiers. A semi-instrumentation amplifier can be realized with three OpAmps and a trimmed bridge circuit for obtaining a high CMRR. If we do not want to use trimming for obtaining a high CMRR, we must rely on current-source isolation. To that purpose, differential voltage-to-current converters can be used to realize instrumentation amplifiers with a high CMRR at the input as well as at the output. The instrumentation amplifier circuit can further be simplified by using matching of the input and output differential V-I converters. In this way, a bipolar and CMOS instrumentation amplifier have been presented having input negative-rail sensing capability. Finally, a simplified circuit diagram and a symbol of an instrumentation amplifier with V-I converters is given.

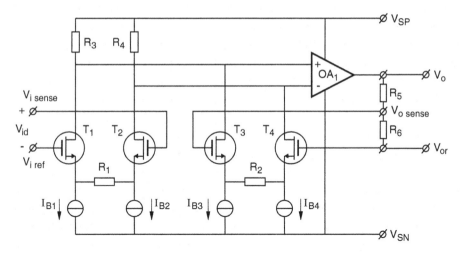

Fig. 9.20 Simplified diagram of an instrumentation amplifier with differential current-to-voltage converters and ideal transistors

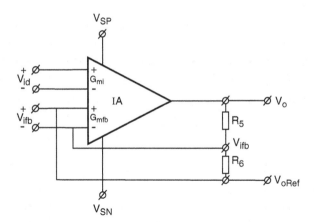

Fig. 9.21 General symbol for an instrumentation amplifier (IA) showing gain setting resistors

9.5   Universal Class-AB Voltage-to-Current Converter Design Using an Instrumentation Amplifier

We have seen that a high-quality instrumentation amplifier can be realized by differential voltage-to-current converters. Now we can use an instrumentation amplifier (IA) to realize a high quality universal voltage-to-current converter. Universal means that signals with a physical positive and negative polarity can be processed. Particularly if we want a class-AB solution, a universal voltage-to-current-converter implementation with an instrumentation amplifier is preferable over that with an OFA. This will later be shown in Sect. 9.7.

Before using the real instrumentation amplifier, we will firstly use the semi-instrumentation amplifier and see its limitations.

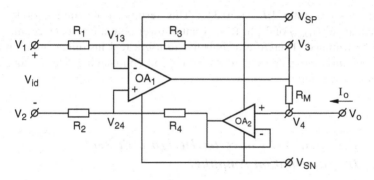

Fig. 9.22 Universal voltage-to-current converter implemented with a bridge-type semi-instrumentation amplifier

9.5.1 Universal V-I Converter Design with Semi-instrumentation Amplifier

A universal voltage-to-current converter can be implemented using a semi-instrumentation amplifier, see Fig. 9.22. A measuring resistor R_M is connected in series with the output of the instrumentation amplifier and the load. The differential output sense terminals V_3 and V_4 are connected across R_M. A buffer amplifier OA_2 is used to isolate output currents from bridge currents. The differential input sense terminals V_1 and V_2 are driven with the input voltage V_{id}.

The voltage-to-current transfer is:

$$G_M = I_o/V_{id} = -A_v/R_M \qquad (9.23)$$

with: $A_v = -R_3/R_1$.

The common-mode output conductance is:

$$G_{OCM} = \frac{1}{R_{OCM}} = \frac{I_o}{V_o} = -\frac{V_3 - V_4}{V_4}\frac{1}{R_5} = -\frac{1}{H_R}\frac{1}{R_M} \qquad (9.24)$$

with: $1/H_R = -\Delta R_B/R_B(-A_{VR} + 1)$
$\Delta R_B/R_B = 1 - R_3 R_2/R_4 R_1$
$A_{VR} = -R_1/R_3$.

$$1/H_R = -\Delta R_B/R_B(-A_{VR} + 1) \qquad (9.25)$$

When we want to use the output to function in a large part of the supply-voltage range, the voltage across the resistor R_5 should be relatively low, say 0.5 V at a supply voltage of 5 V. If the input voltage V_{id} is chosen 1 V, the voltage gain A_v should be set at $-1/2$, and hence the reverse voltage gain A_{VR} is -2. At a nominal output current of 1 mA, the current-measuring resistor R_5 must be 500 Ω, and the voltage-to-current transfer I_o/V_{id} is 1 mS. At a bridge imbalance $\Delta R_B/R_B$ of 1 %, the reverse crosstalk ratio $1/H_R$ is 0.3 %. This results in a common-mode output conductance of G_{OCM} of 3 μS. If we want a lower output conductance, the bridge has to be trimmed. Therefore, it is better to use a real instrumentation amplifier whose common-mode crosstalk does not depend on matching. This is done in the following paragraph.

9.5.2 Universal V-I Converter Design with Real Instrumentation Amplifier

A high quality voltage-to-current converter can be realized with a real instrumentation amplifier. This is shown in Fig. 9.23 with a simplified circuit diagram for an instrumentation amplifier with differential V-I converters, and shown in Fig. 9.24 with a general model. To this purpose, we connect a measuring resistor R_M in series with the output of the instrumentation amplifier and the load. The differential output sense and reference terminals are connected across R_M. The input voltage is connected to the differential input terminals.

The transconductance of the voltage-to-current converters of Figs. 9.23 and 9.24 is:

$$G_M = A_v/R_M \qquad (9.26)$$

In the case of $R_2/R_1 = 1$, as was used with the differential input amplifiers of Figs. 9.17, 9.18, and 9.19, the G_M equals $1/R_M$.

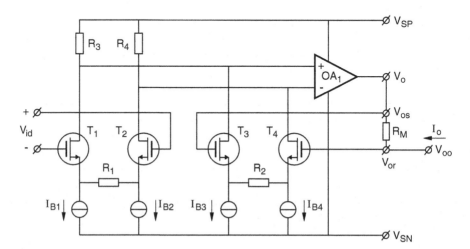

Fig. 9.23 Universal voltage-to-current converter with a simplified schematic diagram for an instrumentation amplifier (IA)

Fig. 9.24 Universal
voltage-to-current converter
with a general model for an
instrumentation
amplifier (IA)

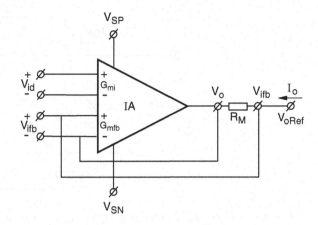

The common-mode output conductance is:

$$G_{OCM} = 1/HR_M \qquad (9.27)$$

The common-mode rejection ratio H for the output differential sense amplifier
has been calculated with Eq. 9.14. The common-mode output resistance
$R_{OCM} = 1/G_{OCM}$ can easily be a factor $H = 10^4$ larger than that of the measuring
resistor. This is a much better result than with the semi-instrumentation amplifier of
Fig. 9.22 without trimming.

For the realization all designs of real instrumentation amplifiers can be used of
Figs. 9.16, 9.17, 9.18, and 9.19.

9.5.3   Conclusion

We have shown in this section that the instrumentation amplifier with differential
V-I converter can be used to realize high-quality universal voltage-to-current
converters which function for physically positive as well as for negative input
and output signals.

9.6   Universal Class-A OFA Design

In this chapter the design of universal class-A OFAs will be treated. The main goal is
to design an OFA with output currents obeying the relation $I_{01} + I_{02} = 2I_{0bias} + 2I_{01}/$
$H_0 = 0$, or $I_{02} = -I_{01}$. This can be achieved by a floating, i.e., isolated, output stage.
On a chip it is realistic to obtain the above by current-source isolation.

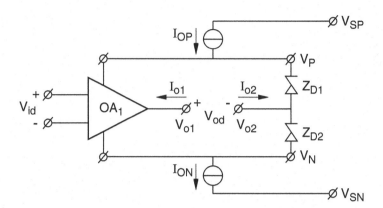

Fig. 9.25 Universal class-A OFA realization with an OpAmp and floating Zener-diode supply

9.6.1 Universal Class-A OFA Design with Floating Zener-Diode Supply

A first approach could use current sources and Zener diodes to create a kind of an internal floating battery [7]. This can be seen as a rough realization of the idea of a floating power supply presented in Fig. 9.1. This approach is shown in Fig. 9.25.

If we take the upper supply current source I_{OP} equal to the lower one I_{ON}, a floating supply voltage $V_S = V_{SP} - V_{SN}$ will result across the Zener diodes Z_{D1} and Z_{D2}. The diodes will take the excess current which is not needed by the OpAmp. The result is that the output bias current I_{ob} is nearly zero.

$$I_{o1} + I_{o2} = I_{ON} - I_{OP} = 2I_{ob} + 2I_{o1}/H_o \approx 0 \qquad (9.28)$$

$$I_{o2} = -I_{o1} + 2I_{ob} + 2I_{o1}/H_o \approx -I_{o1} \qquad (9.29)$$

The Zener diodes divide the total supply-voltage range into one across the OpAmp for V_{o1} and one what remains in between the supply rails and the Zeners. For this reason we lose at least a factor 2 in the total output voltage range [7].

9.6.2 Universal Class-A OFA Design with Supply Current Followers

When we want to avoid the voltage loss of the fixed voltages across the Zener diodes, we may exchange the Zener diodes for current followers as depicted in Fig. 9.26.

These current followers catch and send the supply current of the OpAmp into the second output terminal. The output supply-voltage range can now be extended close into the rails. Moreover, the voltage swings across the total supply current sources

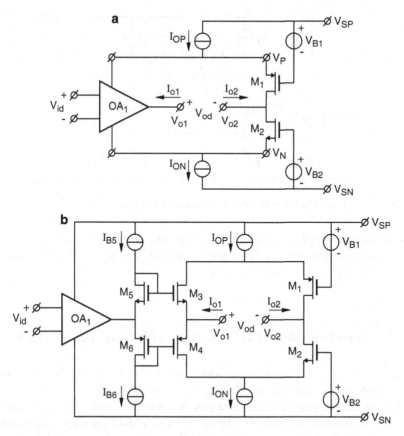

Fig. 9.26 (**a**) Universal class-A OFA realized with an OpAmp and supply current followers (CF). (**b**) Universal class-A OFA realized with a VF output stage of an OpAmp and current followers (CF)

I_{OP} and I_{ON} are now regulated constant by the cascodes M_1 and M_2, which helps to make these currents signal independent and which enlarges the common-mode output impedance.

Not all OpAmp supply currents need to be provided by the total upper and lower supply current sources, only the current for the output transistors. A first approach with a voltage follower output stage is shown in Fig. 9.26b.

The formulas become very clean now:

$$I_{o1} + I_{o2} = I_{OP} - I_{ON} = 2I_{ob} + 2I_{ol}/H_o \approx 0 \tag{9.30}$$

$$I_{o2} = -I_{o1} + 2I_{ob} + 2I_{ol}/H_o \approx -I_{ol} \tag{9.31}$$

Not only the supply-voltage range has been used efficiently now, but also the bias currents I_{OP} and I_{ON} are used as efficiently as can be in class-A. If the total

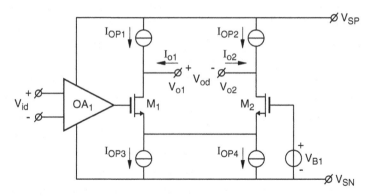

Fig. 9.27 Universal Class-A OFA realized with a long-tailed-pair output stage

positive bias current I_{OP} is used for $-I_{o1}$, the total negative bias current I_{ON} will automatically be used for I_{o2}. This supposes that M_3 and M_4 are properly biased in class-AB by V_{B3} and V_{B4}, see Fig. 5.8a,b.

9.6.3   Universal Class-A OFA Design with Long-Tailed-Pairs

An alternative output stage to the one with the current follower arises when we use a long-tailed-pair as output stage [8]. This is shown in Fig. 9.27.

The long-tailed pair functions as a pair of communicating vessels for currents between the two output terminals. The output bias current I_{ob} and the output current rejection ratio H_o determine the equality of the output currents I_{o1} and $-I_{o2}$:

$$I_{o1} + I_{o2} = I_{OP1} + I_{OP2} - I_{OP3} - I_{OP4} = 2I_{ob} + 2I_{o1}/H_o \approx 0 \qquad (9.32)$$

$$I_{o2} = -I_{o1} + 2I_{ob} + 2I_{o1}/H_o \approx -I_{o1} \qquad (9.33)$$

The way one output current is guided towards the other can clearly be seen. If we follow the output signal current I_{o1}, we see that it is blocked by the current source I_{OP1} from flowing into the positive supply. I_{o1} is forced to flow into M_1 and M_2 to the other output terminal as $-I_{o2}$. On its way there is no escape, because I_{OP3}, I_{OP4}, and I_{OP2} block its way toward the supply rails. The result is a very high output common-mode current rejection ratio H_o. The feedback through external connections, like the current follower of Fig. 1.6b, take the OFA into its functional working mode.

It is interesting that the accuracy of the output stage is even maintained when we use bipolar transistors. We can see this when we go through the long-tailed pair. Going from the left to the right output, we add the base current $1/\beta$, at emitter of Q_1 and lose $1/\beta_2$ at the base of Q_2 [8].

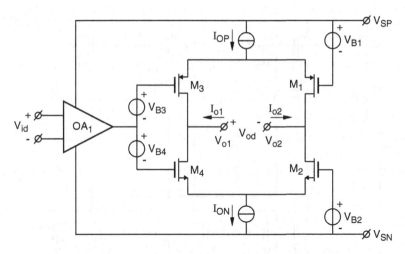

Fig. 9.28   Universal class-A OFA realized with two push–pull complementary long-tail pairs

A disadvantage of the single long-tailed pair is still that we can only use a half of the total output bias current $I_{OP1} + I_{OP2}$ and $I_{ON1} + I_{ON2}$ for one output current I_{o1} and I_{o2}. A more efficient use of the bias currents can be made if we use a P pair and an N pair together as depicted in Fig. 9.28.

The main design problem is the realization of the floating voltage sources V_{B3} and V_{B4}. These have to follow the supply voltages. We will present two approaches to this problem. In the first one we will drive the lower output transistors directly, and then cross the supply voltage and drive the upper output transistors indirectly. In the second approach we will use a similar mesh-drive circuit as we did for the R-R output stage of Fig. 5.27a, b.

A first complete CMOS class-A OFA design is presented in Fig. 9.29. The lower output transistors M_{31} and M_{32} are fed by a lower current source transistor M_{35}. The upper output transistors M_{33} and M_{34} are fed by an upper current source transistor M_{36}. The output transistors function as cascodes for these upper and lower current-source transistors for a high CM output impedance. A model bias stage accurately equates the currents through the upper and lower current source transistors. It is scaled down in current by a factor $1/N$ in regard to the output stage, with transistors M_{41} through M_{46}. The model bias circuit also has the task of differentially driving the upper output transistors in opposite direction in regard to the lower output transistors. In this way the total output bias current is optimally used as maximum positive and negative output currents. Miller capacitors C_{M1} and C_{M2} take care of frequency compensation. The input stage is followed by a folded cascode stage. This cascode stage with transistors M_{21} through M_{28} provides a high internal voltage gain A_{vi}. In the order of 10^4 this is equal to the g_m of the input pair multiplied by the differential output resistance r_{d2} of the folded cascode stage.

This provides the OFA with an overall transconductance G_m which is equal to the internal voltage gain A_{vi} multiplied by the transconductance g_{m4} of the lower output long-tailed pair transistors and doubled by the upper output long-tailed pair transistors as they are driven by the model bias stage.

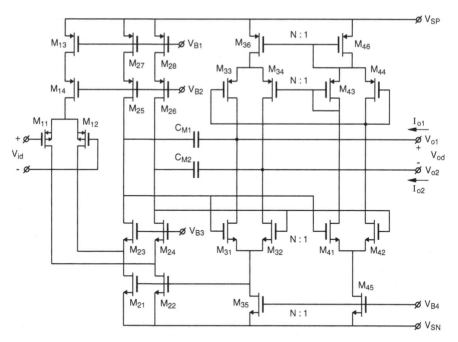

Fig. 9.29 Universal Class-A OFA design with a push–pull complementary pair of long-tailed pair output stages and a model bias stage

The results are:

$$G_m = 2A_{vi}g_{m4'} \qquad (9.34)$$

with

$$A_{vi} = g_{ml}r_{d2} \qquad (9.35)$$

A feedback loop through the lower output transistors and the gates of M_{21} and M_{22} equates the upper and lower common-mode currents in the folded cascode-stage. The differential output impedance may not be that high due to the common-source connections of the long-tailed output pairs. But the series connection of the feedback around the output takes the output impedance to high levels, see Sects. 1.3 and 1.4. A completely alternative design is presented in Fig. 9.30. The output transistors are driven in opposite phases by the meshes M_{51}, M_{52} and M_{53}, M_{54} as explained with Fig. 5.27a, b. However, the output transistors do not have their sources connected to the supply rails but to the current sources M_{35} and M_{36}. Resistors R_{64}, R_{69}, and R_{68} in the diode bias chains M_{63}-M_{68} provide for the extra head room for the current sources. The current sources give the output transistors a class-A bias instead of class-AB bias. The positive and negative current sources through M_{36} and M_{35} are equalized by the model bias circuit M_{41} through M_{56}.

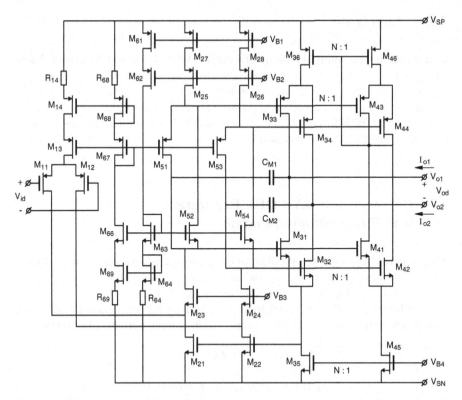

Fig. 9.30 Universal Class-A OFA design with a push–pull complementary pair of long-tailed pair output stages and a model bias stage

The rest of the explanation is similar to that of Fig. 9.29. The advantage of this alternative design over the previous one is a slightly better HF behavior because the upper output transistors are in parallel driven with the lower output transistors by the meshes. Whereas in the previous design, the upper output transistors are indirectly driven through the model bias stage.

The output voltages can approach the rail voltages except for one diode and one saturation voltage. The common-mode current rejection ratio at the output H_o is easily higher than 10^4. That determines the dynamic equality of the output currents. The output bias current I_{ob} is of the order of 0.5 % of the bias current sources I_{BP} and I_{BN}. This determines the offset between the two output currents. The bandwidth can easily be several tens of megahertz.

The OFA of Fig. 9.30 has some similarities with the differential OpAmp of Fig. 8.5. In that case the two output voltages were regulated equally but opposite to each other in regard to a reference voltage. There we needed two equal resistors to regulate the output voltages equally and the equality depends on the matching of the two resistors. With the OFA the output currents are regulated equally but opposite to each other. This does not require matching of elements, apart from offset, because of the nature of the OFA.

9.6.4 Conclusion

In this section the designs of high-quality class-A OFAs have been described. These OFAs can be used in precision instrumentation applications where a free output signal current is desired at a high output impedance, with a signal-to-noise ratio of 120 dB in the audio frequency range, while a bandwidth can be obtained of several tens of megahertz. When used for the composition of gyrator filters, a quality factor Q higher than 1000 can be obtained [8].

9.7 Universal Class-AB OFA Realization with Power-Supply Isolation

A first approach to solve the problem of designing a class-AB output stage, that is isolated from the supply-power ground, is the use of a power supply that is isolated from ground, i.e., floating [9]. This solution is depicted in Fig. 9.31.

If the input current can be disregarded, no output current can leak out of the mesh through the output stage and power supply, so the output currents must obey:

$$
\begin{aligned}
I_{o2} &= I_{SP} - I_{SN} = -I_{o1} \\
I_{o2} &= -I_{o1} \\
I_{o1} + I_{o2} &= 2I_{obias} + 2I_{o1}/H_o = 0
\end{aligned}
\tag{9.36}
$$

This means that the output current I_{o1}, which sources out of one mode, must be accurately equal to the output current $-I_{o2}$, which sinks into the other mode,

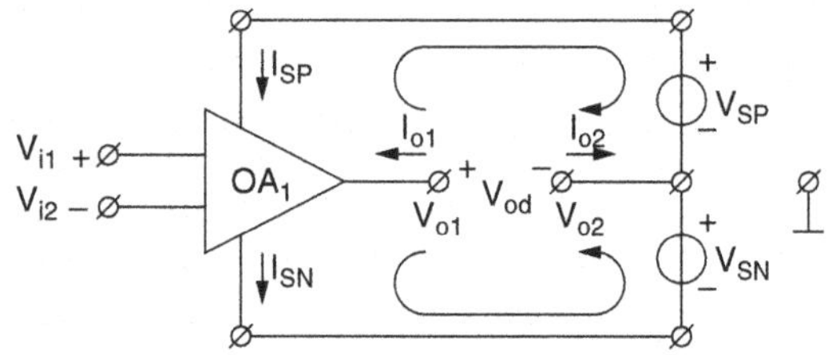

Fig. 9.31 Universal Class-AB OFA realization with one OpAmp and a power-supply source that is isolated from ground, i.e., floating

Fig. 9.32 Realizing a floating power supply source by the method of "flying" capacitors

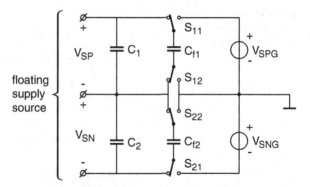

without depending on accurately matched components. If the output stage of the OpAmp is biased in class-AB, the whole OFA is biased in class-AB. A shortcoming of the circuit of Fig. 9.31 is that the maximum differential output voltage V_{od} is limited to half the total supply voltage $V_S = V_{SP} + V_{SN}$. This can be avoided by using a fully differential amplifier as presented in Chap. 8.

9.7.1   Universal Floating Power Supply Design

In this section, the problem of realizing an isolated or floating output port has been shifted into the problem of realizing a power-supply source which is isolated from ground, or floating with regard to ground. This is a problem in itself. But it is realizable. Even on a chip, one can for example use the method of "flying" capacitors, see Fig. 9.32.

When no external capacitors may be used, relatively small on-chip capacitors allow for a small supply current. In some extraordinary cases an external battery or solar cell could be used.

9.7.2   Conclusion

Power-supply isolation is an escape to shift the problem of a floating output stage into the supply of an OpAmp circuit. Realizing a floating supply source is often not easy and may be expensive. For low output currents the method with flying capacitors can be used on chip.

9.8   Universal Class-AB OFA Design

The final task in designing universal OFAs is to provide the output stage with class-AB biasing without using a floating power supply. We will see in this section that class-AB biasing is not easy for OFAs. We will approach the problem

Fig. 9.33 General push–
pull configuration of an
OFAs output stage

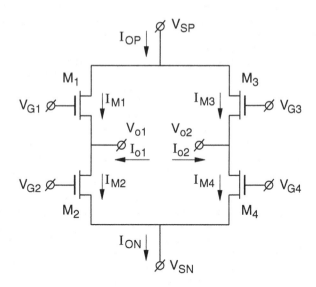

systematically. In Fig. 9.33 the situation is depicted by the currents in four output
transistors of an OFA.

Class-AB biasing means that the total positive and negative output currents I_{OP}
and I_{ON} are strongly changing between their quiescent value I_{OQ} and their maximum value I_{OM}. This changing must be allowed while the OFA requires that I_{o1}
remains equal to I_{o2}. This can be written-out into three equations:

$$I_{ON} - I_{OP} = 2I_{ob} + 2I_{o1}/H_o \approx 0 \tag{9.37}$$

$$I_{M2} - I_{M1} + I_{M4} - I_{M3} = 2I_{ob} + 2I_{o1}/H_o \approx 0 \tag{9.38}$$

$$I_{o1} + I_{o2} = 2I_{ob} + 2I_{o1}/H_o \approx 0 \tag{9.39}$$

(see for I_{ob} and H_o Sect. 2.4).

Each of these three equations gives rise to a class-AB OFA realization. In the
following designs we start from the first formula, then the second one, and finally
the third one.

9.8.1   Universal Class-AB OFA Design with Total-Output-Supply-Current Equalization

In the first realization we use, for example, a mesh of two operational amplifiers,
measure the total positive and negative supply currents and equalize them. That
situation is drawn in Fig. 9.34.

The diodes M_1 and M_2 measure the positive and the negative total supply current
I_{OP} and I_{ON}, respectively. These currents are reproduced in M_3 and M_4, respectively,

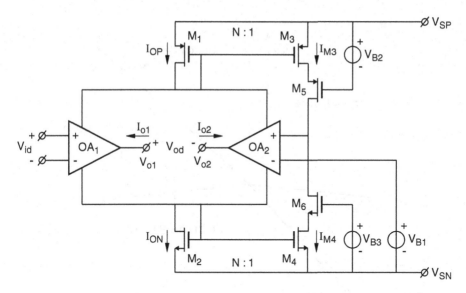

Fig. 9.34 Universal class-AB OFA realization with total-output-supply-current equalization

at a reduced level by a ratio of 1/N. The second OpAmp OA_2 senses the differences in I_{M3} and I_{M4} at its positive input terminal and drives the output in such a way that I_{M3} and I_{M4} are equalized, and so are I_{OP} and I_{ON}, as we hope.

An advantage of the use of this class-AB OFA is that the output bias current $2I_{ob} = I_{o1} + I_{o2}$ is low, when the output currents themselves are low. Also the output bias noise current is low in that case. So the dynamic range of a voltage-to-current converter with such a universal class-AB OFA may be high, i.e., more than 140 dB.

However, a problem arises by the fact that the transistors M_1 through M_4 are not perfectly scaled, or have different drain-source voltages, or different threshold voltages. This results in a nonlinear current transfer as shown in Fig. 9.35. Say that the current I_{OP} through M_1 is $\delta = 1\%$ smaller than I_{ON} through M_2, and that we disregard the bias currents of the OpAmps, then the output current $-I_{o2}$ is 1% smaller than I_{o1} at a positive value of I_{o1}, but $-I_{o2}$ is 1% larger than I_{o1} at a negative value of I_{o1}. This means a kink in the transfer of $-I_{o2} = f(I_{o1})$ at zero current.

We can explain this also in a different way:

$$I_{o2} = -I_{o1} + \delta |I_{o1}|, \quad \text{and} \quad 1/|H_o| = \delta/2 \qquad (9.40)$$

When I_{o1} is physically positive, it is measured and processed by I_{ON}. At the same moment I_{o2} is physically negative and is processed by I_{OP}. If the polarity is reversed the elements that process each of these currents change position. For this reason positive I_{o2} currents are processed in another relation with I_{o1} than negative I_{o2} currents are.

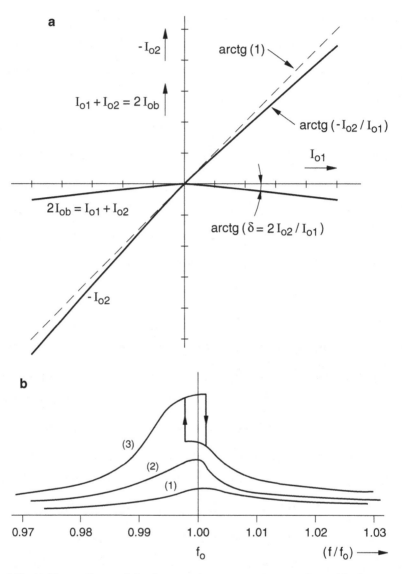

Fig. 9.35 (a) The nonlinear relation between the two output-currents I_{o1} and I_{o2} because of a mismatch of δ between the current-determining transistors. (b) Amplitude characteristics may become instable, while using frequency filters of higher quality factors (1), (2), and (3) while using class-AB OFAs based on the topology of Fig. 9.33 with total-output-supply-current equalization

The result is a strongly nonlinear signal transfer at zero current. If such an OFA is used for audio signals, a noticeable distortion will be heard. If such an OFA is used for filter applications, such as gyrator filters, see Fig. 3.1, then an instability of the filter characteristic will be found at quality factors Q larger than $1/\delta$. This effect is depicted in Fig. 9.35a, b.

The nonlinearity will be partly masked for small signal currents within the class-A-biased range of the OpAmp in which the total supply current does not change much.

We may, of course, trim the transistor matching of the class-AB OFA of Fig. 9.33. If we do this, we can combine the advantage of a high dynamic range with a high linearity. But trimming is costly. We can first try to find a better way.

9.8.2   Universal Class-AB OFA Design with Current Mirrors

The second approach is to find a circuit realization functioning according to the Eq. 9.38 in which the currents in I_{M2} and I_{M4} are equalized with the currents in I_{M1} and I_{M3} [10]. We are able to do so by diagonally connecting current mirrors, as shown in Fig. 9.36. We may expect errors in the signal transfer of the order of 0.5 % due to inaccurate matching of the inaccurate mirror transistor characteristics and due to voltage dependent early effects. This also gives rise to nonlinear distortion of the order of 0.5 %.

Looking to the OFA circuit of Fig. 9.36 we might ask ourselves if we could not avoid mirroring the current signals twice. When we only mirror once, another functional block arises: the operational mirrored amplifier, or OMA [11–14]. The circuit is shown in Fig. 9.37.

The OMA can, as well as the OFA, be used for universal voltage-to-current converters. The output currents are now equal

$$I_{o2} + I_{ol} = \delta|I_{ol}|, \quad \text{and} \quad 1/|H_o| = \delta/2 \qquad (9.41)$$

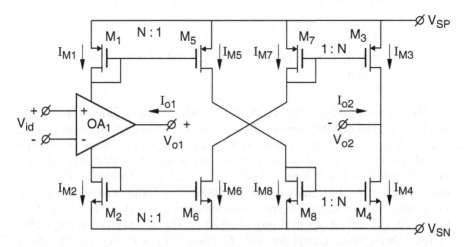

Fig. 9.36 Universal class-AB OFA realization with diagonally connected current mirrors

Fig. 9.37 Universal
operational mirrored
amplifier (OMA) realization

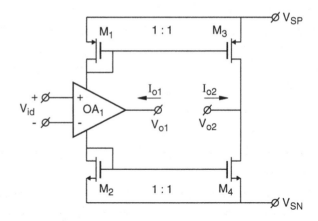

The mismatch δ between the two currents is again caused by transistor mismatch, which gives errors and nonlinearities in the order of 0.5 %.

Though the polarity of the two currents are equal in the OMA, while the OFA has currents with opposite polarity, the OMA can nearly be used in all OFA applications. The advantages and disadvantages of the current-mirror-matching methods are equal to the supply-current-matching methods. This means that a high signal-to-noise ratio for the output current signals can be obtained, but that a strong nonlinearity kink occurs in the zero current point, and that filters may become instable at high quality factors. Only trimming can alleviate these problems for class-AB biasing.

The nonlinearity will be partially masked within the class-A region of the class-AB biasing for small current signals.

Notice that we still have to carefully cascode the current mirrors to avoid significant voltage modulation errors in the current transfer.

9.8.3 Universal Class-AB OFA Design with Output-Current Equalization

The third approach uses the first Eq. 9.37 by which the output currents are directly measured and equated. The way we can approach this is to insert current-measuring resistors R_{M1} and R_{M2} in the output terminals and measure the difference between the voltages V_{M1} and V_{M2} across these resistances, and finally equate these voltages. The differential voltages across these resistors can only be measured with an instrumentation amplifier, having a high CMRR for the differential input voltages, while being able to accurately process the difference, see Figs. 9.16, 9.17, 9.18, 9.19, 9.20, and 9.21. A practical way to depict the situation is given in Fig. 9.38.

The instrumentation amplifier with T_1 through T_4 and OA_2, see Fig. 9.23, measures the voltage difference across a measuring resistor R_{M1} in series with the

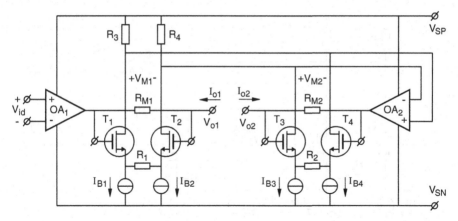

Fig. 9.38 Universal class-AB OFA realization with output-current equalization by means of an instrumentation amplifier

output of an OpAmp OA_1. The instrumentation amplifier is connected as a voltage-to-current converter with a measuring resistor R_{M2} in series with its output amplifier OA_2. The two end terminals of R_{M1} and R_{M2} give shape to the two output terminals of the universal class-AB OFA with V_{o1} and V_{o2}. The output currents I_{o1} and I_{o2} obey the third OFA equation (Eq. 9.39).

Very essential now is that the same physical elements R_1, R_{M1}, and R_2, R_{M2} that measure the output currents in one polarity also measure the output current if their polarity reverses. This means that the current relation between I_{o1} and I_{o2} is basically linear. Only a linear scaling factor α occurs.

For this we may write:

$$I_{o1} = -\alpha I_{o1} = -I_{o1} - (\alpha - 1)I_{o1} = -I_{o1} + \delta I_{o1} \tag{9.42}$$

$$I_{o1} + I_{o2} = \delta I_{o1} = 2I_o = 2I_{ob} + 2I_{o1}/H_o, \quad \text{and} \quad 1/H_o \approx \delta/2 \tag{9.43}$$

We have a linear output current relation. There is no kink in the current transfer, nor an instability in filter characteristics at high quality factors. With these OFAs we can have a signal-to-noise ratio of the output currents in the order of 100 dB. The output current rejection ratio H_o is half the mismatch of the resistors.

9.8.4 Universal Class-AB Voltage-to-Current Converter with Instrumentation Amplifier

One simplification is obvious, though. If we need the OFA for realizing the function of a voltage-to-current conversion, we do not need the full OFA construction of Fig. 9.38, but the connection of an instrumentation amplifier with a current-

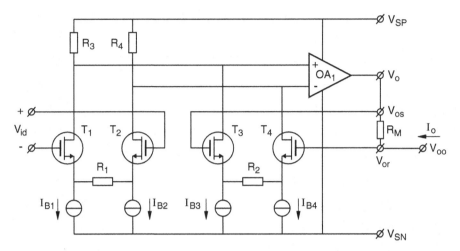

Fig. 9.39 Universal class-AB voltage-to-current converter with instrumentation amplifier

measuring resistor R_M in series with the output, as shown in Fig. 9.23, satisfies our need. It is shown again in Fig. 9.39.

With an instrumentation amplifier, a high-quality universal class-AB voltage-to-current converter of Fig. 9.39 can be obtained. Without trimming, an output impedance of more than $10^4 R_M$ can easily be obtained, at a nonlinearity error of better than 10^{-4}, a bandwidth of several tens of megahertz and a signal-to-noise ratio of 120 dB.

9.8.5 *Conclusion*

We have seen in this section on universal class-AB OFAs that the combination of class-AB biasing and nonlinearity cannot easily be met. Based on the way the output currents can be equalized, three possibilities are open.

Supply-current equalization and current-mirror equalization both show a nonlinear kink in the current transfer at zero current. This gives rise to audible distortion when these circuits are used to process audio signals. Instability may occur when used in filters with a high quality factor. By trimming the nonlinearity can be reduced. For small current signals the nonlinearity is masked in the class-A region of the class-AB biasing. The signal-to-noise ratio of the relation between the output currents can be high. Current-mirror equalization leads to simple circuits.

Output-current equalization using an instrumentation amplifier achieves a high quality and highly linear result. A simplification can be made here. If we only need the function of a voltage-to-current conversion, we do not need the full OFA architecture, but an instrumentation amplifier connected as a voltage-to-current converter can do a high-quality job.

9.9 Problems

9.9.1 Problem 9.1

For the instrumentation amplifier in Fig. 9.19, calculate the common-mode rejection ratio CMRR if the input transistors M_{11} and M_{12} can have a processing-induced threshold voltage difference $\Delta V_{th} = 3$ mV. The circuit is biased at $I_{D31} = I_{D32} = I_{D18} = I_{D28} = 50$ μA, $I_{quies41} = 100$ μA, and the devices in the signal path are sized with $W/L_{11} = W/L_{12} = W/L_{21} = W/L_{22} = 100$ μ/2 μ, $W/L_{37} = W/L_{38} = 40$ μ/2 μ, $W/L_{35} = W/L_{36} = 30$ μ/1 μ, $W/L_{41} = 3.5 W/L_{42} = 70$ μ/1 μ. Source degeneration resistors are all equal to $R_{11} = R_{12} = R_{21} = R_{22} = 5$ kΩ. Transistor parameters are $V_{THN} = 0.5$ V, $V_{THP} = -0.6$ V, $K_N = 56$ μA/V^2, $K_P = 16$ μA/V^2, $\lambda_N = \lambda_P = 0.1$ V^{-1}.

9.9.1.1 Solution 9.1

According to Eq. 4.16, the CMRR limit due to disbalancing of an input stage is

$$CMRR = \frac{\mu^2}{\Delta \mu} \qquad (9.44)$$

Because the mirror M_{37}-M_{38} repeats I_{D11} effects in I_{D38} as well as in I_{D37} there is no common-mode to differential crosstalk in the folded cascode stage, so the input stage voltage gain μ is the only gain factor which degrades CMRR. The variation of μ for the input stage due to the given ΔV_{th} under the assumption of a small threshold voltage change from M_{11} to M_{12} is

$$\Delta \mu = \frac{d\mu}{dV_{th}} \Delta V_{th} \qquad (9.45)$$

In the equation above, an expression for the voltage gain μ as a function of V_{th} is needed to calculate the finite differences. The voltage gain of the input stage is given by its equivalent G_m and its equivalent G_{ds}

$$\mu = \frac{G_m}{G_{ds}} \qquad (9.46)$$

where the conductances are given by

$$G_m = \frac{1}{R_{11} + \dfrac{1}{\sqrt{2K_P \dfrac{W}{L_{11}} I_{D11}}}}$$

$$\frac{1}{G_{ds}} = R_{ds} = \mu_{11} R_{11} + \frac{1}{\lambda I_{D11}} \qquad (9.47)$$

The variation of μ for a small change in V_{th} is given by

$$\frac{d\mu}{dV_{th}} = \frac{dG_m}{dV_{th}}R_{ds} + \frac{dR_{ds}}{dV_{th}}G_m \tag{9.48}$$

The derivative of G_m related to V_{th} will be

$$\frac{dG_m}{dV_{th}} = \frac{dG_m}{dI_{D11}}\frac{dI_{D11}}{dV_{th}} \tag{9.49}$$

Combining $G_m(I_{D11})$ and $I_{D11}(V_{th})$ for transistors working in saturation region, Eq. 9.49 becomes

$$\frac{dR_{ds}}{dV_{th}} = \frac{1}{2}K_P\frac{W}{L_{11}}(V_{SG11} + V_{THP})$$

$$\times \frac{1}{\left(R_{11} + \dfrac{1}{\sqrt{2K_P\dfrac{W}{L_{11}}I_{D11}}}\right)^2} \frac{1}{\sqrt{2K_P\dfrac{W}{L_{11}}}\sqrt{I_{D11}^3}} \tag{9.50}$$

The derivative of R_{ds} related to V_{th} will be

$$\frac{dR_{ds}}{dV_{th}} = \frac{dR_{ds}}{dI_{D11}}\frac{dI_{D11}}{dV_{th}} \tag{9.51}$$

The expression of R_{ds} as a function of I_{D11} gives a derivative

$$\frac{dR_{ds}}{dV_{th}} = -K_P(V_{SG11} + V_{THP}) \times \left(\frac{1}{2}\frac{\sqrt{2K_P\dfrac{W}{L_{11}}I_{D11}}}{\lambda} \frac{1}{\sqrt{I_{D11}^3}}\right) \tag{9.52}$$

Numerically, the parameters involved in calculating CMRR are

$$G_m = \frac{1}{R_{11} + \dfrac{1}{\sqrt{2K_P\dfrac{W}{L_{11}}I_{D11}}}} = 100 \ \mu A/V$$

$$R_{ds} = \mu_{11}R_{11} + \frac{1}{\lambda I_{D11}} = 800 \ k\Omega$$

$$\frac{dG_m}{dV_{dt}} = 200 \ \mu A/V^2 \tag{9.53}$$

$$\frac{dR_{ds}}{dV_{th}} = 3.2 \ M\Omega/V$$

$$\mu = 80$$

$$\Delta\mu = \frac{d\mu}{dV_{th}}\Delta V_{th} = 1.44$$

Replacing these numbers in CMRR expression gives

$$CMRR = \frac{\mu^2}{\Delta\mu} = 73 \text{ dB} \tag{9.54}$$

It is worth noting the strong degradation of common-mode rejection due to an apparently small variation in threshold voltage. In a real implementation this number will be further degraded by additional geometrical mismatches and by the finite impedance of the biasing current source.

9.9.2 Problem 9.2

For the voltage-to-current converter in Fig. 9.22, calculate the typical, maximal and minimal transconductance G_M and the highest common-mode output conductance for resistors sized as $R_1 = R_2 = 10 \text{ k}\Omega$, $R_3 = R_4 = 30 \text{ k}\Omega$ and $R_M = 10 \text{ k}\Omega$. The resistors are matched within $M = 2\%$ and their absolute tolerance is $\Delta R = 10\%$.

9.9.2.1 Solution 9.2

Equation 9.23 gives the expression of this circuit transconductance

$$G_M = A_v \frac{1}{R_M} = \frac{R_3}{R_1} \frac{1}{R_M} \tag{9.55}$$

The matching of resistors R_3 and R_1 affects the voltage gain A_v while the tolerance affects the value of R_M. Thus the limits of G_M become

$$G_{Mmax} = \frac{R_3}{R_1}(1+M)\frac{1}{R_M(1-\Delta R)} = 340 \text{ }\mu A/V$$

$$G_{Mtyp} = \frac{R_3}{R_1}\frac{1}{R_M} = 300 \text{ }\mu A/V \tag{9.56}$$

$$G_{Mmin} = \frac{R_3}{R_1}(1-M)\frac{1}{R_M(1+\Delta R)} = 267 \text{ }\mu A/V$$

It is easy to note that most of this error is induced by the high tolerance of R_M.

For the worst-case common-mode conductance, i.e., the highest possible, Equation set (Eq. 9.24) shows

$$G_{OCMmax} = -\frac{1}{H_{Rmax}R_M(1-\Delta R)} - \frac{1}{H_{Rmax}}$$

$$= \left(1 - \frac{R_3}{R_1}\frac{R_2}{R_1}(1-M)^2\right)\left(1 + \frac{R_1}{R_3}(1+M)\right) = 0.0134 \tag{9.57}$$

The equations above give a maximal common-mode output impedance of

$$G_{OCMmax} = 1.5 \ \mu A/V \tag{9.58}$$

9.9.3 Problem 9.3

The universal voltage-to-current converter shown in Fig. 9.24 uses the instrumentation amplifier depicted in Fig. 9.16, built using a conductance $G_1 = 2G_2 = 100 \ \mu A/V$ and current sources have conductances $G_{IB} = 0.5 \ \mu A/V$ matched within $\Delta G_{IB}/G_{IB} = 3 \ \%$. All operational amplifiers are considered to have $CMRR = 80$ dB, and the conversion resistor is $R_M = 10$ kΩ. Calculate the differential transconductance and the common-mode output conductance, G_M and G_{OCM}.

9.9.3.1 Solution 9.3

Combining Eqs. 9.25 and 9.19, the differential voltage gain and transconductance can be calculated.

$$
\begin{aligned}
A_v &= \frac{G_1}{G_2} = 2 \\
G_M &= \frac{A_v}{R_M} = 200 \ \mu A/V
\end{aligned}
\tag{9.59}
$$

The common-mode rejection ratio of the instrumentation amplifier's input stage limits the value of the common-mode output conductance, according to Eq. 9.26.

$$G_{OCM} = \frac{1}{H}\frac{1}{R_M} \tag{9.60}$$

The CMCR factor $1/H$ can be calculated based on Eq. 9.14

$$\frac{1}{H} = \frac{G_{IB}}{G_1} \frac{\Delta G_{IB}}{G_{IB}} + \frac{1}{H_{OA1}} + \frac{1}{H_{OA2}} = -69 \ \text{dB} \tag{9.61}$$

and G_{OCM} becomes

$$G_{OCM} = \frac{1}{H}\frac{1}{R_M} = 35 \, nA/V \tag{9.62}$$

References

1. D.D.H. Tellegen, On nullators and norators. IEEE Trans. Circ. Theory **CT-13**, 466–468 (1966)
2. J.J. Ebers, J.J. Moll, Large-signal behavior of junction transistors. Proc. IRE **42**, 1761–1772 (1954)
3. Philips Semiconductors Application Note NE 5230
4. G.J.A. van Dijk, A. Bakker, J.H. Huijsing, Low-power CMOS transadmittance amplifier with extended common-mode input range for a smart thermocouple interface, in *Proceedings of ProRisc Workshop*, Mierlo, The Netherlands, 27–28 Nov 1997, pp. 143–147
5. A.P. Brokaw, P.M. Timko, An improved monolithic instrumentation amplifier. IEEE J. Solid-St. Circ. Sci **10**(6), 417–423 (1975)
6. B.J. van den Dool, J.H. Huijsing, Indirect current feedback instrumentation amplifier with a common-mode input range that includes the negative rail. IEEE J. Solid-St. Circ. **38**(7), 743–749 (1993)
7. E.H. Nordholt, Extending op amp capabilities by using a current-source power supply. IEEE Trans. Circ. Syst. **CAS-29**, 411–414 (1982)
8. J.H. Huijsing, J. de Korte, Monolithic nullor – a universal active network element. IEEE J. Solid-St. Circ. **SC-12**(1), 59–64 (1977)
9. J.H. Huijsing, Design and application of the Operational Floating Amplifier (OFA): the most universal operational amplifier. Analog. Integr. Circ. Signal Process **4**(2), 115–129 (1993)
10. A.S. Sedra, The current conveyer: history and progress. Proc. Int. Symp. Circ. Syst. **3**, 1567–1570 (1989)
11. J.H. Huijsing, C.J. Veelenturf, Monolithic operational mirrored amplifier (OMA). Elect. Lett. **17**(3), 119–120 (1981)
12. J.W. Haslett, M.K.N. Rao, A high quality controlled current source. IEEE Trans. Instrum. Meas. **IM-28**(2), 132–140 (1979)
13. B.L. Hart, R.W.J. Barker, Universal operational-amplifier converter technique using supply-current sensing. Electron. Lett. **15**(16), 496–497 (1979)
14. F.J. Lidgey, C. Toumazon, Accurate current follower. Electron. Wireless World **91**(1590), 17–19 (1985)

Chapter 10
Low-Noise and Low-Offset Operational and Instrumentation Amplifiers

Abstract This chapter gives an overview of techniques that achieve low offset, low noise, and high accuracy in CMOS operational amplifiers (OA or OpAmp) and instrumentation amplifiers (IA or InstAmp). Auto-zero and chopper techniques are used apart and in combination with each other. Frequency-compensation techniques are described that obtain straight roll-off amplitude characteristics in the multi-path architectures of chopper-stabilized amplifiers. Therefore, these amplifiers can be used in standard feedback networks. Offset voltages lower than 1 μV can be achieved. Instrumentation amplifiers with capacitive coupled chopper inputs are described. They facilitate CM input voltage ranges outside the supply voltages for applications of beyond the rail current sensing.

10.1 Introduction

The simplest type of amplifier that can be made with a low-offset voltage V_{os} and high common-mode rejection ratio (CMRR) is the operational amplifier (OA). But this amplifier does not have a well-determined gain. The gain of an OA is normally so high that feedback around the OA must be applied to produce an accurate result [1]. This situation is depicted in Fig. 10.1.

However, feedback destroys the CMRR because the feedback network may have unbalance and is connected to a ground reference. Therefore, other types of amplifiers have to be found to combine an accurate voltage gain, a low offset, and a high CMRR.

Instrumentation amplifiers (IA), on the other hand, can have the combination of accurate gain, low offset voltage V_{os}, and high common-mode rejection ratio. But they are more difficult to implement than operational amplifiers. A general symbol for an instrumentation amplifier is given in Fig. 10.2.

This chapter discusses the design of low-noise and low-offset operational amplifiers and instrumentation amplifiers:

1. Introduction
2. Application of IA
3. Three-OpAmp IA

© Springer International Publishing Switzerland 2017

J. Huijsing, *Operational Amplifiers*, DOI 10.1007/978-3-319-28127-8_10

Fig. 10.1 Operational amplifier (OA) in feedback network, $V_{id} = 0$, $I_{id} = 0$, $I_{ic}(CM) = 0$, input CMRR = Low

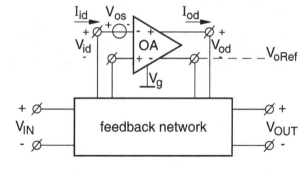

Fig. 10.2 Instrumentation amplifier (IA), $V_{id} \neq 0$, $I_{id} = 0$, $I_{ic}(CM) = 0$. $V_{od} = A_v V_{id}$, input CMRR = High

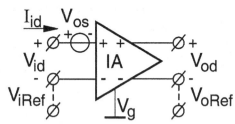

4. Current-feedback IA
5. Auto-zeroing
6. Chopping
7. Chopper stabilization
8. Chopping + AZ or chopper-stabilized
9. Ripple-reduction loop (RRL)
10. Capacitive coupled input
11. Gain accuracy of IA
12. Summary low offset

10.2   Applications of Instrumentation Amplifiers

All applications of an IA use the combination of accurate gain and high CMRR. The first application example is a general one: to overcome a ground loop. This occurs when we want to transfer a voltage signal referred to a different ground potential V_{sRef} than that of the destination potential V_{oRef}. The situation is depicted in Fig. 10.3.

This is the case, for instance, when an instrument has to interface a sensor, like a thermocouple, that is connected to a remote ground. The small output voltage of the thermocouple requires a low offset voltage of the amplifier, while the remote ground can have a large potential difference in regard with the ground of the sensing instrument. This requires a high CMRR.

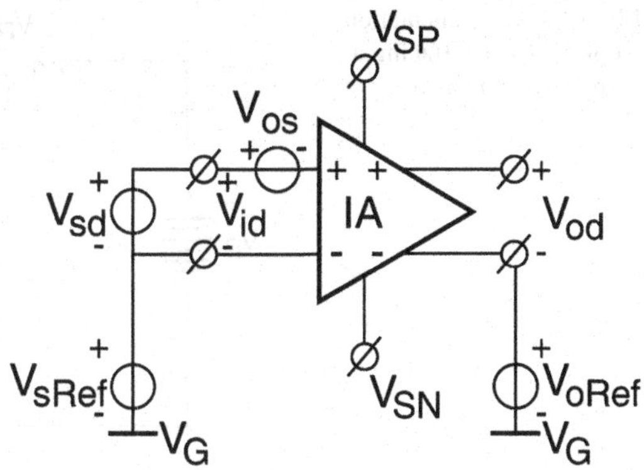

Fig. 10.3 Instrumentation
amplifier bridging the
common-mode voltage
between $V_{s\,ref}$ and $V_{o\,ref}$

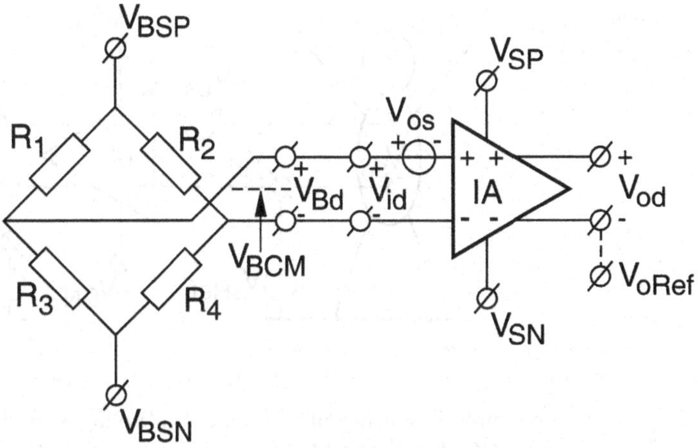

Fig. 10.4 Instrumentation amplifier for the readout of a sensor bridge

A second common application is the interfacing of the differential output voltage V_{Bd} of a sensor bridge that has a large common-mode voltage V_{BCM}, as shown in Fig. 10.4. Accuracy and low offset of the measurement in this application are of high priority.

A third application example is monitoring the voltage V_{Rsd} across a current-sense resistor R_s in supply lines of battery-powered systems like cell phones and laptops (see Fig. 10.5). Power management and battery life make this application rapidly more important.

A high dynamic range is required for the current-sense application, as we want to be able to measure high as well as low supply currents reasonably accurately, and do not want to spill a large amount of power across the sense resistor at high currents. This means that the sense voltage must be small and that the IA or "current-sense" amplifier needs to have a low-offset voltage under high CM input voltages. The CM input voltage range may be even far beyond its supply voltage. This thoroughly complicates the design of the IA.

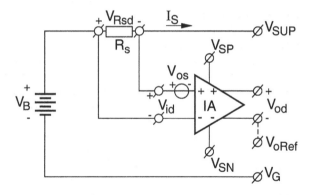

Fig. 10.5 Instrumentation amplifier for interfacing a current-sense resistor

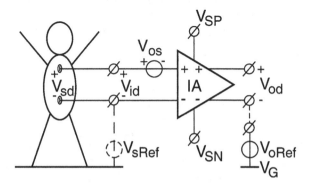

Fig. 10.6 Instrumentation amplifier for interfacing medical electrodes

A final application example is sensing of differences in voltages of skin electrodes for measuring an ECG, EEG, or EMG of a person (see Fig. 10.6). These differential voltages are in the order of 10 μV to 100 mV, while the person is capacitive coupled to large CM voltages from mains operated sources on the order of 10–100 V. High CMRR and patient safety are main requirements here.

10.3 Three-OpAmp Instrumentation Amplifiers

The most common approach to an IA is the three-OpAmp topology as shown in Fig. 10.7 (see Fig. 3.4).

The actual IA consists of an OA that is feedback by a resistor bridge network R_{11}, R_{12}, R_{13}, and R_{14}. If the bridge is in balance, the gain for differential signals is:

$$A_d = -R_{12}/R_{11} \approx -R_{14}/R_{13} \tag{10.1}$$

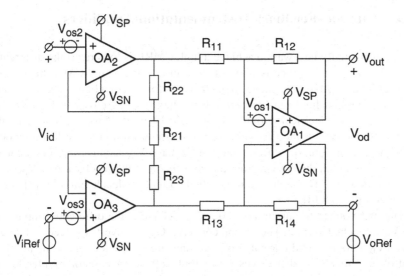

Fig. 10.7 Three-OpAmp instrumentation amplifier with resistor-bridge feedback and input buffer amplifier

To achieve a high input impedance, buffer amplifiers OA_2 and OA_3 have been placed in front of the bridge resistors. These amplifiers are connected in a non-inverting gain configuration with R_{21}, R_{22}, and R_{23}. Their extra gain is:

$$A_{d2} = (R_{21} + R_{22} + R_{23})/R_{21} \qquad (10.2)$$

The total voltage gain is:

$$A_V = -(R_{21} + R_{22} + R_{23})R_{12}/(R_{11}R_{21}) \qquad (10.3)$$

The main problem of the three OpAmp approach is the CMRR. In this topology the CMRR is dependent on the matching of the feedback bridge resistors, as explained in Sect. 3.2:

$$CMRR = (R/\Delta R)A_V \qquad (10.4)$$

in which $\Delta R/R$ is the relative error in one of the bridge resistors in regard to its ideal value if the bridge were balanced [2]. For instance:

$$\Delta R_{11}/R_{11} = 1 - R_{12}R_{13}/(R_{11}R_{14}) \qquad (10.5)$$

Another shortcoming of the three-OpAmp approach is that the input CM range can not include the negative nor positive supply rail voltage. This is the consequence of the feedback connection from the output of the input buffer amplifiers OA_2 and OA_3 to their inputs. Only when a level shift is built-in in the positive input modes of these amplifiers one of the rail voltages can be reached.

10.4   Current-Feedback Instrumentation Amplifiers

The fundamentally best way to achieve a high CMRR is to convert the differential input signal V_{id} into a type of signal that is insensitive to the CM voltage V_{iCM}. Such a signal could be a magnetic signal in a transformer, or a light signal between a light-emitting and light-sensing diode. But when we stay closer to the electrical domain, also an electrical current signal could be used, if we can make it sufficiently insensitive for the CM voltage. For a circuit on a chip the last method is preferable. Therefore, the differential input voltage V_{id} is converted into a current and compared with the current from the conversion of the feedback part V_{fb} of the output voltage V_o. The architecture is called current-feedback amplifier [3], and is shown in Fig. 10.8 [4].

The first voltage-to-current converter G_{m21} converts the differential input voltage V_{id} into a first current. The second converter G_{m22} converts the feedback output signal V_{fb} into a second current. Both currents are subtracted and compared by a control amplifier G_{m1} that drives the output voltage. A resistor divider R_2, R_1 determines the part V_{fb} of the output voltage V_o that is fed back. The gain of the whole amplifier will be:

$$A_V = (G_{m21}/G_{m22})(R_2 + R_1)/R_1 \qquad (10.6)$$

Often we can not easily make the transfer of G_{m21} and G_{m22} accurately different. But we can make G_{m21} and G_{m22} accurately equal. In that case the gain of the amplifier simplifies to:

$$A_V = (R_2 + R_1)/R_1, \quad \text{while} : G_{m21} = G_{m22} \qquad (10.7)$$

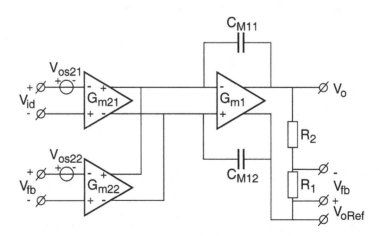

Fig. 10.8 Current-feedback instrumentation amplifier

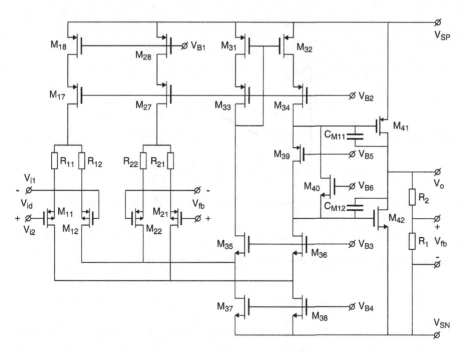

Fig. 10.9 Simple circuit-diagram of an current-feedback instrumentation amplifier

The CMRR is now not determined by matching of main elements but just by the ratio of the Gm and small parasitic conductances, which keep the CMRR large.

A simple example of an current-feedback InstAmp is given in Fig. 10.9.

The InstAmp is Miller compensated by the capacitors C_{M11} and C_{M12}.

The input and feedback VI converters are as simple as possible. They can be degenerated to increase the differential input voltage range if needed. Their linearity is not good in itself, but they match quite well for gain accuracy. The input CM voltage range may include the negative supply-rail voltage V_{SN}. This allows the output voltage V_o being referenced to V_{SN}. The input stages are followed by folded cascodes with a current mirror at their upper end. The push-pull output transistors are biased in class-AB by a class-AB mesh composed from M_{39} and M_{40} and proper bias voltages V_{B5} and V_{B6} (see Fig. 5.27).

A general symbol for an current-feedback IA is given in Fig. 10.10. It shows that inside the IA there are two G_m stages: one for the input G_{mi} and one for the feedback G_{mfb}.

It is interesting that the output as well as the input has a high CMRR. This means that we can connect the output reference voltage V_{oRef} terminal to any voltage as shown in Fig. 10.11. The voltage across the measuring resistor R_M and the current through R_M are not influenced by the voltage on V_{oRef}. Hence, we obtain a voltage controlled current source at the V_{oRef} terminal. The whole topology of Fig. 10.11 act as an accurate general-purpose V-I converter with a transconductance of $1/R_M$. Hence, $I_o = -V_{id}/R_M$.

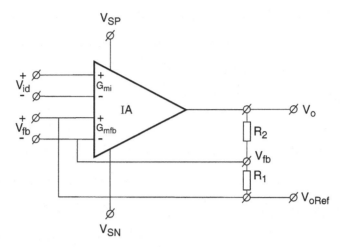

Fig. 10.10 Symbol for an current-feedback instrumentation amplifier

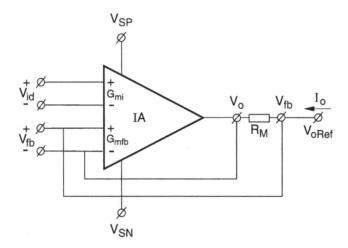

Fig. 10.11 Universal voltage-to-current converter with an current-feedback instrumentation amplifier

10.5 Auto-Zero OpAmps and InstAmps

In Sect. 10.2 we have seen several applications that need low offset. Auto-zeroing and chopping are the main tools to obtain low offset.

In this paragraph we start with auto-zeroing. Firstly, we will apply auto-zeroing to an OA in order to reduce its offset. Out of the many ways to implement auto-zeroing we firstly have chosen the simple method with switched capacitors at the input as shown in Fig. 10.12.

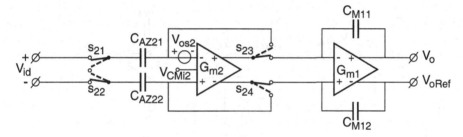

Fig. 10.12 Switched-cap auto-zero OpAmp. Vos = 100 μV

The auto-zero OA consists of an auto-zeroing input stage G_{m2} with input CM control and a Miller compensated output stage G_{m1}.

Auto-zeroing has two phases. In phase 1 the forward path is broken, and G_{m2} is being fully fed back, so that its offset appears at its input. The auto-zero capacitors C_{AZ21} and C_{AZ22} store this offset voltage as their inputs are short-circuited together. In phase 2 G_{m2} is connected straight forward, and the auto-zero capacitors are connected to the input. Their stored offset voltage now compensates for the offset of G_{m2}. Therefore, G_{m2} shows no offset in phase 2.

An improved auto-zero topology with storage capacitors at the output is shown in Fig. 10.13a. When the input switches S_{21} and S_{22} are short-circuited, and the auto-zero switches S_{23} and S_{24} are in auto-zero position, the output current of G_{m2} charges the capacitors C_{31} and C_{32} at its output until the correction amplifier G_{m3} compensates this current. The output of G_{m2} is CM controlled at its output.

The advantage of this topology is that the capacitors can store the offset independent of the input signal. This means that the capacitors and conductance G_{m3} can be taken 10× smaller for the same kT/C noise and that the compensation voltage on the storage capacitors can be taken 10× larger. The offset of G_{m3} is not of interest because it is automatically taken into account in the capacitive stored voltage.

A further improvement can still be made if we replace the passive integrator capacitors for an active integrator as shown in Fig. 10.13b. The switches S_{23} and S_{24} do not need to be switched back and forth anymore between the virtual ground at the input of the output amplifier G_{m1} and the stored compensation voltage on the capacitors C_{31} and C_{32}, but between the two virtual grounds at the input of G_{m1} and G_{m3}. This reduces the charge injection of the switches. For simplicity most following auto-zero circuits have been drawn with the simple circuit of Fig. 10.12.

Very important is that the auto-zero action removes offset and 1/f noise. But extra noise V_{naz} is added in the frequency range below $2f_{AZ}$ due to noise folding back from the bandwidth BW of the local auto-zero feedback loop. This is depicted in Fig. 10.14. Hence, it is of importance to keep this BW not too much above $2f_{AZ}$. If the duty-cycle of the auto-zero loop is ½, the noise level is at least raised by a factor $2^{1/2}$.

a

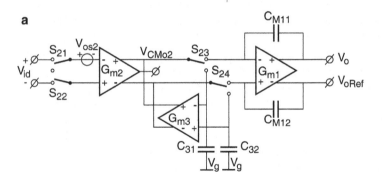

b

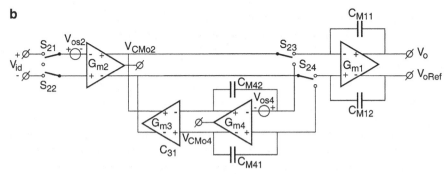

Fig. 10.13 (**a**) Auto-zero OpAmp with storage capacitors C_{31} and C_{32} at the output and correction amplifier G_{m3}. Vos $= \sim20$ μV. (**b**) Auto-zero OpAmp with further improved auto-zero storage by an active integrator. Vos $= \sim10$ μV

Fig. 10.14 Noise densities with and without auto-zeroing

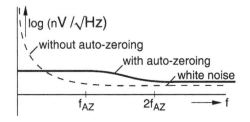

$$V_{naz} = V_n(white) \; BW^{1/2}/f_{az}^{1/2} + 2^{1/2}V_n(white) \qquad (10.8)$$

A problem is that the auto-zero OA has no continuous-time transfer. This means that when the output has to follow a ramp, a staircase with steps at the clock frequency is the result. Moreover, a factor $2^{1/2}$ must be added to the noise as the amplifier is only used half of the time effectively. To overcome these problems the ping-pong auto-zero [5] concept of Fig. 10.15 has been invented.

In Fig. 10.15 two auto-zero input stages G_{m21} and G_{m22} alternately are connected between the input and the output stage in order to obtain a continuous-time solution.

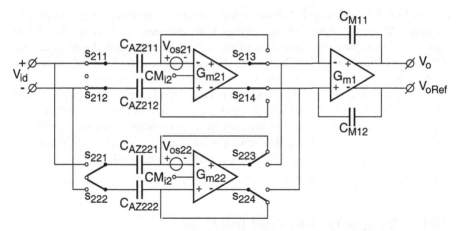

Fig. 10.15 Ping-pong auto-zero OpAmp. Vos = ~100 μV

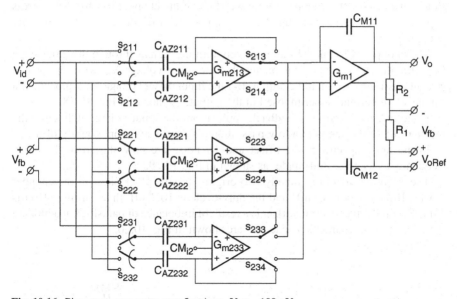

Fig. 10.16 Ping-pong-pang auto-zero InstAmp. Vos = 100 μV

The stage that is not connected gets time to auto-zero itself. This allows the OA to be generally used in continuous-time feedback configurations.

We can extend the principle of ping-pong to ping-pong-pang in order to obtain a suitable InstAmp topology, as shown in Fig. 10.16 [27].

In Fig. 10.16 three auto-zero input stages G_{21}, G_{22}, and G_{23} are used. Sequentially, two stages are connected to the output stage G_{m1}, while one stage is in auto-zero mode. In this way a continuous-time IA is shaped while its offset and 1/f noise is strongly reduced by auto-zeroing.

The limitation of offset reduction is due to parasitic capacitors of capacitors and switches. When the input switches change from auto-zero mode to transfer mode and vice versa, parasitic capacitors to ground are charged and discharged. Any unbalance in this charge will change the offset voltage stored on the AZ capacitors. Offline auto-zero as in Fig. 10.13a or b would therefore be preferable.

In practice the offset can maximally be reduced by a factor on the order of 100 or 1000 with auto-zeroing, reducing the offset from 10 mV to 100 μV or 10 μV.

It is very interesting to see that not only the offset voltage is reduced by the AZ function, but any CM induced differential input voltage at frequencies lower than the AZ frequency. This means that also the CMRR is drastically increased.

10.6   Chopper OpAmps and InstAmps

Before we discuss the chopper IA we will look at the chopper OA [6]. This OA is depicted in Fig. 10.17. We suppose a 6σ input offset of 10 mV for G_{m2} without chopping.

The choppers Ch_2 and Ch_1 alternatively turn the signals through the input stage G_{m2} straight and reverse. This means that the input voltage V_{id} will appear as a continuous-time current at the output. But the input offset voltage V_{os2} appears as a square wave current, superimposed in the output, as shown in Fig. 10.18.

If the OA is placed in a feedback application, the input voltage will show the residual offset voltage with a low-pass filtered square wave ripple on top of it.

In the noise spectra of the offset and 1/f noise are now shifted to the clock frequency f_{cl} as noise and ripple, as shown in Fig. 10.19.

The resulting offset has mainly two origins: Firstly, clock skew in the chopper clocks. If the offset is 10 mV and the duty-cycle is 10^{-4} off, the resulting offset is 1 μV. Secondly, the resulting offset is a result of imbalance of parasitic capacitors in the choppers. The parasitic capacitors are shown in Fig. 10.20.

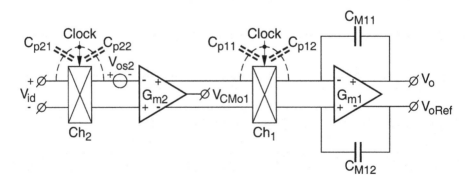

Fig. 10.17 Chopper OpAmp with continuous-time transfer. Vos = ~10 μV, Vrip = ~10 mV

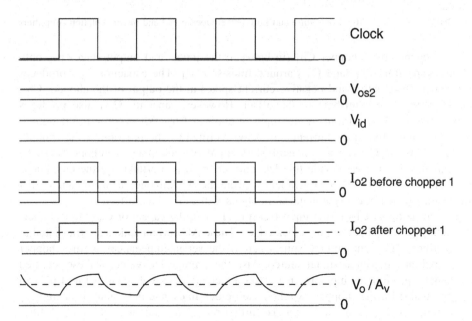

Fig. 10.18 Voltage and current signals as function of time in a chopper amplifier

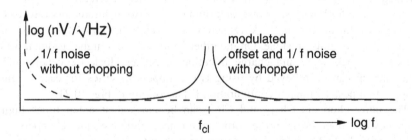

Fig. 10.19 Noise densities in an amplifier with and without chopping

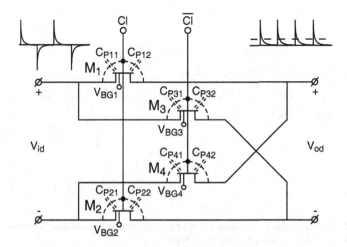

Fig. 10.20 Charge injection current in C_{p11} of Chopper Ch_1 gets rectified at output

Suppose that chopper Ch_1 (in between the input- and output stage) has only the capacitors C_{p11} and C_{p12} around transistor M_1. The capacitor C_{p12} produces alternative positive and negative current spikes at the output of the chopper Ch_1. This does not contribute to the offset. However, capacitor C_{p11} also produces alternative spike currents at the input of chopper Ch_1. When going to the output, these alternative spike currents are being rectified by the function of the chopper Ch_1 into a DC current. An equivalent input offset voltage component can be calculated when we divide this DC current by G_{m2}. Fortunately, the chopper is fully balanced. Hence, charge injection from the clock in one transistor cancels that of the other. But every imbalance in layout will cause a net offset.

For chopper Ch_2 at the input (see Fig. 10.17), the capacitor C_{p22} injects alternating current spikes on the clock edges. These AC current spikes are translated in rectified DC input voltage spikes across the series impedances of the chopper switch and the input signal source. Also the average DC value of these rectified spikes appear at the input as a net offset. Practical offset voltage to below 1 µV can be obtained if the chopper switches, their clock lines, and the signal lines are very carefully balanced in the layout. To further reduce the influence of the clock lines on the amplifier circuit is to layout both clock lines of a chopper within a fully shielded cable on the chip from the chopper to the clock generator in a corner.

In our quest for low offset, noise, and ripple we see two contradictory effects. On the one hand, the higher the clock frequency, the smaller the ripple at the output and the lower the 1/f noise residue. On the other hand, we see a higher residual offset caused by clock skew and charge injection at higher clock frequencies. This contradiction can be relieved by using two choppers in series for each original chopper in a nested chopper configuration [7] according to Fig. 10.21.

The inner choppers Ch_{211} and Ch_{11} can be clocked at a ten times higher frequency Cl_H than the 1/f noise corner frequency, while the outer choppers C_{221} and C_{12} are clocked at a ten times lower frequency Cl_L to take away the residual offset by the charge injection of the inner choppers. This architecture can lead to

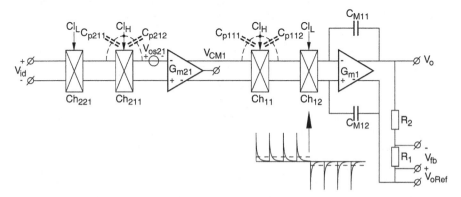

Fig. 10.21 Nested-chopper operational amplifier with better compromise between 1/f noise, ripple, and offset. $Vos = \sim0.1$ µV, $Vrip = \sim100$ µV

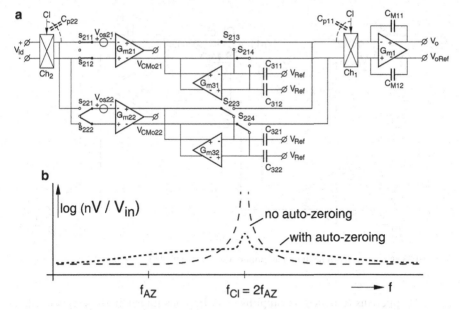

Fig. 10.22 (**a**) Operational chopper amplifier with ping-pong auto-zero input stages. Vos = ~2 μV, Vrip = ~10 μV (**b**) Noise in an operational chopper amplifier with ping-pong auto-zero input stages

offset voltages as low as on the order of 0.1 μV. But a small ~100 μV filtered input-referred ripple at Cl_H still remains due to the original offset, and an even smaller ripple at Cl_L due to charge injection of Ch_{11}.

An other way to reduce the ripple is to combine an auto-zeroed amplifier in a ping-pong fashion with a chopper amplifier in order to obtain a low-ripple continuous-time signal transfer [8]. The block diagram is shown in Fig. 10.22a.

The choppers Ch_1 and Ch_2 chop the signal alternately positive and negative through the whole set of two ping-pong auto-zeroing amplifiers G_{m21} and G_{m22}. The switches S_{211} through S_{222} and S_{213} through S_{224} sequentially switch the amplifiers G_{m21} and G_{m22} in a transfer or auto-zero mode in a full clock cycle.

The capacitors C_{311} through C_{322} differentially store the auto-zero correction voltages. The transconductances G_{m31} and G_{m32} correct the amplifiers G_{m21} and G_{m22} for their offsets, respectively. The auto-zero switches S_{213} through S_{224} switch the outputs of G_{m21} and G_{m22} between the stored voltages on the auto-zero capacitors and the input offset voltage of the output stage. This causes some extra charge injection. The amplifier achieves an offset of 2 μV and an input referred ripple on the order of 10 μV. The noise of the auto-zero amplifier is now transposed by the choppers to the clock frequency, which keeps the low frequencies cleaner, as shown in Fig. 10.22b.

An advantage of the ping-pong continuous-time topology is the simplicity of the frequency compensation. It is restricted to one set of Miller-compensation capacitors.

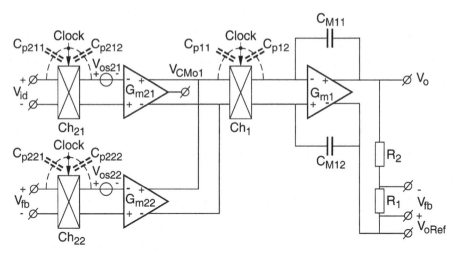

Fig. 10.23 Chopper instrumentation amplifier. Vos = ~20 µV, Vrip = ~20 mV

A chopper instrumentation amplifier can be constructed if we use two input stages G_{m21} and G_{m22}, each preceded by a chopper, Ch_{21} and Ch_{22}, respectively. This situation is shown in Fig. 10.23.

The gain is:

$$A_v = ((R_1 + R_2)/R_1)(G_{m21}/G_{m22}) \qquad (10.9)$$

The accuracy of the instrumentation amplifier fully depends on the equality of G_{21} and G_{22}. In Sect. 10.11 we will discuss ways to increase the accuracy of G_m stages. Even with an ordinary differential pair in weak inversion, and well matched tail currents, an accuracy better than 1 % can easily be achieved without trimming.

The CMRR is also strongly increased by the chopper function for frequencies below the clock frequency. Easily 60 dB can be added to the CMRR by chopping. The improvement is limited, firstly, by the clock skew in the chopper clocks, and secondly, by unequal modulation of the charge injection spikes in the choppers as a function of the CM voltage. The resulting offset can be as low as 20 µV, which is twice that of the chopper OpAmp, and an input-referred ripple of 20 mV, which is twice of that of the OpAmps. The factor 2 is a rough simplification, and results from the fact that there are two parallel input stages.

To improve the offset and ripple, we may also apply the nested-chopper [7] principle to the chopper instrumentation amplifier, as shown in Fig. 10.24. By this a better compromise of chopper ripple and 1/f noise on the one hand and residual offset on the other hand can be achieved as explained with Fig. 10.21. An offset on the order of 0.2 µV can be achieved and a residual ripple on the order of 200 µV.

Finally, just like the chopper OA with ping-pong input stages, the chopper IA can be devised with auto-zero ping-pong-pang input stages, as explained with Fig. 10.16.

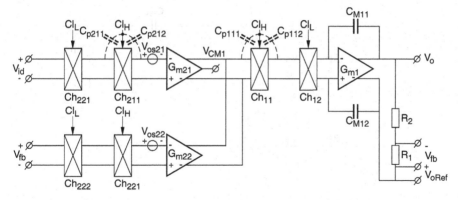

Fig. 10.24 Nested chopper instrumentation amplifier with better compromise between 1/f noise, ripple, and offset. Vos = ~0.2 μV, Vrip = ~200 μV

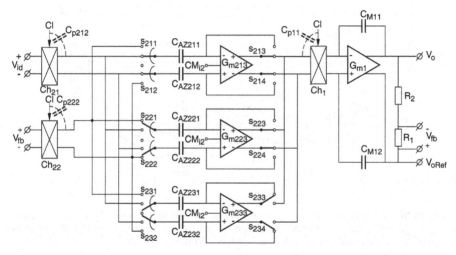

Fig. 10.25 Chopper instrumentation amplifier with ping-pong-pang auto-zero input stages. Vos = ~4 μV, Vrip = ~20 μV

This results in the circuit of Fig. 10.25. It roughly can obtain an offset Vos of 4 μV, and a ripple Vrip of 20 μV. Besides reducing offset the 2-step approach also increases the CMRR.

10.7 Chopper-Stabilized OpAmps and InstAmps

The output ripple from a chopper amplifier invites us to search for ways to reduce it. The chopper-stabilized amplifier is one of the best approaches [9]. A basic multipath nested Miller compensated OA topology is suited to incorporate chopper stabilization, as shown in Fig. 10.26a.

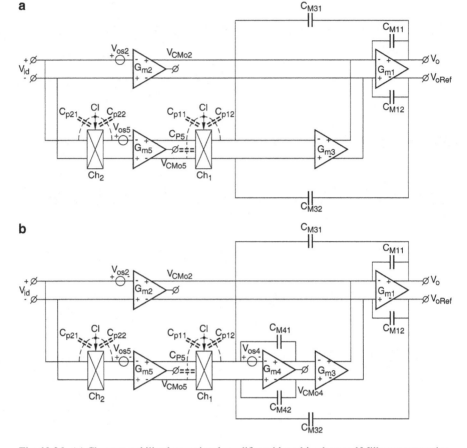

Fig. 10.26 (**a**) Chopper-stabilized operational amplifier with multipath nested Miller compensation. Vos = ~10 μV, Vrip = ~ 10 mV. (**b**) Chopper-stabilized operational amplifier with multipath hybrid-nested Miller compensation. Vos = ~ 10 μV, Vrip = ~ 100 μV

The basic OA is composed of two stages G_{m1} and G_{m2}. The output stage G_{m1} is differentially Miller compensated by C_{m11} and C_{m12}. The input stage G_{m2} forms the "high-frequency" path. The input stage G_{m2} has an offset V_{os2}. When the OA is placed in a feedback loop, the offset V_{os2} appears at the input.

This input error voltage V_{id} is now measured and corrected by the chopper amplifier's "gain" path. This path starts with an input chopper Ch_2 that translates the input error voltage V_{id} into a square wave. The sense amplifier G_{m5} produces a square-wave output error current proportional to V_{id} together with a DC output current due to its own DC offset V_{os5}. The chopper Ch_1 chops the square-wave error current back to a DC error current, while the DC offset current is changed into a square-wave current. The square-wave current due to offset of G_{m5} is filtered and reduced by the Miller integrator capacitors C_{M11} and C_{M12}, while the DC error current as a function of the input error voltage V_{id} is amplified by the DC gain of the intermediate amplifier G_{m3}. Finally the output current of G_{m3} is being added to the output current of the input amplifier G_{m2} in order to compensate its offset. It should be noted that the output CM levels of G_{m2} and G_{m5} have to be controlled to a CM level.

We have now obtained a two-path amplifier: a high-frequency low-gain path through G_{m2}, and a low-offset low-frequency high-gain path through G_{m5} and G_{m3}. The offset can only be reduced to the extent that the high-gain path has a higher gain than the low-gain path.

The above circuit has two shortcomings: Firstly, the gain of G_{m3} is only roughly 20 % of that of G_{m2}, as this amplifier should be able to correct the offset of G_{m5} while not adding too much noise. This means that the offset is not reduced so much. Secondly, the integration function of the outer Miller capacitances C_{M31} and C_{M12} is not strong, particularly not at low overall closed loop gains where the overall closed bandwidth through the low-frequency path of G_{m5} is larger than the clock frequency. In that case the square-wave ripple at the input has nearly the full size of the initial offset of the offset sense amplifier G_{m5}.

For the above reasons it is much better to step on to the multipath hybrid nested Miller compensated OA of Fig. 6.28a which has an extra integrator in the "gain path." This circuit with chopper stabilization is given in Fig. 10.26b. The extra integrator, firstly, is able to strongly reduce the ripple, and secondly, provides much more gain. The square-wave current due to offset of G_{m5} is strongly filtered out by the integrator G_{m4}. The integrator time constant can be chosen freely by the value of the integrator capacitors C_{M41} and C_{M42}. The residual ripple at the output of integrator G_{m4} is further reduced by the relative weak G_{m3}. The DC error current as a function of the input error voltage V_{id} is integrated and strongly amplified by the DC gain of the integrator G_{m4}. The integrated error voltage at the output of G_{m4} is added to the output current of the input amplifier G_{m2} through G_{m3} in order to compensate the offset of G_{m2}.

One of the old struggles with chopper-stabilization is that the two poles in the gain path lead to a non-straight 6 dB per octave role-off, as shown in Fig. 10.27.

This problem can be solved in practice by applying the principle of hybrid nesting as described in Sect. 6.2 [10]. To that end we connect two differential hybrid-nested Miller capacitors C_{M31} and C_{M32} from the final output to the input of the integrator G_{M4}.

If we choose the bandwidth of the two-stage Miller-compensated HF amplifier path equal to the bandwidth of the four-stage hybrid-nested Miller loop, the overall frequency characteristic becomes straight form very low frequencies to the bandwidth of the OA. Therefore, we choose $G_{m2}/C_{M1} = G_{m5}/C_{M3}$ with $C_{M1} = C_{M11} = C_{M12}$ and $C_{M3} = C_{M31} = C_{M32}$. The result is a straight frequency characteristic, as shown in Fig. 10.27.

The low-frequency behavior, and thus the offset of the whole amplifier is determined by that of the chopper loop. That means that we have to carefully balance the parasitic capacitors C_{p11} and C_{p22} of the choppers Ch_1 and Ch_2, respectively, and their lay-out. Also the duty-cycles of the chopper clocks

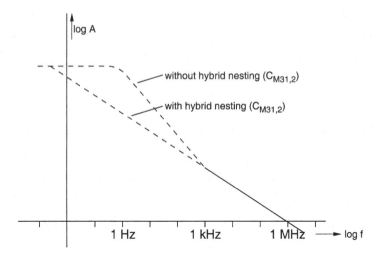

Fig. 10.27 Amplitude characteristic of a chopper-stabilized amplifier with and without hybrid-nested Miller capacitors C_{M31} and C_{M32}

determine the offset. If the duty-cycle is 10^{-4} off from 50 %, and the offset of the amplifier is 10 mV, an offset of 1 μV is resulting.

There is one more source of offset we have to watch for. That is caused by a combination of the parasitic capacitor C_{p5} between the outputs of G_{m5} and the offset Vos_4 of the integrator amplifier. The chopper Ch_1 chops this offset voltage up and down each chopper period on C_{p5}, while the chopper rectifies the associated charge current into a DC value I_{p5} at the input of the integrator equal to:

$$I_{p5} = 4 \; Vos_4 C_{p5} f_{cl} \tag{10.10}$$

This current cannot be distinguished anymore from the DC output current of the chopper sense amplifier that is also presented at the input of the integrator. The resulting input offset V_{osi} is:

$$V_{osi} = I_{p5}/G_{m5} = 4 \; Vos_4 C_{p5} f_{cl}/G_{m5} \tag{10.11}$$

The resulting offset is smaller than 1 μV referred to at the input, only if we take measures to make C_{p5} small, i.e., in the order of 0.1 pF. We can always chopper-stabilize or auto-zero-stabilize the integrator amplifier to further reduce this offset term.

The input referred ripple has now been reduced by a factor 100 from a square wave of about 10 mV in the chopper amplifier into a triangular wave of about 50 μV in the chopper-stabilized amplifier. If we want to decrease the ripple further, we can auto-zero the chopper amplifier [11], as shown in Fig. 10.28.

We have now a combination of a chopper-stabilized amplifier in which the chopper amplifier is auto-zeroed. In this way the ripple can further be reduced to

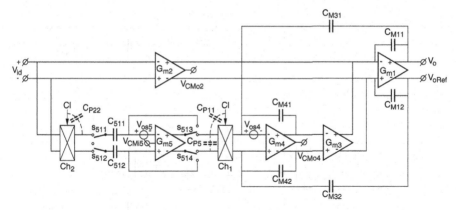

Fig. 10.28 Ch-stab. OpAmp with auto-zero G_{m5}. Vos $= \sim 1\,\mu V$, Vrip $= \sim 10\,\mu V$

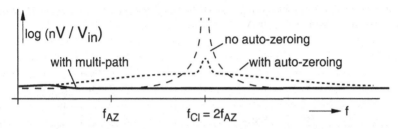

Fig. 10.29 Noise densities of a chopper-stabilized multi-path instrumentation amplifier with and without auto-zeroing

the $1\,\mu V$ level. The noise spectrum of such an amplifier is shown in Fig. 10.29. It still suffers from noise folding and a factor $2^{1/2}$ from a duty-cycle of 50 %. But at higher frequencies the HF path through G_{m2} takes over and the noise reaches it thermal floor.

An interesting alternative way to reduce the ripple is using a sample-and-hold after the integrator [12], as shown in Fig. 10.30.

In this design two passive integrators have been connected as a ping-pong sample and hold with C_{41}, C_{42}, and C_H. The design is simple and elegant and has an offset of $3\,\mu V$, while the ripple is on the order of $10\,\mu V$.

Now, the step has to be made to an instrumentation amplifier. Therefore, the chopper-stabilized OA must be transformed into the current-feedback IA architecture [13]. The circuit is shown in Fig. 10.31.

The IA has a HF path through G_{m21} and G_{m22} and a LF gain path through G_{m51} and G_{m52}. The LF gain path not only determines the offset and CMRR, but also sets the gain accuracy at low frequencies.

The gain at low frequencies is:

$$A_{VL} = (G_{m51}/G_{m52})(R_1 + R_2)/R_1, \qquad (10.12)$$

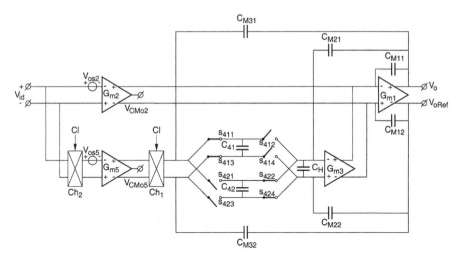

Fig. 10.30 Chopper-stabilized OpAmp with passive integrator and sample and hold. (Rod Burt),
Vos = ~ 3 μV, Vrip = ~ 20 μV

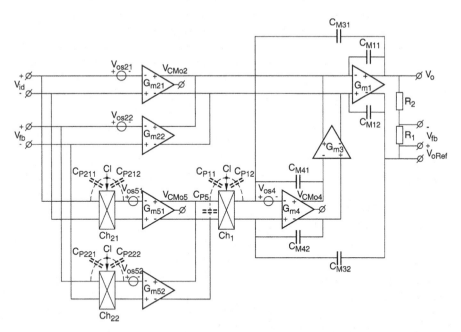

Fig. 10.31 Chopper-stabilized InstAmp with multipath hybrid-nested Miller comp.
Vos = ~ 20 μV, Vrip = ~ 200 μV

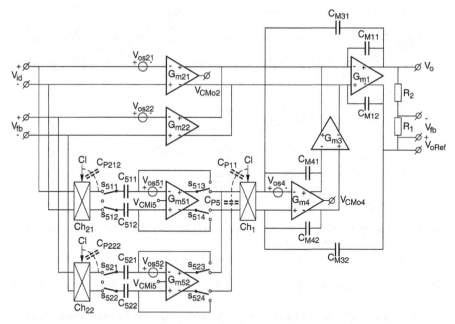

Fig. 10.32 Chopper-stabilized InstAmp with auto-zero sense amplifiers. Vos = $\sim 2\,\mu V$, Vrip = $\sim 20\,\mu V$

and at high frequencies:

$$A_{VH} = (G_{m21}/G_{m22})(R_1 + R_2)/R_1 \qquad (10.13)$$

An offset in the order of 20 μV and an output ripple of 200 μV can be obtained. The offset and ripple is a factor $2^{\frac{1}{2}}$ larger than in the OA case because we have two input stages in parallel in both the HF and LF gain path. Also, the noise is $2^{\frac{1}{2}}$ times larger than in the OA case.

If we want to further reduce offset and ripple the chopper amplifiers can be auto-zeroed as in the OA case [13]. The resulting block diagram is shown in Fig. 10.32.

This topology may result in an input-referred offset voltage lower than 2 μV and a ripple lower than 20 μV.

10.8 Chopper-Stabilized Chopper OpAmps and InstAmps

The smooth continuous-time chopper amplifier is the best approach to low offset. However, a 0.01 % error in duty-cycle of the clock multiplied by an initial 6σ offset voltage of 10 mV of the first stage of a CMOS amplifier presents a lower limit to the residual offset on the order of 1 μV. Moreover, the initial offset voltage on the order of 10 mV at 6σ results in an input-referred chopper square wave ripple of 10 mV.

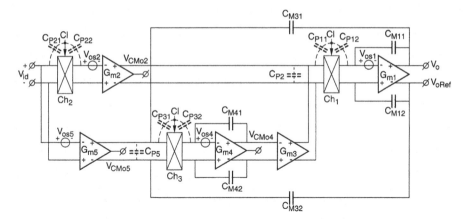

Fig. 10.33 Chopper-stabilized chopper OpAmp with multipath hybrid-nested Miller compensation. $V_{os} = \sim 1\,\mu V$, $V_{rip} = \sim 50\,\mu V$

Hence, the ripple and resulting offset caused by the input amplifier must be further reduced.

The next step of improvement is to chopper-stabilize the chopper amplifier [14]. The topology is shown in Fig. 10.33.

If an amplifier has a high loop gain the differential input voltage becomes zero, except for the input offset voltage. This means in the case of the chopper-stabilized chopper amplifier of Fig. 10.33 that the right-hand side of chopper Ch_2 sees V_{os2}. Hence, the left-hand input side carries a square wave voltage equal to V_{os2}. This allows us to directly connect the correction amplifier G_{m5} to the input without extra chopper. We do not need to discuss the chopper-stabilizer loop anymore, because we already discussed this at Fig. 10.26. However, there are major differences.

Firstly, the first stage of the main amplifier now determines the noise at low and high frequencies, while the correction loop determines the noise and ripple at the clock frequency.

Secondly, the hybrid nested capacitors C_{M31} and C_{M32} are not anymore connected to the input of the integrator, but to the input of chopper Ch_3, in order to maintain continuous negative feedback in the loop including Ch_1 [10]. This means that the parasitic capacitor C_{p5}, at the output of the sense amplifier, is now in parallel increased by the series connection of C_{M31} and C_{M32}. This parallel combination of capacitors is now charged and discharged by the offset voltage V_{os4} of the integrator G_{m4}. To avoid the extra offset of this parallel combination of capacitors in combination with V_{os4}, either the offset V_{os4} has to be reduced, or C_{M31} and C_{M32} in parallel with C_{p5} can be reduced by connecting them through a (folded) cascode at the output of G_{m5} to chopper Ch_3.

Thirdly, the parasitic capacitor C_{p2} before chopper Ch_1 is now charged and discharged to the offset voltage V_{os1} of the output stage G_{m1}. This causes spikes at the output through the first set of Miller capacitors C_{M11} and C_{M12} at the size of

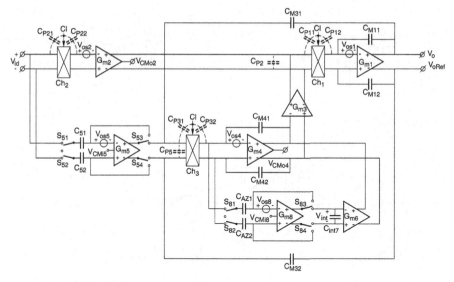

Fig. 10.34 Chopper-stabilized chopper OpAmp with multipath hybrid-nested Miller compensation, auto-zero G_{m5} and G_{m4}. $Vos = \ \sim 0.1\,\mu V$, $Vrip = \ \sim 10\,\mu V$

$V_{os1}C_{p2}/C_{M1S}$. with $C_{M1S} = C_{M11}\ C_{M12}/(C_{M11} + C_{M12})$. Therefore, the parasitic capacitor C_{p2} at the output of G_{m2} and G_{m3} has to be kept small.

The offset of G_{m5} causes a triangular ripple at the output of the integrator and a saw-tooth like ripple through Ch_1 at the output. This can be eliminated if the offset of the sense amplifier G_{m5} is auto-zeroed similar to the chopper-stabilized amplifier of Fig. 10.28. To further reduce the offset caused by the parasitic capacitor C_{p5} in combination with the offset of the integrator amplifier G_{m4} this amplifier can also be auto-zero stabilized by an extra loop around it [14]. These features are shown in Fig. 10.34. In this way an offset of 0.1 µV can be achieved with a ripple lower than 10 µV. Nanosecond chopper spikes of several millivolts can still be observed at the output.

A chopper-stabilized chopper instrumentation amplifier appears when the HF and LF amplifier paths are doubled [15] according to Fig. 10.35. In contrast to the chopper-stabilized IA of Sect. 10.7, the gain in a chopper IA is not set by the ratio of G_{m51} and G_{m52} of the correction loop, but by the ratio of G_{m21} and G_{m22} of the main amplifier in cooperation with the feedback network.

$$A_v = \ G_{m21}(R_1 + \ R_2)/G_{m22}R_1 \qquad (10.14)$$

The reason that the sense amplifiers G_{m51} and G_{m52} do not determine the gain by their ratio is because their influence is shifted to the clock frequency by the choppers around the main amplifiers G_{m21} and G_{m22}. G_{m52} is sensing the feedback ripple as a result of the offset of G_{m21} and G_{m22}. The output current of G_{m52} is

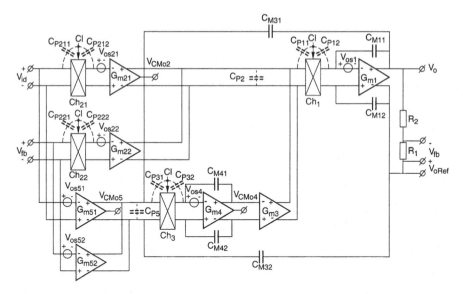

Fig. 10.35 Chopper-stabilized chopper InstAmp with multipath hybrid-nested Miller compensation. Vos = 2 μV, Vrip = ∼ 200 μV

rectified by chopper Ch_3 and amplified by the integrator G_{m4} and coupled by G_{m3} to the output of G_{m21} and G_{m22} in order to compensate the offset of G_{m21} and G_{m22} in the main chopper path. The feedback signal-dependent part at the input of G_{m52} is compensated for by the signal-dependent part at the input of G_{m51}. Therefore, the signal does not interfere with the offset cancellation.

The offset of the correction amplifiers G_{m51} and G_{m52} is chopped into a square wave by chopper Ch_3. The integrator does not amplify this square wave, but reduces it to a small triangular wave. Referred to the input it is translated by an attenuation of G_{m3}/G_{m21} and a chopper Ch_{21}. This means that the shape at the input results in a small saw-tooth at the double clock frequency.

The next step to reduce the saw-tooth ripple is to auto-zero the sense stages G_{m51} and G_{m52} [15]. This is shown in Fig. 10.36.

The most important offset contribution of the chopper-stabilized chopper instrumentation amplifier that is left, comes from the combination of the parasitic capacitance C_{p5} at the output of G_{m5} in combination of the offset voltage V_{os4} at the input of G_{m4}, see Eq. 10.11 This is particularly important as the hybrid nested Miller capacitors C_{M31} and C_{M32} are connected in parallel to the parasitic capacitor C_{p5} at the output of G_5. To further reduce this offset component also G_{m4} is auto-zeroed too, as shown in Fig. 10.36. In this way the final offset can be reduced to values well below 0.2 μV with a ripple lower than 20 μV.

It has to be kept in mind that the voltage gain of the correction loop G_{m5}, G_{m4}, G_{m3} must be taken 10^5 times larger than the voltage gain of G_{m2} in order to reduce its offset from 20 mV to 0.2 μV.

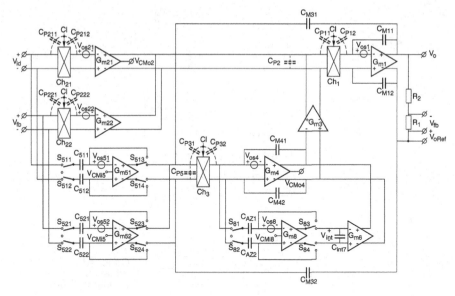

Fig. 10.36 Chopper-stabilized chopper InstAmp with multipath hybrid-nested Miller comp. and auto-zero G_{m5} and G_{m4}. Vos $=$ 0.2 μV, Vrip $=$ $\sim$ 20 μV

10.9   Chopper Amplifiers with Ripple-Reduction Loop

The chopper-stabilized chopper amplifiers of Sect. 10.8 combine low offset, low ripple, and a straight 6 dB/octave frequency characteristic. The latter was obtained by hybrid nested Miller compensation with capacitors C_{M31} and C_{M32}, as explained before. An interesting simplification can be made if we do not make use of this basic frequency compensation technique, but if we select a ripple-reduction notch filter to reduce the chopped offset. The notch filter will also take away signals in a small band around the clock frequency. But if we do not care about the notch for signals, for instance because we are only interested in a frequency band below the clock frequency, we can allow ripple reduction by a notch filter.

A chopper amplifier with a feedback ripple-reduction loop (RRL) [17] as a notch filter is sketched in Fig. 10.37a. The circuit senses the ripple at the output by the sense capacitors C_{M31} and C_{M32}. The ripple currents through these capacitors are rectified by synchronous detection to DC by chopper Ch3. The DC current is integrated by integrator G_{m4} on C_{M41} and C_{M42}, and fed back to the output current of G_{m2} through an amplifier G_{m3}.

Waveforms of unsettled chopper ripple and sense current in the circuit of Fig. 10.37a are shown in Fig. 10.37b. The ripple at the output approaches a square wave if the bandwidth of the closed loop gain of the OpAmp is larger than the clock frequency f_c. But if the bandwidth of the closed loop gain is smaller than f_c, the output ripple looks more similar to a triangle wave.

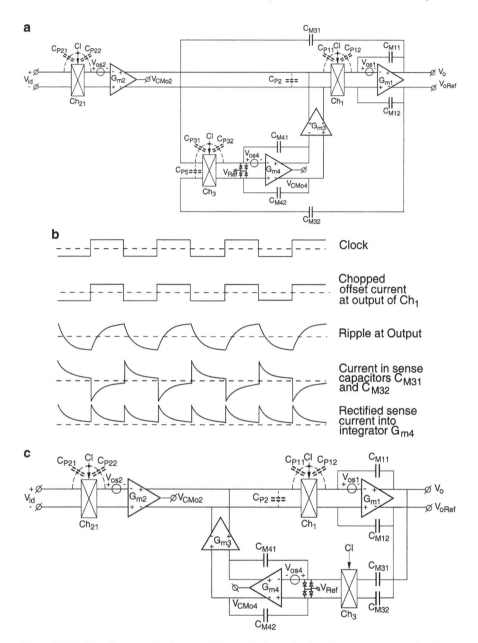

Fig. 10.37 (**a**) Chopper OpAmp with a ripple reduction loop as a notch filter. $V_{os} = 10\,\mu V$(at input), $Vorip = \sim 10\,mV$(at output). (**b**) Waveforms in Fig. 10.37. (**c**) Chopper OpAmp with reverse drawn ripple reduction loop as a notch filter. $Vos = 10\,\mu V$(at input), $V_{orip} = \sim 10\,mV$(at output)

Assuming that the polarity in the feedback RRL is correct and that the loop gain of the RRL is large enough, then the offset voltage V_{os2} of G_{m2} will be compensated and the ripple at the output will be reduced to nearly zero. The ripple reduction factor is equal to the DC loop gain of the ripple-reduction loop. The part of the loop through the integrator G_{m4}, and through G_{m3} amplifies DC offset correction signals. The part of the loop between the chopper Ch_1 and chopper Ch_3 through the output amplifier G_{m1} carries AC ripple signals at the clock frequency. Chopper Ch_1 together with the output amplifier G_{m1} and Miller capacitors C_{M11} and C_{M12} can be regarded as a modulating switched-capacitor transimpedance amplifier. This converts the DC current at the input of chopper Ch_1 into a ripple voltage at the clock frequency at the output with a transimpedance of $Y_{M1} = 2f_c C_{M1}$, with $C_{M1} = C_{M11} = C_{M12}$. Sense capacitors C_{M31} and C_{M32} in combination with chopper Ch_3 can be regarded as demodulating switched-capacitor impedances $Z_{M3} = 1/(2f_c C_{M3})$, with $C_{M3} = C_{M31} = C_{M32}$. They convert the output voltage ripple into a DC current that is being integrated on C_{M4}. The resulting voltage at the output of integrator G_{m4} thus represents the average rectified output ripple. The gain from the average output ripple voltage to the integrated DC voltage at the output of G_{m4} is limited to the finite DC gain A_{O4} of G_{m4}. The output voltage of G_{m4} is converted into an offset-compensating current by G_{m3}. Hence, the DC loop gain A_{LO} of the ripple-reduction loop and reduction factor R_r is:

$$R_r = A_{L0} = A_{o4} G_{m3} / 2f_c C_{M1} \tag{10.15}$$

If the factor part $G_{m3}/2f_c C_{M1}$ is estimated at 1, the integrator G_{m4} needs a DC gain $A_{04} = 10^4$, to obtain a ripple reduction factor R_r of 10^4.

The bandwidth B_L of the notch filter at the clock frequency equals twice the frequency f_L where the AC loop gain A_L of the ripple-reduction loop is 1. The AC loop gain is:

$$A_L = (C_{M3}/C_{M4})(G_{m3}/2\pi f_c C_{M1}) \tag{10.16}$$

Hence, the bandwidth B_L is:

$$B_L = 2A_L = (C_{M3}/C_{M4})(G_{m3}/\pi C_{M1}) \tag{10.17}$$

In practice the bandwidth of the ripple-reduction notch filter is several kilohertz.

When we compare Fig. 10.37a with Fig. 10.33 we see that the only difference is that in Fig. 10.37a the input sense amplifier G_{m5} has been eliminated. And hence, the capacitors C_{M31} and C_{M32} do not need to obey the rule for hybrid nesting anymore. They can be optimized for ripple sensing.

The functioning of the ripple-reduction loop (RRL) can be depicted in a simpler way if it is drawn reversed as shown in Fig. 10.37c. It clearly shows that the RRL measures the output ripple and feeds the correction signal back to correct the offset of the input transconductance G_{m2}.

The ripple-reduction loop cancels the ripple originating from the offset V_{os2} of G_{m2}. There is another ripple source, which originates from the offset of V_{os4} of the integrator amplifier G_{m4}. This is limiting the ripple reduction. The ripple originating from V_{os4} can be analyzed as follows. The chopper Ch_3 switches the offset voltage V_{os4} each clock cycle back and forth on the sense capacitors C_{M31} and C_{M32}. The resulting alternating charge spikes through these capacitors are rectified by chopper Ch_3 and fed back into integrator G_{m4}. At the output a ripple will occur with an average square-wave or triangle-wave AC voltage V_{orip} equal to the offset voltage V_{os4} of G_{m4}:

$$V_{orip} = V_{os4} \qquad (10.18)$$

Referred to the input this roughly results in an equivalent input ripple V_{rip} of:

$$V_{rip} = V_{os4}2f_cC_{M1}/G_{m2} = V_{os4}f_c/\pi f_0, \qquad (10.19)$$

in which f_0 is the bandwidth of the amplifier.

One good solution to reduce this ripple is to auto-zero the integrator. One of the many circuits that can auto-zero the integrator is shown in Fig. 10.38. For simplicity, the same auto-zero circuit is used as was used in the chopper-stabilized chopper opamp of Fig. 10.34. In fact, the only difference with the whole circuit of Fig. 10.34 is that the auto-zeroed sense amplifier G_{m5} has been omitted including its sampling capacitors.

From the above chopper OpAmps with ripple-reduction loop a chopper instrumentation amplifier with RRL can easily be derived. When both sense amplifiers G_{m51} and G_{m52} of Fig. 10.35 are omitted the InstAmp of Fig. 10.39a, b appears.

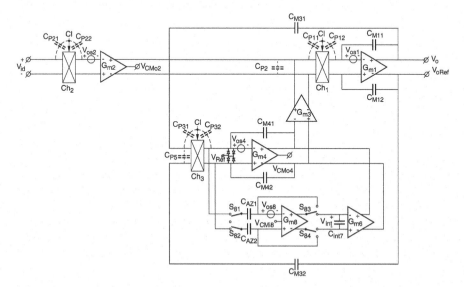

Fig. 10.38 Chopper OpAmp with ripple-reduction loop of which the integrator is auto-zeroed. Vos $= 1\,\mu V$, $V_{orip} = \ \sim 10\,\mu V$ (at output)

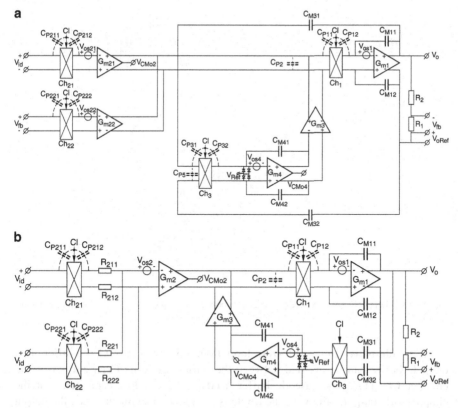

Fig. 10.39 (**a**) Chopper InstAmp with ripple-reduction loop. Vos = 20 μV, V_{orip} = ~ 10 mV (at output). (**b**) Chopper InstAmp with reverse drawn ripple-reduction loop. Vos = 20 μV, V_{orip} = ~ 10 mV (at output)

A disadvantage of the current feedback instrumentation amplifier (CFIA) of Fig. 10.39a is that the gain is determined by the ratio of G_{m21} and G_{m22}. As these G_m's are normally realized by simple differential pairs the accuracy may not be better than 1 % without trimming. How to improve the gain accuracy automatically is described in Sect. 10.11.

A simple way to improve the accuracy is to go back to an OpAmp with a resistor bridge around it like described with Fig. 3.4. Without chopping, the CMRR would be not better than the inverse of the inaccuracy of the bridge multiplied by the gain setting of the bridge. But if we chop the bridge as shown in Fig. 10.39b, the average CMRR is high, while the ripple is taken away by the RRL. The limitation on the CMRR is in differences of the series resistances of the choppers. These differences are not chopped away. These differences can be seen as differences in signal source resistances which are loaded by the bridge.

Applying auto-zeroing of the integrator the simplification in Fig. 10.40 over Fig. 10.36 is even clearer, as two auto-zeroed sense amplifiers G_{m51} and G_{m52} are omitted.

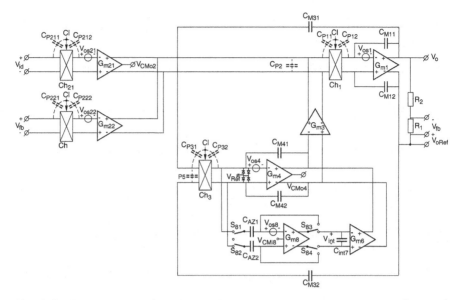

Fig. 10.40 Chopper InstAmp with ripple-reduction loop. $V_{os} = 2\,\mu V$, $V_{orip} = \sim 20\,\mu V$(at output)

As an alternative to the auto-zero loop around the integrator G_{m4} a differential cascode buffer can be inserted between the sense capacitors C_{M31} and C_{M32} and chopper Ch_3 [17]. The cascode buffer needs to have a low capacitance at the chopper-side output; otherwise the ripple is not reduced so much. Also, the output offset current of the cascode needs to be made low; otherwise Ch_3 will modulate the offset current to the second harmonic of the chopping frequency, and an other type of ripple will appear at the output.

It is essential that the choppers Ch_1, Ch_2 and Ch_3 are precisely synchronized, so that there is not much delay in the amplifier stages. Otherwise the compensation does not work precise and the ripple and resulting offset is larger.

When a signal step occurs the sense capacitors $C_{m31,32}$ will punch the RRL slightly out of balance. The RRL will work to return to balance. But during this time a reducing ripple is seen at the output. When we want to get rid of this effect, one can built-in a step-sense circuit at the output, and if the step size is larger than a prescribed value, momentarily block the current through the sense capacitors by short-circuit and open-circuit switches.

When we want to eliminate the notch in the frequency characteristic further, and further reduce the chopper spikes of the chopper amplifier with RRL, one can embed it one level down in a chopper-stabilized OpAmp topology, as shown in Fig. 10.41a. The feed-forward amplifier stage G_{m2} now bypasses the notch and the ripple at the higher frequencies.

The RRL can be reversely drawn for clarity, as shown in Fig. 10.41b.

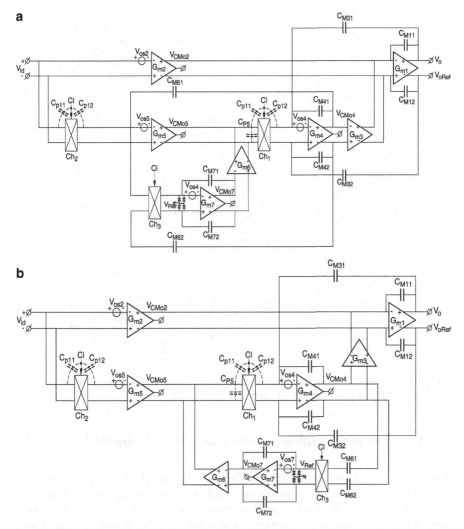

Fig. 10.41 (a) Chopper-stabilized OpAmp with RRL. $V_{os} = 10\,\mu V$, $V_{orip} = \sim 100\,\mu V$ (at output).$V_{os} = 1\,\mu V$, $V_{orip} = \sim 10\,\mu V$(at output) when G_{m7} is auto-zeroed. (b) Chopper-stabilized OpAmp with reversed drawn RRL. $V_{os} = 10\,\mu V$, $V_{orip} = \sim 100\,\mu V$(at output). $V_{os} = 2\,\mu V$, $V_{orip} = \sim 20\,\mu V$ when G_{m7} is auto-zeroed

In order to lower the offset and ripple of the OpAmp to the level of $V_{os} = 1\,\mu V$ and $V_{orip} = \sim 10\,\mu V$ (at output), respectively, the integrator G_{m7} has to be auto-zeroed likewise shown in Fig. 10.2 [20].

Finely, a low-offset, low-ripple InstAmp without notch in the frequency characteristic can be devised from the above OpAmps. This is drawn in Fig. 10.42. To lower the offset and ripple to the level of $V_{os} = 2\,\mu V$, $V_{orip} = \sim 20\,\mu V$ (at output) the integrator G_{m7} has to be auto-zeroed. This is a better InstAmp compared with Fig. 10.32 in terms of offset, ripple, and noise [20].

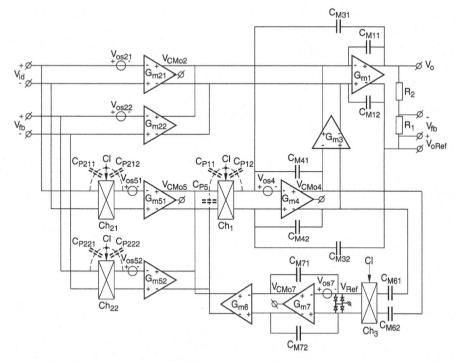

Fig. 10.42 Chopper-stabilized instrumentation amplifier with reversed drawn RRL. $V_{os} = 2\ \mu V$, $V_{orip} = \sim 20\ \mu V$ (at output) when G_{m7} is auto-zeroed (not drawn)

10.10   Chopper Amplifiers with Capacitive-Coupled Input

There is an increasing request for interfacing high input CM voltages: Firstly, in high-voltage current-sense applications, such as power management in laptops; secondly, in solenoids of smart electro motors, for instance in electric or hybrid cars; thirdly, in biomedical sensors that make direct electrical contact to the body, as in skin electrodes or implanted devices. Of course, high-voltage transistors can be used. But often the voltage requirements are higher than the values the transistors allow. And in the case of biomedical electrodes it might be the patient safety that forbids direct contact with a transistor's gate. Therefore, it is useful to see how we can transfer signals through on-chip metal–oxide–metal capacitors.

Figure 10.43a shows a chopper-stabilized OpAmp like that described at Fig. 10.26 with an additional chopper-capacitor-chopper combination as a chopped-capacitor coupled input. At the input the signal voltage is firstly chopped by Ch_2, next coupled through the capacitors C_{21} and C_{22} and finally chopped back by Ch_1 and Ch_3. The input chopper Ch_2 can be built from low-voltage CMOS transistors in an isolated N-well. Its clock can be differentially driven by two small capacitors from a grounded clock. The CM level of the clock at the chopper side has

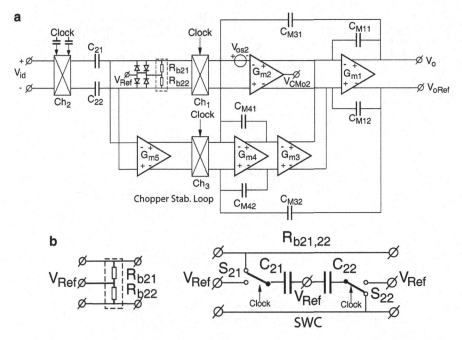

Fig. 10.43 (a) Chopper-stabilized OpAmp with chopped-capacitor input. (b) Bias Resistors Rb_{21} and Rb_{22}, on the *left-hand side* as normal resistors, and on the *right-hand side* as switched capacitors connected to V_{Ref}

to be established by diodes coupled to the inputs. Hence, the input chopper Ch_2 can be fully capacitive isolated from ground. To avoid unbalanced signal attenuation in the capacitors C_{21} and C_{22} by parasitic capacitances to ground on their bottom plates, these plate sides should be connected to the input chopper Ch_2. Unbalance in combination with offset will produce an extra ripple.

At the right-hand side of the couple capacitors the CM voltage level must be established at a certain internal level V_{Ref}. Transistors connected as head-to-tail diodes serve to limit the CM swing at the input of the input amplifier G_{m2}. The high resistances of these diodes at zero V_{DS} softly serve to establish the level V_{Ref}. A firm reference level, which can cope with leakage currents and does not allow large offset, can be established either by regular resistors or by switched-capacitor resistors R_{b21} and R_{b22}. The switched-capacitor resistors are depicted in Fig. 10.43b. On the left hand we see the resistors, and at the right hand side the switched capacitors. For a capacitor value of 5 pF, and a clock frequency of 10 kHz, we find a CM resistance of $R_{b21} = R_{b22} = 1/(5pF. 2.10kHz) = 10$ MOhm.

The head-to-tail connected diodes serve to limit the CM swing at the input of the input amplifier G_{m2}. The chopper-stabilization loop with sense amplifier Gm5, Ch3, integrator Gm4, and correction amplifier Gm3 has been described with Fig. 10.26.

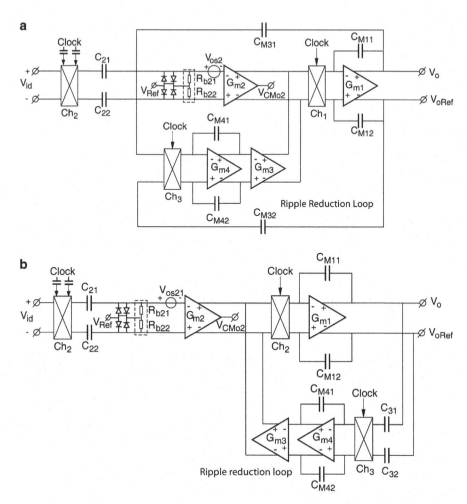

Fig. 10.44 (**a**) Chopper OpAmp with chopped-capacitor input and ripple-reduction loop. (**b**) Chopper OpAmp with chopped-capacitor input and ripple-reduction loop drawn reverse

For lowering the 1/f noise and offset it is better to include the input amplifier G_{m2} in between the two choppers Ch_2 and Ch_1. In this way a Chopper OpAmp arises, as shown in Fig. 10.44a with chopped-capacitor input. The ripple of the chopper amplifier could have been reduced by a stabilization loop like the chopper amplifier of Fig. 10.34. But then we need to also capacitive couple the sense amplifier G_{m5} to the input, for instance as it was done in Fig. 10.35. To avoid this extensive circuitry we rather completely eliminate the sense amplifier and keep only a ripple-reduction loop as explained with Fig. 10.37a. We have to keep in mind, though, that elimination of the sense amplifier produces a notch in the frequency response at the clock frequency.

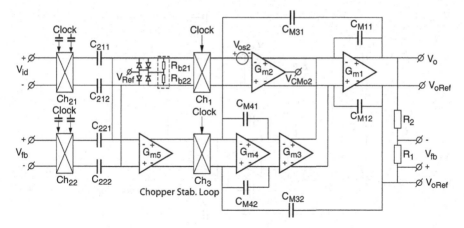

Fig. 10.45 Chopper-stabilized InstAmp with chopped-capacitor bridge input

The signal through the coupling capacitors C_{21} and C_{22} will be attenuated by their load with parasitic capacitors like in the circuit of Fig. 10.43a. To keep the attenuation minimal and balanced, we should connect the bottom plates of the oxide isolated coupling capacitors to the input side, and take care that the parasitic capacitors of the diodes are small and balanced enough. To further reduce the ripple we can take all measures as explained in Sect. 10.9.

It is just interesting to draw Fig. 10.44a little different with the ripple-reduction loop from the output side back to the input of chopper Ch_1. This is shown in Fig. 10.44b. It looks simpler to understand.

After the discussion of OpAmps with chopped-capacitor input coupling, it is easy now to make the step to instrumentation amplifiers. Figure 10.45 shows a chopper-stabilized InstAmp with chopped-capacitor input coupling. The input capacitors C_{211}, C_{212}, C_{221}, and C_{222} together with the input choppers Ch_{21} and Ch_{22} make up a chopped-capacitor bridge InstAmp like the resistor-bridge InstAmp of Fig. 3.4. The closed-loop gain can be set either by the ratio of $(C_{211}, C_{212})/(C_{221}, C_{222})$ or by the feedback attenuator $R_1/(R_1 + R_2)$. Unbalance in the input capacitors will not so much deteriorate the CMRR, as it did in the bridge InstAmp of Fig. 3.4, because the whole capacitor bridge is being chopped. The input impedance of this InstAmp will not be very high as a result of the chopped-capacitor bridge at the input. For C_{211} and C_{212} of 10 pF, and a clock frequency of 10 kHz, the input impedance R is $1/(10pF.2.10kHz) = 5$ MOhm. As this value is not so high an unbalance of the impedances in the input signal source attenuated by the input impedance of the amplifier may cause a slightly lower CMRR.

In line with the OpAmp of Fig. 10.44a it is better to include the input amplifier G_{m2} in-between the two choppers Ch_2 and Ch_1 for lowering its 1/f noise and offset. In this way a chopper InstAmp arises, as shown in Fig. 10.46a with chopped-capacitor input. Like in Fig. 10.44b ripple-reduction loop has been used to lower the

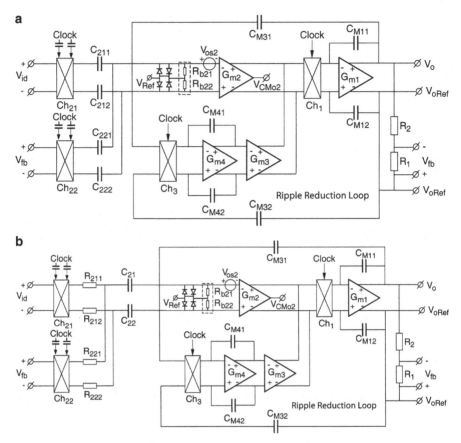

Fig. 10.46 (**a**) Chopper InstAmp with chopped-capacitor bridge input coupling and ripple-reduction loop. (**b**) Chopper InstAmp with capacitive coupled chopped-resistor bridge input and ripple-reduction loop

chopper ripple. The CMRR will remain high by the chopping of the whole capacitor bridge, and only restricted by the loading of possible unbalanced source resistances as in the circuit of Fig. 10.45.

It is interesting to see that we can replace the chopped-capacitor bridge by a chopped-resistive bridge followed by capacitive coupling, as shown in Fig. 10.46b to obtain the same goal. The coupling capacitors C_{21} and C_{22} isolate the CM voltage of the bridge mid points from ground. Now the transfer function of the bridge between input and output is determined by resistors. This topology can be used to our advantage if the resistors can be made more accurate or more linear than capacitors. A further advantage is that the capacitive peak charge currents with full capacitive coupling are absent. A disadvantage is that the choppers now are loaded by a resistive bridge. The chopper series resistances may deteriorate the bridge accuracy. Moreover, the resistive bridge creates more noise. The CMRR remains high due to the chopping of the whole bridge.

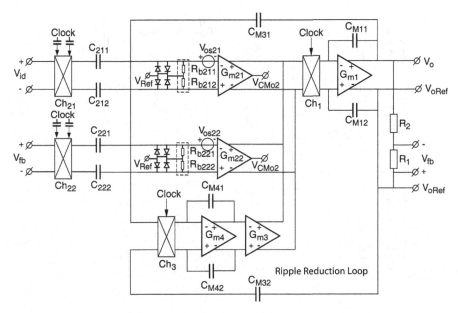

Fig. 10.47 Current-feedback chopper InstAmp with chopped-capacitor coupling and ripple-reduction loop with high-impedance input

To increase the input impedance it is better to step away from the bridge-type InstAmp and go to the current-feedback InstAmp. This topology is presented in Fig. 10.47 in combination with chopped-capacitor input coupling. The input impedance of in the InstAmp of Fig. 10.47 will now be made up by the chopped parasitic ground-plate capacitances at the chopper side of the input capacitances C_{211}, C_{212}, C_{221}, and C_{222}, and the parasitic ground capacitors of the elements following. If we suppose that the input capacitances have a value of 10 pF, and the parasitic ground capacitances 1 pF, we can expect an input chopped-capacitor resistance of $1/(1pF.2.10\,kHz) = 50\,MOhm$.

A disadvantage of the circuit of Fig. 10.47 is that the accuracy can be deteriorated, firstly, by inequality of G_{m21} and G_{m22} (cures for this inaccuracy are presented in the next paragraph), secondly, if the parasitic ground capacitances that load the coupling capacitances are not well matched.

The last circuit that will be discussed here is the Chopper InstAmp with Chopped-Capacitor input and Ripple-Reduction Loop and with Biomedical Electrode offset voltage compensation of Fig. 10.48. A basic problem of biomedical electrodes is that they may generate a DC offset of several 100 mV, while AC voltages of the order of μV have to be measured. A natural solution would be to couple these electrodes by capacitors. However, if we want to measure signals from a well determined frequency of 1 Hz and higher, and the input resistance of the amplifier can be reliably made 100 MΩ, then we still need 10 nF input coupling capacitors. And those can not be easily integrated on chip. Therefore, an additional integrator loop following the output is made with a large time-constant. It uses a

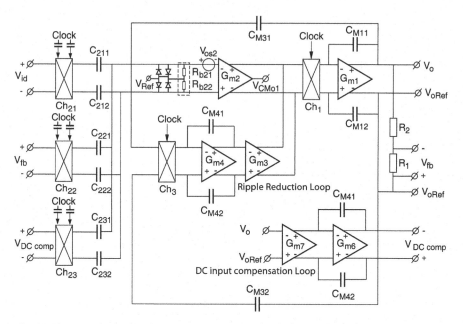

Fig. 10.48 Chopper InstAmp with chopped-capacitor input and ripple-reduction loop with biomedical electrode offset voltage compensation

special chopped-capacitor integrator [19] depicted as G_{m7} and G_{m6}. At the output of that integrator a compensation voltage V_{DCcomp} will appear that will increase proportionally to the offset. This compensation voltage will be fed through a third input chopper Ch_{23} into C_{231} and C_{232} to compensate the DC offset at the input. As a basis the InstAmps of Fig. 10.46a or Fig. 10.47 can be used. For simplicity we have chosen Fig. 10.46a. The offset that can be cancelled is limited by the supply voltage.

10.10.1   Wide-Band Chopper Amplifiers with Capacitive-Coupled Input

The aforementioned chopper amplifiers with the capacitive-coupled input of Figs. 10.43, 10.44, 10.45, 10.46, 10.47, and 10.48 all show a notch in their amplitude characteristic at the chopping frequency. This notch occurs because an input signal at the chopping frequency is down-converted to DC by the input chopper and subsequently blocked by the high-pass filter of the coupling capacitors connected to the bias resistors behind them. This not only affects the flat frequency characteristic but also results in a slow-settling ripple at the chopping frequency as a response to an input signal step. Furthermore, as all these amplifiers are used in feedback loops, an insufficiently controlled chopper ripple up-swing can occur due to the zero loop gain at the notch frequency.

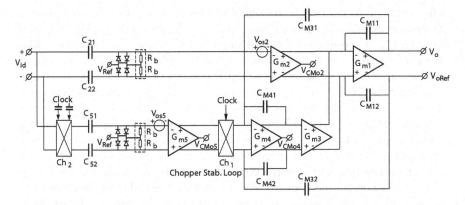

Fig. 10.49 Capacitive-coupled chopper-stabilized operational amplifier with a flat frequency characteristic extending beyond the chopping frequency, combining a HF path with a precision DC path

Consequently, these amplifiers are only suitable for applications where the signal bandwidth is lower than the chopping frequency and where there are no fast settling requirements. However for wide-band applications, where fast settling is required, another approach is necessary which maintains a flat frequency response over the entire band of interest including the chopping frequency. This type of approach calls for dual signal paths.

In the first approach, shown in Fig. 10.49, an operational amplifier combines a high-frequency (HF) path, including the chopper frequency, with a precision DC path [23, 28].

The circuit of Fig. 10.49 can be seen as being derived from the chopper-stabilized operational amplifier of Fig. 10.26b. The difference here is that coupling capacitors have been inserted both directly into the HF path, and into the DC path behind Ch_2. Also, chopper Ch_2 has been replaced by a fully floating chopper with a wide CM input voltage range able to follow the CM range of the input signal. Its driver clock has been coupled in a capacitive way. This will be dealt with in the next section on fully floating capacitive-coupled input choppers. Chopper amplifier G_{m5} has yet to be auto-zeroed, chopper-stabilized, or given a ripple-reduction loop to reduce the output ripple caused by the offset V_{os5} of G_{m5}. The hybrid-nested Miller compensation scheme (see Fig. 6.28a) with G_{m2} and C_{M11}, C_{M12} and G_{m5} and C_{M31}, C_{M32} ensures a flat frequency characteristic, under the condition that the ratios of G_{m2}/C_{M1} and G_{m5}/C_{M3} are taken equally, with C_{M1} equal to the series connection of C_{M11} and C_{M12}, likewise C_{M3} equal to the series connection of C_{M31} and C_{M32}.

The CM input level of the amplifiers G_{m2} and G_{m5} is set by four large resistors R_b to a reference voltage V_{Ref}. Head-to-tail connected diodes across these resistors prevent excess voltages at the inputs of G_{m2} and G_{m5} during the edges of large input voltage steps. The input offset voltage V_{os2} of G_{m2} is compensated by the output current of G_{m3} at the current summing point of G_{m2} and G_{m3}, which is at the output of the DC path. The reduction factor by which the offset is reduced is determined by the voltage gain ratio of $A_{m5} \times A_{m4} \times G_{m3}/G_{m2}$. The transconductance G_{m3} is normally taken five times lower than G_{m2} to prevent it from increasing the noise of G_{m2} too much, while still allowing it to compensate an offset related to 20 % in the full output current range of G_{m2}. Therefore, if we want to decrease the offset of G_{m2} by a factor of 10^4 we have to be sure that the voltage gain $A_{m5} \times A_{m4}$ is more than 5×10^4.

A disadvantage of the circuit in Fig. 10.49 is that due to the capacitive coupling there is no overall DC feedback on the offsets V_{os2} and V_{os5}. Hence, these offsets are amplified open loop, and independent of the closed loop gain. For V_{os2} the chopper stabilization loop through G_{m5} will take care, as we have already seen. However, the output ripple caused by V_{os5} is only suppressed by the integrator G_{m4} and the Miller compensation capacitors around G_{m1}. The ripple still has to be reduced by auto-zeroing or chopper-stabilize G_{m5}, or with a ripple-reduction loop around G_{m5}, as shown in [23]. A simple yet effective way to turn the open loop gain of these offsets into a closed loop gain, and thus to further reduce the offset and ripple, is shown in the second approach of Fig. 10.50 [24].

A third chopper Ch_3 has been inserted between the high-frequency and low-frequency paths behind the coupling capacitors. This chopper, firstly, restores a DC coupling between the circuit input and the input of G_{m2} by down-converting the signal from the up-converting chopper Ch_2 and the coupling capacitors C_{51}, C_{52}. Secondly, it measures the output chopper ripple through the HF coupling capacitors C_{21}, C_{22} and rectifies the ripple synchronously to DC, compensating the offset voltage V_{os5} that causes the ripple. To reduce the output ripple further, G_{m5} still has to be either auto-zeroed (see Fig. 10.28) or chopper-stabilized, otherwise a ripple-reduction loop can be applied.

From the operational amplifier of Fig. 10.50 a capacitive-coupled bridge-type instrumentation amplifier can be made simply by adding two sets of input coupling capacitors C_{31}, C_{32} and C_{41}, C_{42} along with an input chopper Ch_4. The circuit is shown in Fig. 10.51 [24].

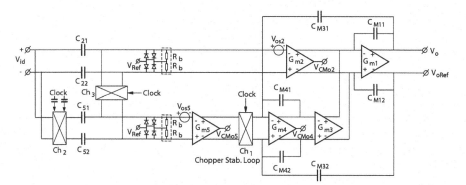

Fig. 10.50 Capacitive-coupled chopper-stabilized operational amplifier with a flat frequency characteristic extending beyond the chopping frequency, and with reduced chopper ripple

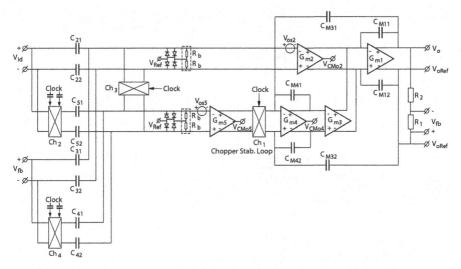

Fig. 10.51 Capacitive-coupled chopper-stabilized bridge-type instrumentation amplifier with a flat frequency characteristic extending beyond the chopping frequency, and with reduced chopper ripple

Now, however, this version has two capacitive bridges at its input: C_{21}, C_{22} and C_{311}, C_{32} for higher frequencies, and C_{41}, C_{42} and C_{51}, C_{52} for low frequencies. The high-frequency bridge functions as the resistive bridge in the instrumentation amplifier of Fig. 10.7. The input impedance is equal to the inverse of the product of the input capacitance and four times the clock frequency. Furthermore, this bridge restricts the CMRR in the high-frequency path to the product of gain $A_v = C_{31}/C_{21} = C_{32}/C_{22}$ and the inverse relative imbalance $C/\Delta C$ of the bridge, as expressed in Eq. 10.4. For the low-frequency bridge on the other hand, the imbalance is averaged out by the choppers Ch_2 and Ch_4 at its input, which results in the CMRR also being restored.

To overcome these disadvantages, an option is to isolate the CM voltage of the input path from that of the feedback path. This can be achieved by giving each path with their coupling capacitors its own G_m stage with a good CMRR. This results in the capacitive-coupled current-feedback instrumentation amplifier of Fig. 10.52. Giving each path with their coupling capacitors a G_m stage also drastically improves the input impedance for both the input path as well as for the feedback path. However, differences in the nonlinearity of the two sets of G_m's may lead to more harmonic distortion. Some measures, as discussed in Sect. 10.11, may help to make the G_m's more equal.

Besides the class of chopper-stabilized amplifiers, a class of chopper-stabilized chopper amplifiers also exists, as we have seen in Sects. 10.7 and 10.8, respectively. Chopper-stabilized chopper amplifiers have a lower offset, since these amplifiers already have a chopper amplifier core which is further enhanced by chopper stabilization. The same holds for the capacitive-coupled versions. In Fig. 10.53 a capacitive-coupled chopper-stabilized chopper amplifier is shown.

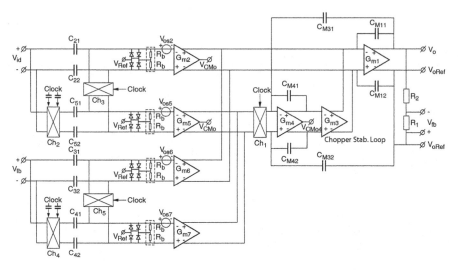

Fig. 10.52 Capacitive-coupled chopper-stabilized current-feedback instrumentation amplifier with a flat frequency characteristic extending beyond the chopping frequency, with reduced chopper ripple, a high CMRR, and high input impedance

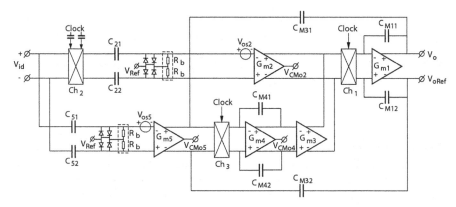

Fig. 10.53 Capacitive-coupled chopper-stabilized chopper operational amplifier with a flat frequency characteristic extending beyond the chopping frequency

Similarly to the circuit of Fig. 10.49, the offsets of G_{m2} and G_{m5} are processed open loop to the output. Therefore, we also insert a chopper Ch_4 between the HF and LF coupling capacitors, deriving Fig. 10.54.

The high-frequency path with amplifier stage G_{m2} is now inserted between the input and output choppers Ch_2 and Ch_1, respectively. The offset voltage V_{os2} of G_{m2} produces an offset current at its output that is transformed into a square-wave ripple current by chopper Ch_1. This ripple is integrated by the Miller-compensated output stage G_{m1} and presented at its output in the shape of a triangular voltage

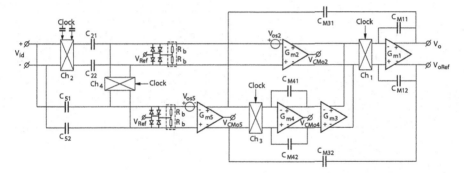

Fig. 10.54 Capacitive-coupled chopper-stabilized chopper operational amplifier with a flat frequency characteristic extending beyond the chopping frequency, with reduced chopper ripple

ripple. An external feedback loop feeds the output voltage ripple into the input. Amplifier stage G_{m5} in the DC path senses this voltage ripple at the input and transforms it into a ripple current at its output. Chopper Ch_3 synchronously detects this ripple current of G_{m5} and transforms it into a DC error current. Integrator G_{m4} integrates the error current and builds up an error voltage at its output. Finally, the correction amplifier G_{m3} corrects the offset current of G_{m2} by adding its output current to that of G_{m2}. In this way the offset of G_{m2}, from which the ripple originates, is sufficiently corrected, depending on the loop gain of this process. The transconductance G_{m3} is chosen small enough, for instance $G_{m2}/5$, that it does not add much noise to that of G_{m2}, but large enough that it is able to correct the worst case offset of G_{m2}.

There are still two remaining obstacles. Firstly, the offset V_{os5} of G_{m5} is chopped by Ch_3 into another ripple. Although this ripple is suppressed by integrator G_{m4} and the Miller-compensated output stage G_{m1}, we still want to suppress it further. This can be done by auto-zeroing or chopper-stabilizing of G_{m5}, which has already been shown in the circuit of Fig. 10.34. Secondly, the offset of G_{m4} limits the effectiveness of the ripple reduction because that offset appears as a square wave at the input of chopper Ch_3. There it charges and discharges all present parasitic capacitors including the hybrid nesting capacitors C_{M31}, C_{M32} (see Fig. 6.28a). This square wave cannot be effectively distinguished from the output voltage ripple, as these two are related by C_{M31}, C_{M32}. Therefore, we want to lower the offset voltage of G_{m4}. This can be done by auto-zeroing or chopper stabilization, as also explained together with Fig. 10.34.

As with the chopper-stabilized amplifier, we can transform the chopper-stabilized chopper amplifier into an instrumentation amplifier by adding two capacitive bridges between its input and feedback path. This is shown in Fig. 10.55. The negative-feedback bridge restricts the input impedance to the inverse of the product of the input capacitance and 4 times the clock frequency. Moreover, the bridge now restricts the CMRR in the low-frequency path to the product of gain C_3/C_2 and the inverse relative imbalance $C/\Delta C$ of the bridge, as expressed in Eq. 10.4.

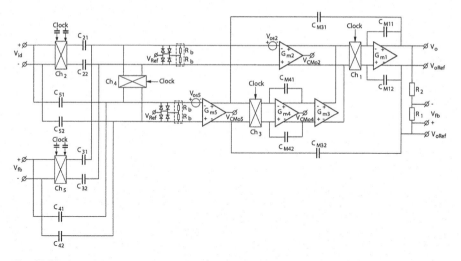

Fig. 10.55 Capacitive-coupled chopper-stabilized bridge-type chopper instrumentation amplifier with a flat frequency characteristic extending beyond the chopping frequency, and with reduced chopper ripple

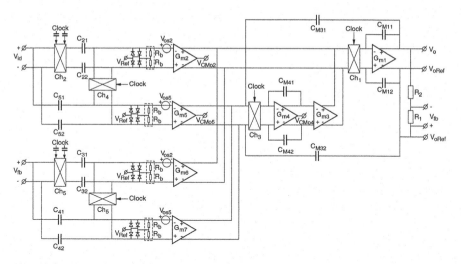

Fig. 10.56 Capacitive-coupled chopper-stabilized current-feedback chopper instrumentation amplifier with a flat frequency characteristic extending beyond the chopping frequency, with reduced chopper ripple, a high CMRR, and high input impedance

To improve the input impedance and CMRR both the input and feedback paths with their coupling capacitors can be given their own subsequent G_m stage. In this way a capacitive-coupled chopper-stabilized current-feedback chopper instrumentation amplifier arises, as shown in Fig. 10.56. However, differences in the nonlinearity of the two sets of G_m's may result in more harmonic distortion. Some measures, as discussed in Sect. 10.11, may help to make the G_m's more equal.

10.10.2   Fully Floating Capacitive-Coupled Input Choppers

In current-sensing applications for battery management in laptops and motor management in electric cars, we need to sense small voltages of several to tens of millivolts across a current-sense or shunt resistor at CM voltages of 10–100 V. For electric motors the CM voltage could even become negative by inductive kick back. In the previous section we saw how capacitive coupling would allow these applications. We only have to build fully floating choppers in front of the coupling capacitors which can accommodate the large input CM-voltage range reaching beyond the positive and negative supply-voltage rails [25]. To realize these fully floating choppers we have three requirements to fulfill: (1) The chopper transistors need to be floating within a large CM-voltage range beyond the supply-rail voltages. (2) A chopper clock has to be delivered to the chopper transistors in a floating way, preferably by-on chip high-voltage capacitors. (3) The coupling capacitors at the signal output should be able to isolate large voltages between their plates and be linear. Figure 10.57 shows a basic circuit that can facilitate these requirements.

The four N-channel transistors are each built in a P-well. Each P-well is isolated from the P-substrate by an N-buried layer. Each P-well is connected to its source and driven from an input. The P-substrate is connected to ground. The N-buried layers are left floating to facilitate large (beyond the supply rails) positive and negative voltages on the P-wells. Potential latch-up issues associated with the parasitic NPNP (N-source, P-well, floating N-buried layer, P-substrate) structure of the NMOS transistors can be circumvented by two measures. Firstly, a connection between source and P-well will prevent the upper NP-diode from conducting. Secondly, a HV clamping diode between the positive supply and the floating N-buried layer will prevent this N-pocket from dropping below the P-substrate, which means that the lower NP-diode cannot turn on [26].

When the CM input voltage rises to higher positive levels, the CM voltage of the N-MOS transistors also rises to higher positive levels together with their connected P-wells. The P-wells take up the N-buried layers by the PN diodes between them. Hence, the IC process should allow a high voltage between the N-buried layer and the P-substrate. When the CM input voltage sinks to levels below that of the substrate the voltage on the P-wells sinks below the N-buried layers. This means that the IC process should allow a high negative voltage between the P-wells and the N-buried layers. The N-buried layers are prevented from sinking below the P-substrate by the HV diode clamps connected to the positive supply, as explained before.

The floating driver clock is realized by connecting small (0.1 pF) coupling capacitors between the chopper clock and the gates of the chopper transistors. High-voltage and linear signal-coupling and clock-coupling capacitors can be realized in many IC-processes by an interwoven metal finger layout with oxide isolation in between.

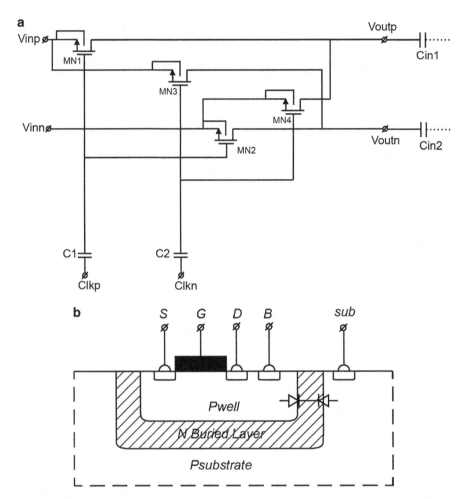

Fig. 10.57 Principle of a floating chopper with four Floating low-voltage N-MOS transistors each in their own P-well that are isolated from the substrate by floating N-buried layers, and with capacitive-coupled signal output and driver clock, accommodating large positive and negative CM input voltages

The CM levels of the gates have to be fixed in relation to those of their sources. The simplest approach to fix the CM levels is using sets of diode clamps. This is shown in Fig. 10.58.

Reverse clamp diodes D_{r11} through D_{r41} prevent the gate voltages of the 4 chopper transistors from going lower than one diode voltage below their source voltages. This determines the lower clock voltage level on their gates. Forward chains of clamp diodes D_{f11}–D_{f1x} through D_{f41}–D_{f4x} prevent the gate voltages from going higher than two or more diode voltages above their source voltages. If we choose three diodes in the forward chain, then we can have a clock amplitude of four diode voltages, providing an effective drive voltage over the threshold voltage of

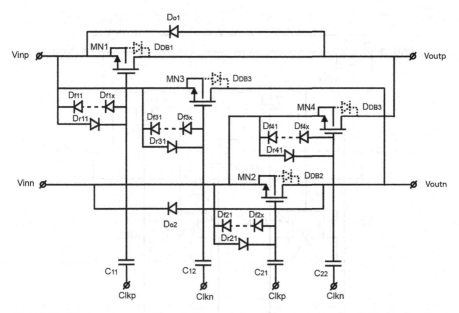

Fig. 10.58 Floating chopper with capacitive-coupled driver clocks having CM levels fixed and protected by diode clamps

the chopper transistors of two diode voltages. The peak-to-peak clock amplitude should always be kept lower than the sum of the reverse and forward diode voltages. This rule should also be obeyed over the ambient temperature range. If the clock amplitude becomes too large, the diodes conduct at the clock edges, and injection-current peaks occur at the input terminals.

At sudden positive input surges the built-in back-gate diodes D_{DB} (dotted) of the chopper transistors protect the chopper transistors against large source–drain voltages. Incidentally, these back-gate diodes limit the maximum differential input voltage to one diode voltage. For higher differential voltages, some will conduct. Sudden negative surges require the insertion of the diodes D_{o1} and D_{o2} between input and output to limit the drain–source voltage of the low-voltage chopper transistors.

A disadvantage of using passive clamping diodes, besides all these diodes having parasitic capacitors, is that we always lose one additional diode voltage. Therefore, it is better to use active clamps instead of passive diode clamps so that no diode voltage will be lost. A basic circuit for an active clamp is a latch. A basic floating chopper circuit that uses active clamps is shown in Fig. 10.59. The active clamp transistors $M_{N5}-M_{N8}$ are also made with floating N-wells so that they are capable of beyond-the-rail operation.

When the Clk_p edge goes high and Clk_n edge low, the NMOS latch transistors M_{N6} and M_{N8} start to conduct. They shortcut the gates of the cross-chopper transistors M_{N3} and M_{N4} to their sources so that these transistors are opening. At the same time the latch transistors M_{N5} and M_{N7} become non-conducting and

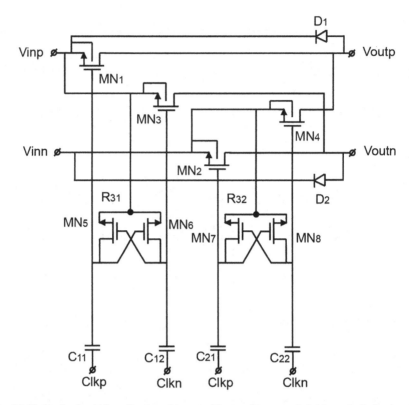

Fig. 10.59 Basic circuit for a floating chopper with capacitive-coupled driver clocks having CM levels fixed by active latches

their drain–source voltages become high. This activates the gates of the straight chopper transistors M_{N1} and M_{N2} so that they turn on. The chopper now conducts the input signal straight forward. We see that the latches are quite helpful. First of all, they facilitate a fast and simultaneously positive and negative clock switching behind the coupling capacitors. Secondly, they position the negative voltage levels of the clock and clock-bar behind the coupling capacitors at the input level so that the full clock amplitude is efficiently used to open and close the chopper switches. Another advantage of using the latches is that the amplitude of the clocks can now be freely chosen on the basis of the optimal voltage to switch on the chopper transistors. The amplitude is not restricted by diode clamps. The coupling capacitors chosen for the clock should preferably be as small as possible, but still large enough that the gate capacitors of the chopper switches do not reduce the clock amplitude too much. Practical values in the order of 0.1 pF have been used.

Still many precautions have to be taken to protect the low-voltage chopper switches. This is shown in Fig. 10.60.

Besides the capacitive coupling and active latches the floating chopper of Fig. 10.60 shows other measures to protect the chopper transistors. Again the

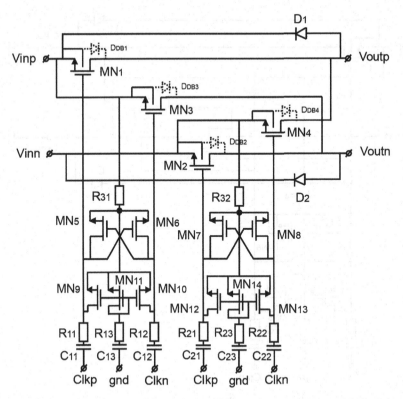

Fig. 10.60 Full circuit for a floating chopper with capacitive-coupled driver clocks having CM levels fixed by active latches, and protected by diode clamps

built-in back-gate-to-drain diodes D_{DB1} through D_{DB4} protect the chopper switches when a sudden upward input CM surge occurs. Additional diodes D_1 and D_2 between input and output are inserted for protection during a sudden downward input CM surge. Incidentally, these diodes limit the maximum differential input voltage to that of a diode voltage. The drain-to-back-gate diodes of M_{N5} through M_{N8} protect the chopper gates during a sudden positive CM input surge. A series chain of diodes between the drains and sources of M_{N5} through M_{N8} could protect the chopper gates at a sudden negative CM input surge, even if the clock signal is absent and the latches are inactive. However, in our circuit we have chosen to protect the chopper gates with small active circuits. If a negative CM input transient occurs, both the capacitors C_{13} and C_{23}, which are equal to the other driver capacitors, and diodes M_{N11} and M_{N14} "feel" the CM movement on the driver capacitors and mirror the associated current through M_{N10}, M_{N12}, M_{N13}, and M_{N15} back into the driver capacitors so that the gate side of the driver capacitors are accordingly following down. Lastly, relatively small series resistors have been inserted in the lines connecting to the inputs in order to reduce clock spikes in the input signals. Clock spikes at the input can also be kept small if we choose the

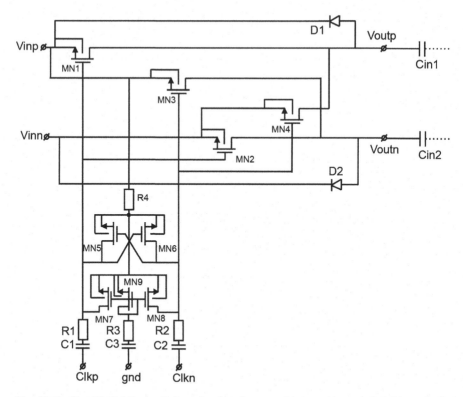

Fig. 10.61 Simplified full circuit for a floating chopper with capacitive-coupled driver clocks having CM levels fixed by active latches, and protected by diode clamps

driver clocks flash, meaning that the ascending edge of one clock line occurs simultaneous with the descending edge on the complementary clock-bar line. In that way the positive charge injection and negative charge injection in the chopper switches largely compensate each other.

We can simplify the circuit of Fig. 10.60 to that of Fig. 10.61 by using only one input as a reference for all input chopper transistors. Of course this brings imbalance to the on-voltages of the chopper switches. What is more, the chopper response may show more second harmonics. However, when the differential input voltage is small the imbalance is low, and the spurious responses might be negligible, while keeping the circuit simpler.

In some cases, such as in an instrumentation amplifier or in an ADC, we need to chop differential input voltages larger than a few hundred mV. The built-in diodes in the chopper transistors between their P-well back-gates and N-doped sources and drains prohibit the previously described floating choppers from processing differential voltages larger than several hundreds of mV. Nonetheless, if we are able to keep the back-gates at any moment on the lowest voltage present in the chopper, this limitation is eliminated. To that end a minimum selector with MNS1 and MNS2 (see Fig. 10.62), has been inserted between the input terminals with its minimum

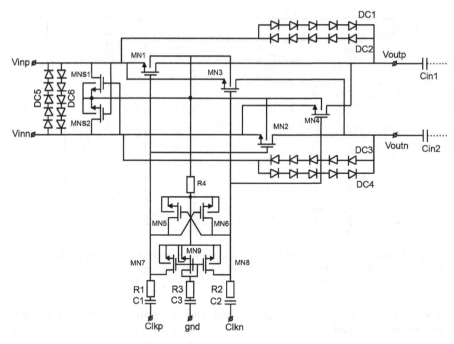

Fig. 10.62 Full circuit for a floating chopper with capacitive-coupled driver clocks having CM levels fixed by active latches, and protected by diode clamps for 2 V differential input voltages

point connected to all back-gates. If the voltage at the positive input terminal is higher than that at the negative one, MNS2 conducts and connects the minimum point to the negative input terminal. If the voltage on the negative input terminal is higher than that of the positive one, MNS1 conducts and connects the minimum point to the positive input terminal. In any case the minimum point is connected to the lowest input voltage.

Figure 10.62 shows the full circuit of a floating chopper with capacitive-coupled driver clocks and protection by active latches and diode clamps that can handle 2 V differential input voltages, limited by the protection diodes. It is further explained in [26]. To protect the floating chopper transistors from large positive and negative CM input surges, two antiparallel diode chains DC1 and DC2 have been placed across the chopper between the positive input and output, and another two antiparallel diode chains DC3 and DC4 have been placed between the negative input and output. For protection against large positive and negative DM surges two antiparallel diode chains DC5 and DC6 have been connected between the two input terminals.

If we want to chop even larger differential voltages, the limitation by the back-gate diodes can be removed by using two anti-series-connected transistors for each chopper transistor. A basic sketch for this floating chopper is given in Fig. 10.63. If MN1, MN11 and MN2, MN12 are turned on, the floating chopper connects straight forward. Even if the differential signal is +5 V, the switched-off transistors

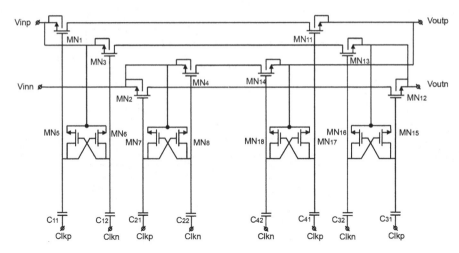

Fig. 10.63 Sketch of a floating chopper with capacitive-coupled driver clocks and having CM levels fixed by active latches for 5 V differential input signals (protection circuitry has still to be added)

MN13 and MN4 block the cross path. Their back-gates are not turned on. In each straight and cross path there are always two transistors that block a large positive or negative differential voltage. It stands to reason that the chopper transistors must be able to carry 5 V between their drains and sources. Protection circuitry still has to be added around the high-differential-voltage floating chopper of Fig. 10.63.

10.11 Gain Accuracy of Instrumentation Amplifiers

The inaccuracy of current-feedback instrumentation amplifiers is proportional to the relative difference in G_m of the input stage and feedback stage. The input stages are drawn in Fig. 10.64. For accurate matching the transistors need to be large. For high CMRR and for a high signal-to-noise ratio the input transistors need to have the highest possible G_m. This means they have to be biased in weak inversion with a relative large width/length ratio. Also the current sources I_{T1} and I_{T2} need to be matched well, as the G_m of the differential input transistor pairs is dependent on their tail current. Therefore, the current-source transistors need to be made less sensitive to differences in their gate-source threshold voltages by biasing them in strong inversion using long transistors, or in weak inversion in combination with degeneration resistors.

The most used input stage is of the P-Channel type for two reasons: Firstly, P-Channel transistors always have an isolated back gate. This makes it possible to bootstrap the back gates with the source voltages. By this measure the input G_m is much less dependent on the input CM voltage. This raises the CMRR from about 50 to 80 dB. Secondly, P-Channel transistors have less 1/f noise than N-Channel

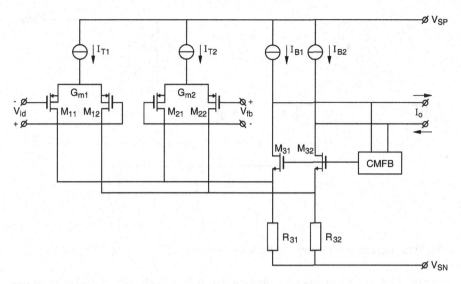

Fig. 10.64 Basic input stages for an current-feedback InstAmp

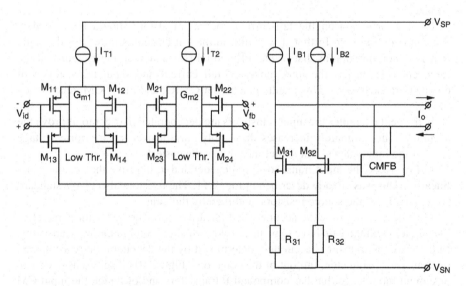

Fig. 10.65 Cascoded input stages for an current-feedback InstAmp

transistors. Folded cascodes behind the input stage allow the input CM voltage to include the negative rail, and increase the voltage gain of the input stage.

To further increase the CMRR of the input stages the input transistors have been cascoded in Fig. 10.65. This can be easily done if transistors are available with two different threshold voltages. For the input transistors we use the higher threshold transistors, and for the cascode transistors the lower threshold transistors. In that

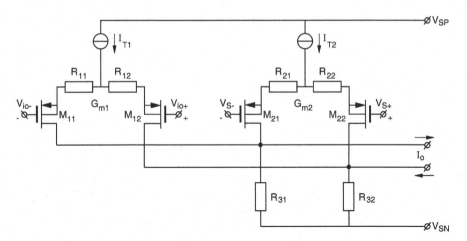

Fig. 10.66 Input stages of current-feedback InstAmp with degenerated input stages

way both transistors function in saturation at high voltage gain. Cascoding of the input transistors also helps to decrease the gain dependency of G_m to the CM voltage.

For the same reason the tail-current source transistors should be cascoded. We can even improve further if we built up the tail-current source with the same transistor combination as that of the differential pair and keep the elements at the same current density, the impedances of tail current and input transistors will compensate each other. This results in a higher CMRR and lower CM dependency of G_m.

In the next examples the input transistors are degenerated in order to improve the accuracy. This ultimately increases the supply current over square noise voltage ratio of the input stages. But sometimes this is the easiest.

In Fig. 10.66 the input transistors are degenerated to improve the accuracy and linearity. This was already described in Fig. 9.11. The resistors also give a standard way to calibrate the source resistors, and thereby the gain.

If we need to improve the accuracy and linearity more, the transistor parameters like G_m and voltage gain have to be increased by using a combination of transistors. In this way the transfer is accurately determined by the degeneration resistors and the transistor parameters fall out of the equations. Figure 10.67 shows how we can design an input stage that has compound P transistors and of which the input CM range includes the negative supply rail voltage V_{SN}. M_{11} and M_{12} are the input transistors. Their drain current is kept constant by M_{13} and M_{14}. These transistors take on the current needed to drive the degeneration resistors R_{11} and R_{12} and feed that current to the output load resistors R31 and R_{32}. M_{15} and M_{16} are folded cascodes to allow the input voltage include the negative rail voltage V_{SN} [18]. A disadvantage of this circuit is that the current over square noise ratio is roughly 8–16 times worse than that of a simple differential input pair.

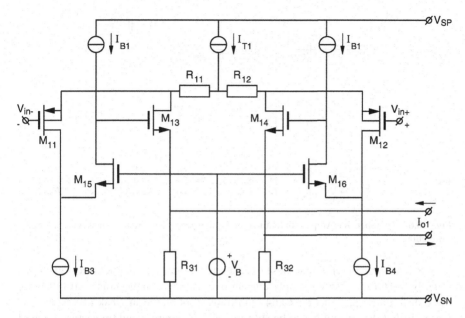

Fig. 10.67 Input stage with degenerated compound P transistors

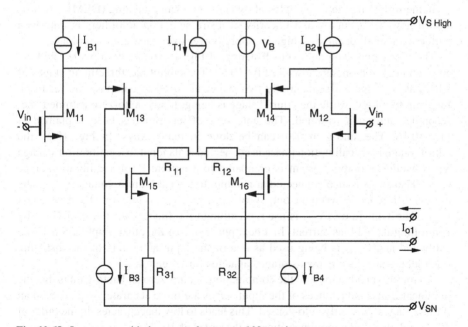

Fig. 10.68 Input stage with degenerated compound N transistors

Figure 10.68 shows an accurate input N transconductance which input CM range includes the positive supply voltage V_{SP}. It has the same functionality as the circuit of Fig. 10.67 [16].

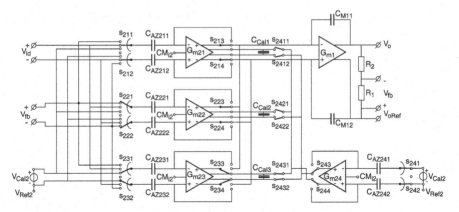

Fig. 10.69 Chopper InstAmp with auto-zeroed ping-pong-pang input stages and auto-gain calibration

We must keep in mind that these degenerated input transconductances result in a roughly 8–16 times higher supply current over square voltage-noise ratio than the basic differential pair of Fig. 10.64 with transistors biased in weak inversion. The reason is that the current has to be split into several sources and transistors, each of which contribute to noise, and that the G_m is lowered by degeneration.

In the following three examples of dynamic element matching (DEM) it will be shown how the accuracy and linearity of instrumentation amplifiers is improved with only a small penalty on higher supply power over square noise voltage.

The ping-pong-pang auto-zero InstAmp of Fig. 10.16 can be provided with an auto-gain calibration, as shown in Fig. 10.69. The circuit has three input stages of which at any moment sequentially two stages are used to compose the feedback InstAmp topology, while the "third" stage is being, firstly, auto-zero trimmed and, secondly, auto-gain trimmed. The auto-zero offset trim has been explained at Fig. 10.16. The auto-gain trim can be done in many ways. In Fig. 10.69 the "third" stage in its calibration phase is connected at its input to a calibration voltage V_{Cal2}, while its output current is compared to a current from a calibration stage G_{m24}. That stage is also connected at its input to the calibration voltage V_{Cal2}. The current difference between the outputs of G_{m23} and G_{m24} is integrated and stored on C_{Cal3}. The voltage on C_{Cal3} is used to calibrate the transconductance of G_{m23} by controlling its tail bias current. In a next phase one of the other amplifiers is being calibrated and G_{m23} is being used as one of the input amplifiers of the InstAmp. Each input stage has its own storage capacitor for its gain trim.

An important advantage of the combination of auto-zero and auto-gain is that the nonlinear transconductances of the input stages are more accurately equalized than if these stages were only auto-zeroed. This leads to low inaccuracies, in the order of 10^{-4}, and also to low nonlinearities, in the order of 10^{-4}.

In the same way the chopper instrumentation amplifier with ping-pong-pang auto-zero input stages of Fig. 10.24 can be auto-gain calibrated. This leads to an

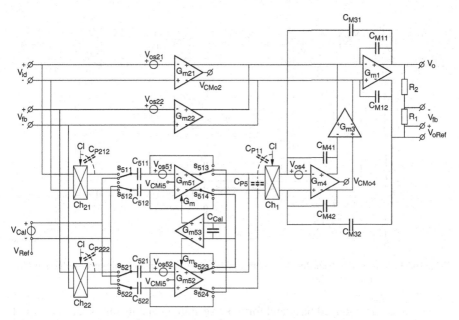

Fig. 10.70 Chopper-stabilized InstAmp with auto-zero sense amplifiers and auto-gain calibration

accurate and low-ripple auto-zero chopper instrumentation amplifier. This has not been shown separately.

The chopper-Stabilized InstAmp with auto-zeroed sense amplifiers of Fig. 10.32 can also be provided with auto-gain calibration [22]. This circuit is sketched in Fig. 10.70.

Calibration of the two offset sense amplifiers G_{m51} and G_{m52} needs only to result in the equality of these two amplifiers. This means that in the auto-calibration phase both inputs can simply be connected to a calibration voltage V_{Cal} and that the output currents can be compared. The difference of the output currents is being integrated and stored on a capacitor C_{Cal}. The voltage on the store capacitor controls through G_{m53} the difference of the tail bias currents and thus the difference of the transconductances. At a large control loop gain the transconductances of G_{m51} and G_{m52} become equal. The auto-zero and auto-calibration can be placed in one main auto-correction phase.

We have to keep in mind that with chopper-stabilized instrumentation amplifiers the stabilization loop accurately controls the gain at low frequencies. At high frequencies the gain is set by differences in the main amplifier input stages, which are not auto-calibrated. Normally, this is not a problem because at high frequencies the gain is not accurate anyway because of lack in overall loop gain.

In the above example the auto-gain calibration was done in a time-discrete way. A time-continuous way to achieve a high accuracy and remove differences in the two input G_m's of a chopper instrumentation amplifier can be obtained by applying dynamic-element matching (DEM) of the two input stages. To that purpose the two

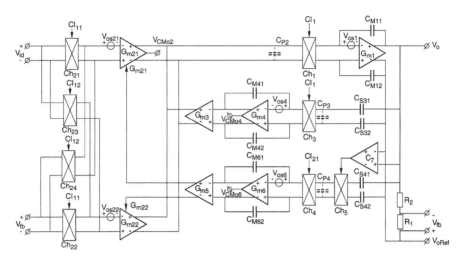

Fig. 10.71 Current-feedback InstAmp with ripple-reduction loop and gain-error-reduction loop (order of gain error is 0.01 %)

input stages are chopped back and forth between input and feedback. As a result of nonequal gains of the input stages an output ripple will arise at the frequency by which we interchange the input stages. If we DEM the input stages, for instance, at half the chopper frequency, a ripple will occur at the output at half the chopper frequency. This gain-error ripple can be reduced by a gain-error reduction loop (GERL) [21] independently of the offset RRL, which runs at the full clock frequency, as in the chopper InstAmp of Fig. 10.39a of the previous paragraph. The resulting circuit is shown in Fig. 10.71.

The gain-error reduction loop is made up from the capacitors C_{S41} and C_{S42}, which sense the output ripple, chopper Ch_4, integrator G_{m6}, and correction amplifier G_{m5}. The last one differentially corrects a small part (for instance 2 %) of the bias currents of the input G_m's. The result is a multiplicative correction on the input signals. If the gain is too large because g_{m21} is larger than g_{m22}, the output ripple is positive in regard to the clock for a positive input signal. But the output ripple will be negative in regard to the clock for a negative input signal. Both situations need a correction in the same direction. Therefore, a comparator C_7 is used to measure the output polarity. A chopper Ch_5 is inserted to multiply the sense signal by the output polarity.

This analog GERL can also be made up in a digital way by using an ADC after Ch_4, a digital integrator, followed by a DAC. That way the digital loop can be provided with long-term memory for the best gain setting in times that the signal is small, and measured when the signal is not so small.

If the CM voltage levels of the input and feedback output are different, the CM voltage levels of the input transistor pairs are alternatively chopped high and low. This has two side effects: Firstly, if the input stages have a G_m that depends on the CM voltage, this dependency is not taken away by the DEM action. Therefore, the overall gain is slightly depending on the difference of the input and feedback CM voltage level. Hence, it is very important to choose input stages with a highly CM-independent G_m, as described in the beginning of this paragraph. Secondly, parasitic capacitors between the back gates and ground will cause large CM current spikes in the input stages. When these CM current spikes are larger than the bias currents the signal transfer is hampered. To lower the CM current spikes, the back gates of the input transistors and the cascodes can be actively bootstrapped to their CM input levels by class-AB source followers [21].

10.11.1 Conclusion

The combination of automated offset and gain calibration leads to an accurate equalization of the nonlinear characteristics of the input stages. This results not only in a low offset, of the order of microvolts, but also in a low inaccuracy, of the order of 10^{-4}, and moreover in a low nonlinearity, of the order of 10^{-4}, as the curved G_m characteristics of input and feedback are accurately matched. The continuous-time chopping and calibration method of simple differential transistor pairs in weak inversion leads to the lowest ratio between supply current and the square input voltage noise.

10.12 Summary Low Offset

Table 10.1 gives an overview of the roughly estimated offset and ripple of the operational amplifiers in the Sects. 10.5–10.9.

Chopping generally can reduce offset by a factor of 10,000. But the ripple stays equal to the offset without other measures. Auto-zeroing reduces the offset by a factor of 100–500, depending whether the AZ store capacitors are placed at the

Table 10.1 Summary of offset and ripple that can be obtained

OpAmps	V_{os} (μV)	Vrip	InstAmps	V_{os} (μV)	Vrip
AZ	20–100		AZ	20–100	
Chopper	10	10 mV	Chopper	20	20 mV
N Chopper	0.1	100 μV	N Chopper	0.2	200 μV
ChSt	10	100 μV	ChSt	20	200 μV
ChSt + AZ	1	10 μV	ChSt + AZ	2	20 μV
Ch + ChSt	1	100 μV	Ch + ChSt	2	200 μV
Ch + ChSt + AZ	0.1	10 μV	Ch + ChSt + AZ	0.2	20 μV
Ch + RRL	1	100 μV	Ch + RRL	2	200 μV
Ch + RRL + AZ	0.1	10 μV	Ch + RRL + AZ	0.2	20 μV

input or at the output. Further improvement can be obtained when we combine chopping and auto-zeroing. Abbreviations used in Table 10.1 are: AZ = auto-zeroing, N = nested, ChSt = chopper-stabilized, Ch = chopping.

References

1. J. Huijsing, *Operational Amplifiers, Theory and Design (Chapter 1)* (Kluwer Academic Publishers, Dordrecht, 2001), p. 456
2. J. Huijsing, *Operational Amplifiers, Theory and Design (Chapter 3)* (Kluwer Academic Publishers, Dordrecht, 2001), p. 456
3. B. van den Dool, J. Huijsing, Indirect current feedback instrumentation amplifier with a common-mode input range that includes the negative rail. IEEE J Solid-St Circ **28**(7), 743–749 (1993)
4. J. Huijsing, *Operational Amplifiers, Theory and Design (Chapter 9)* (Kluwer Academic Publishers, Dordrecht, 2001), p. 456
5. I.E. Opris, G.T.A. Kovacs, A rail-to-rail ping-pong OpAmp. IEEE J Solid-St Circ **31**(9), 1320–1324 (1996)
6. C. Enz, E. Vittoz, F. Krummenacher, A CMOS chopper amplifier. IEEE J Solid-St Circ **22**(3), 708–715 (1987)
7. A. Bakker, K. Thiele, J. Huijsing, A CMOS nested chopper instrumentation amplifier with 100 nV offset. IEEE J Solid-St Circ **35**(12), 1877–1883 (2000)
8. A. Tang, Ping-pong amplifier with auto-zeroing and chopping, U.S. Patent 6,476,671, 11 May 2002. Analog Devices
9. C. Enz, G. Temes, Circuit techniques for reducing the effect of OpAmp imperfections: autozeroing, correlated double sampling, and chopper stabilization. Proc IEEE **84**(11), 1584–1614 (1996)
10. J. Huijsing, J. Fonderie, B. Shahi, Frequency stabilization of chopper-stabilized amplifiers, U.S. Patent 7,209,000, 24 April 2007
11. J.F. Witte, K. Makinwa, J. Huijsing, A CMOS Chopper offset-stabilized OpAmp, 2006, in European solid–state circuits conference, proceedings, pp. 360–363
12. R. Burt, J. Zhang, A micropower chopper-stabilized operational amplifier using a SC notch filter with synchronous integration inside the continuous-time signal path. IEEE J Solid-St Circ **41**(12), 2729–2736 (2006)
13. J.F. Witte, J. Huijsing, K. Makinwa, A current feedback instrumentation amplifier with 5 μV offset for bidirectional high-side current sensing. IEEE solid–state circuits conference 2008, San Francisco, Session 3.5, 4–6 Feb 2008
14. J. Huijsing, J. Fonderie, Chopper Chopper-stabilized operational amplifiers and methods, U.S. Patent 6,734,723, 11 May 2004
15. J. Huijsing, B. Shahi, Chopper Chopper-stabilized instrumentation and operational amplifiers, U.S. Patent 7,132,883, 7 Nov 2006
16. J.F. Witte, K.K.A. Makinwa, J.H. Huijsing, *Dynamic Offset Compensated CMOS Amplifiers* (Springer, Dordrecht, 2009)
17. R. Wu, K.A.A. Makinwa, J.H. Huijsing, A chopper current-feedback instrumentation amplifier with a 1 mHz 1/f noise corner and an AC-coupled ripple-reduction loop. IEEE solid-state circuits conference 2009 8–12 Feb 2009, pp. 322–323, 323a
18. J.H. Huijsing, B. Shahi, Accurate voltage to current converters for rail-sensing current-feedback instrumentation amplifiers, U.S. Patent 7,202,738, 10 April 2007
19. T. Denisson et al., A 2 μW 100 nV/rtHz chopper-stabilized instrumentation amplifier for chronic measurement of neural field potentials. IEEE J Solid-St Circ **42**(12), 2934–2945 (2007)

20. Q. Fan et al., A 21 nV/√Hz (10.5 nV/√Hz) chopper-stabilized multi-path current-feedback instrumentation (operational) amplifier with 2 μV offset. IEEE solid–state circuits conference 2010, San Francisco, 8–11 Feb 2010

21. R. Wu, J.H. Huijsing, K.A.A. Makinwa, A current-feedback instrumentation amplifier with a Gain-Error Reduction Loop (GERL) achieving 0.05 % gain accuracy and 1 ppm/degree C gain drift. IEEE solid-state circuits conference 2011, San Francisco, Feb 20–23, 13.5

22. R.E. Boucher, J.H. Huijsing, Auto-gain correction and common mode voltage cancellation in a precision amplifier, U.S. Patent 7,696,817B1, 7 April 2010

23. Q. Fan, J.H. Huijsing, A capacitively coupled chopper instrumentation amplifier with a ±30 V common-mode range, 160 dB CMRR and 5 μV offset. 2010 I.E. solid–state circuits conference, San Francisco, 19–23 Feb. 2012

24. J.H. Huijsing, Q. Fan et al., Fast-settling capacitive coupled amplifiers. US Patent 9,294,049, March 22, 2016

25. J.H. Huijsing, Q. Fan et al., Fully capacitive coupled input choppers. US Patent 9,143,092, September 22, 2015

26. Long Xu, et al., A 110 dB SNR ADC with ±30 V input common-mode range and 8 μV offset for current sensing applications. 2015 I.E. international solid-state circuits conference, San Francisco, 22–26 Feb 2015

27. J.H. Huijsing, Dynamic offset cancellation in operational amplifiers and instrumentation amplifiers, in *Analog Circuit Design*, ed. by M. Steyaert et al. (Springer, New York, NY, 2009), pp. 99–123

28. Q. Fan, et al., Capacitively-Coupled Chopper Amplifiers, Book, to be published by Springer Science+Business Media B.V. 2016

Author Biography

Johan H. Huijsing was born on May 21, 1938. He received the M.Sc. degree in Electrical Engineering from the Delft University of Technology, Delft, the Netherlands in 1969, and the Ph.D. degree from this University in 1981 for his thesis on operational amplifiers.

He has been an assistant and associate professor in Electronic Instrumentation in the EE Faculty of the Delft University of Technology since 1969. He became a full professor in the chair of Electronic Instrumentation in 1990 and professor emeritus since 2003.

From 1982 through 1983 he was a senior scientist at Philips Research Labs in Sunnyvale, California, USA. From 1983 until 2005 he was a consultant for Philips, Sunnyvale, and since 1998 he is a consultant for Maxim, Sunnyvale, CA.

The research work of Johan Huijsing is focussed on the systematic analysis and design of operational amplifiers, analog-to-digital converters, and integrated smart sensors. He has supervised more than 30 Ph.D. students. He is author or coauthor of some 350 scientific papers, 40 US patents, 15 books, and coeditor of 13 books.

He initiated the international Workshop on Advances in Analog Circuit Design in Europe in 1992. He co-organized it yearly until 2005. He was a member of the program committee of the European Solid-State Circuits Conference from 1992 until 2002. He has been chairman of the Dutch STW Platform on Sensor Technology and chairman of the biennial national Workshop on Sensor Technology from 1991 until 2002.

He is fellow of IEEE for contributions to the design and analysis of analog integrated circuits. He received the title of Simon Stevin Meester by the Dutch Technology Foundation.

October 1, 2015.

© Springer International Publishing Switzerland 2017   415
J. Huijsing, *Operational Amplifiers*, DOI 10.1007/978-3-319-28127-8

Index

A

All-NPN
high-frequency all-NPN OpAmp
with mixed PC and MC, 243–245
LM101 class-AB all-NPN OpAmp
with MC, 246
NE5534 class-AB OpAmp with bypassed
NMC, 247–249
1GHz, all-NPN class-AB OpAmp
with MNMC, 253–255
precision all-NPN class-AB OpAmp
with NMC, 249–250
precision HF all-NPN class-AB OpAmp
with MNMC, 251–255
2 volt power-efficient all-NPN class-AB
OpAmp with MDNMC, 254–255
Applications, 3, 5, 7, 12, 18, 29–55, 72, 81,
112, 117, 118, 132, 140, 157, 167,
175, 191, 215, 228, 242, 292, 295,
300, 314, 336, 340, 342, 351–354,
358, 362, 384, 391, 397
Auto-gain calibration for Instamps, 408, 409
Auto-zero (AZ), 352, 358–362, 365, 370, 371,
373, 375–377, 380, 382, 408, 409
chopper-stabilized OpAmps and InstAmps,
367–373
instrumentation amplifier, 358–362
Opamp, 358–362

B

Back-gate influence, 79
Balancing techniques, 59–63, 71, 73
Basic bipolar R-R-out class-A OpAmp,
229–230

Bias

Bias
FBB, 129–143
FFB, 112–129
generator for constant G_m, 66
offset, bias and drift, 57–69
Biasing for constant transconductance G_m,
66–69
Boosting
input class-AB boosting, 226–227
voltage-gain boosting, 227
Bridge instrumentation amplifier, 33–35

C

Capacitive input chopper amplifiers, 384–390
Cascade
folded-cascode OpAmp, 218–221, 280
telescopic-cascode OpAmp, 221–222
Cascoded input stages, 177, 405
Charge injection in chopper amplifier,
364, 365
Chopper
instrumentation amplifier (IA), 366, 367,
375, 376, 380, 396, 408, 409
operational amplifier (OA/OpAmp), 352,
362–367, 373–376, 380, 381, 395
Chopper amplifiers
with capacitive input, 384–390
with ping-pong auto-zero, 365
Chopper-stabilized chopper InstAmps,
373–376, 409
Chopper-stabilized chopper OpAmps,
373–376, 380, 394, 395
Chopper-stabilized OpAmps, 367–373, 382,
384, 385

© Springer International Publishing Switzerland 2017
J. Huijsing, *Operational Amplifiers*, DOI 10.1007/978-3-319-28127-8

Printed in the United States
By Bookmasters